# Selected Titles in This Series

# JACQUES HADAMARD,

## A Universal Mathematician

Jacques Hadamard

*1865 – 1963*

**History of Mathematics**
**Volume 14**

# JACQUES HADAMARD,

## A Universal Mathematician

**Vladimir Maz'ya**
**Tatyana Shaposhnikova**

**American Mathematical Society**
**London Mathematical Society**

1991 *Mathematics Subject Classification.* Primary 01–02, 01A70, 01A99;
Secondary 11–03, 26–03, 30–03, 35–03, 46–03, 53–03, 73–03, 76–03.

The main part of the text was translated from Russian and French or edited by Peter Basarab-Horwath.

Cover design by Nekod Singer.

A list of photograph and figure credits is included at the beginning of this volume.

---

**Library of Congress Cataloging-in-Publication Data**

Maz'ia, V. G.

Jacques Hadamard , a universal mathematician / Vladimir Maz'ya, Tatyana Shaposhnikova.

p. cm. — (History of mathematics ; v. 14)

Includes bibliographical references and index.

ISBN 0-8218-0841-9 (alk. paper)

1. Hadamard, Jacques, 1865–1963. 2. Mathematicians—France—Biography. 3. Mathematics—France—History—20th century. I. Shaposhnikova, T. O. II. Title. III. Series.

QA29.H18M39   1998

510'.92

[B]—DC21                                                                                 97-36357

CIP

---

*To our son Michael
with love*

# Contents

# Preface and Acknowledgements

We owe a special and deep debt of gratitude to E. Polishchuk (1913-1987), who awakened our interest in Hadamard's life and mathematics when he invited us to participate in writing a book about the great French mathematician. The book "Jacques Hadamard", by Polishchuk and Shaposhnikova written with participation of Maz'ya, was published in 1990, in Leningrad (now Saint Petersburg).

The biographical part of that book was very small, because original documents, letters and publications were inaccessible to us. Since our emigration to Sweden in 1990, we have been able to travel and consult new material concerning Hadamard. His personal papers disappeared during the war, and we have had to piece together the mosaic of his life from the odd fragments dispersed in archives and libraries in many different countries. The first part of the present book is a result of this search.

Professional historians of science have never focused their attention on Hadamard. During the thirty years after his death no attempt was made at writing his comprehensive biography. Meanwhile, there remain only a few of those who could share their reminiscences of the man who was once called the "living legend of mathematics". As J.-P. Kahane remarked in his essay [II.29],[1] "no mathematical library contains the whole mathematical work of Hadamard, because – unlike the work of much less important mathematicians – it was never collected and published in its entirety. No street in Paris bears his name.[2] The legend needs a revival, especially in France."

---

[1] Roman numeration refers to one of four lists of references: I Bibliography of Jacques Hadamard, II Publications about Jacques Hadamard and his work, III General Bibliography, and IV Archival Material.

[2] Also there are no memorial plaques on the houses where Hadamard was born and died, and even Hadamard's gravestone at the Père Lachaise cemetery does not have his name on it. (Auth.)

Although our exposition is documented, it does not pretend to be a deep historical study either. We neither could nor wished to attain this goal. Indeed, our aim was more modest: we merely wanted to tell a story of Hadamard's life for professional mathematicians and undergraduate students with an interest in mathematics. It is Jacques Hadamard, his relatives, teachers, colleagues, friends and pupils who speak in the pages of this book.

In our endeavours to make the account of Hadamard's life intelligible for a wider readership, we tried in the first part to obey the Justinian law dealing with "Malefactors and mathematicians and their like" which says: "The art of mathematics is also forbidden under pain of punishment" [III.88, p. 379], so we only mention some of Hadamard's most important mathematical results, to give an idea why Hadamard is worth writing about.

We have described in more detail Hadamard's contribution to mathematics in the second part of the book. Here the order of chapters reflects mainly the chronological order of his research interests. As before, we tried to address readers with rather modest mathematical background. We hope, nevertheless, that even an expert can find something of interest. Together with new material, Part II contains an extended and revised version of mathematical chapters of the Russian book [II.52]. Sections 11.3, 16.1 and excerpts of Chapter 9 reproduce with small changes the text left by E. Polishchuk, who died at the beginning of the work on [II.52]. Moreover, his notes were helpful when we wrote Chapters 10-12. We gratefully acknowledge his contributions to the present volume.

This book would never have appeared without the generosity of Hadamard's grandson, the physicist Francis Picard (1929-1995). We met him in Paris in 1992, and were overwhelmed by his willingness to help. He put at our disposal interesting material about his grandfather, which had been collected by his aunt, Jacques Hadamard's youngest daughter Jacqueline. Moreover, Francis Picard gave us permission to cite Jacqueline Hadamard's unpublished autobiographical manuscript. He was fascinated by the project and, as his wife Sabine Gayet told us, even during the last days of his life, he dreamed of seeing this book in print. To him, also, we express our profound indebtedness.

We are heavily indebted to Jeremy J. Gray for his generous help at various stages in the preparation of the manuscript. After taking on the burden of analyzing a preliminary version of the book he gave us much welcome criticism and advice on style, the main concepts, and the exposition of particular questions. For this, and his indulging our demands on his time, our deep-felt thanks and admiration.

Our heartfelt gratitude goes to François Murat for reading the preliminary and the final versions of the manuscript and for the infinitely many corrections and suggestions he made. He gave freely of his time in answering our questions about French scientific and public life. It is his achievement that the book contains fewer mistakes in mathematics, French, English, as well as in the life and culture of France.

We are indebted to our friend and colleague Lars-Inge Hedberg, who read drafts of our text. It is due to his erudition and his unalloyed criticisms that many errors have been eliminated and the text improved.

We are deeply obliged to Marie-Hélène and Laurent Schwartz for their encouragement and for sharing their reminiscences about Hadamard with us.

We express our appreciation to Benoît Mandelbrot for his critical comments which led to many improvements, for his permission to incorporate the reminiscences of his uncle, Szolem Mandelbrojt, and some of his own recollections about Hadamard.

We record our warmest thanks to the artist Nekod Singer for designing the cover and illustrating many stories from Hadamard's life with amusing drawings.

Natasha and Alexander Movchan deserve our particular credit for their friendly help in editing a preliminary version of the manuscript.

Our hearty thanks go also to Matelda and the late Gaetano Fichera for reading a draft of this book and for their generous assistance with our search for Hadamard's connections with Italian mathematicians.

We are most grateful to Eva and Lars Gårding for reading a preliminary version of the book and for making useful comments. Lars Gårding kindly helped to date Hadamard's article written in Swedish, which we found in the library of the Accademia dei Lincei.

We are deeply indebted to Ernest Kahane for his permission to include his recollections of Hadamard in the book, and to his son Jean-Pierre Kahane for telling us about his meetings with Hadamard and about French trade unions. He kindly supplied us with some interesting photographs.

We wish to thank Jean-Pierre Puel for explaining the French education system in a detailed lecture which proved to be extremely useful to us.

Our appreciation also extends to Howard Stone for his zeal in reading the text and for his valuable comments.

We unreservedly thank Nan Strömberg for her willingness to take on the extra burden of typing many unreadable handwritten pages of the manuscript.

We wish to express our deepest appreciation to Tjavdar Ivanov for his generous assistance in the technical preparation of the manuscript.

We cordially thank S. Agmon, K. Amaratunga, V.M. Babich, G. Birkhoff, R. Bürger, M. Costabel, S.S. Demidov, V.F. Demyanov, D. Dionisi, T. Ganelius, A. Grigorian, J. Horváth, B. Håkansson, V. Katasonov, J. Král, G. Kresin, R. Kress, N. Kuznetsov, S. Latkovic, P.R. Masani, L. Meister, J.-C. Nedelec, I. Netuka, L. Nikolski, N. Nikolski, O. Oleinik, S. Prössdorf, A.-M. Sändig, Ya. Sinaĭ, Zhen Sun, and J. Veselý for providing us with various materials and information concerning Hadamard.

We are grateful to F. Norstad for his expert help in solving our problems with LaTeX, to B. Edgar and M. Ludvigsen for the prompt answering of our questions in English grammar, and to J. Björn, G. Hsiao, K. Marciniak, P.E. Ricci, and B.O. Turesson for their kind help in translating some documents from Czech, Chinese, Polish, Italian, and Latin.

We also wish to thank those who helped us illustrate this book with photographs and drawings: H. Brezis, F. Dauphragne, N. Ermolaeva, E. Fraenkel, L. Gibiansky, N. Grigorian, S. Johns, J.-P. Kahane, G. Kresin, J. Lützen, W. Lenferink, B. Mandelbrot, J. Mawhin, M. Mendès-France, C. Monod-Broca, F. Murat, L. Nikolski, N. Nikolski, J. Polking, M. Rågstedt, M. Raulin, I. Romanovskaya, Y. Schetz, G. Schmidt, A. Slutskiĭ, H. Stone, Zhen Sun, Ji-Guang Sun, V. Tikhomirov, M. Tucscnak, and I. Verbitsky.

While working on the book, we have enjoyed the friendly atmosphere of the Mathematics Department of Linköping University, which greatly facilitated and stimulated our work.

We are grateful to the staff of the Mittag-Leffler Institute, where we spent January 1991, for providing us with excellent working conditions.

We acknowledge the support given to our work by Linköping University and by the Swedish Natural Science Research Council. Our thanks also go to the Swedish Institute, which offered us the rare opportunity of staying at the Swedish Cultural Centre during our two visits to Paris: its old walls and historical surroundings added much to our inspiration.

We wish to thank the staff of the libraries and archives where we obtained a lot of the information used in this book. We always met a readiness to help and great efficiency in the library of Linköping University, the library of the Mittag-Leffler Institute, the Royal Library in Stockholm, the library of Uppsala University, the library of the *Accademia dei Lincei*, the library of the Massachusetts Institute of Technology, the Woodson Research Center (Rice University), the library of the Göttingen University, the library of Queen's College (Oxford), the *Bibliothèque Nationale* in Paris,

*Bibliothèque de l'Institut de France, Bibliothèque de l'École Polytechnique, Bibliothèque de l'École Normale Supérieure, Bibliothèque de l'Institut National de Recherche Pédagogique*, and in the *Archives Nationales, Archives de l'Académie des Sciences, Archives de l'École Polytechnique, Archives du Collège de France, Archives Municipales de Versailles*, Archive of the Hebrew University of Jerusalem, and Archive of the Russian Academy of Sciences.

Our special thanks go to our sixteen-year-old son Michael who is used to hearing Hadamard's name from infancy. He did his best to help us with English and French, typing, copying and scanning. We owe him so much for his tolerance, understanding and encouragement.

# Photograph and Figure Credits

The AMS gratefully acknowledges the kindness of these individuals, institutions and publishers in granting the following permissions.

H. Brezis

   Photograph on p. 60; Courtesy of H. Brezis.

L. Gibiansky

   Photograph on p. 236; Courtesy of L. Gibiansky.

A. Grenberger

   Drawing of the Mittag-Leffler Institute on p. 153 by A. Grenberger; Courtesy of A. Grenberger.

Tj. Ivanov

   Figures 1-12 and 14; Courtesy of Tj. Ivanov.

J.–P. Kahane

   Photographs on p. 283, 284; Courtesy of J.-P. Kahane.

B. Mandelbrot

   Photograph on p. 186; Courtesy of B. Mandelbrot.

J. Mawhin

   Photograph on p. 69; Courtesy of J. Mawhin.

V. Maz'ya and T. Shaposhnikova

> Photographs on p. 10, 11, 17, 22, 39, 123, 147, 189, 214, 218, 296, portrait of V.I. Smirnov on p. 215, photograph of S.L. Sobolev on p. 216; Courtesy of V. Maz'ya and T. Shaposhnikova.

M. Mendès-France

> Drawing on p. 194 by M. Mendès-France; Courtesy of M. Mendès-France.

C. Monod–Broca

> Photograph on p. 126; Courtesy of C. Monod-Broca.

A. Movchan

> Figures on p. 19, 20, 169; Courtesy of A. Movchan.

F. Murat

> Photograph on p. 32; Courtesy of F. Murat.

L. Nikolski and N. Nikolski

> Photographs on p. 61, 62; Courtesy of L. Nikolski and N. Nikolski.

F. Picard

> Photographs on p. V, 3, 63, 74, 100, 111, 115, 138, 139, 201, 212, 222, 229, 233, 240, 246, 254, 260, 262, 270, 271, 273, 274, 276, 289, 291-294; photograph of L. Hadamard on p. 248; facsimiles of the letters on p. 120, 210, 267, 286, 295, 388, 493; list of Hadamard's awards on p. 20; telegram on p. 234; drawing on p. 273; facsimile of the beginning of Hadamard's copy-book on p. 124; facsimile of the list of Hadamard's papers on p. 511; Courtesy of F. Picard.

I. Romanovskaya

> Photograph of L.V. Kantorovich on p. 216; Courtesy of I. Romanovskaya.

N. Singer

> Cover and drawings on p. 12, 13, 21, 59, 94, 108, 140, 143, 162, 164, 165, 168, 223, 231, 237, 239, 254, 256, 272, 275, and collage on p. 297 by N. Singer; Courtesy of N. Singer.

Ji–Guang Sun

> Photograph on p. 264; Courtesy of Ji–Guang Sun.

*Archives de l'Académie des Sciences*

> Photographs on p. 66, 78, 159, 180, 209, 350, first page of Hadamard's report on p. 217; Courtesy of the *Archives de l'Académie des Sciences.*

*Archives de l'École Polytechnique*

> Hadamard's results for the entrance examinations for the *École Polytechnique* on p. 24; Courtesy of the *Archives de l'École Polytechnique.*

*Archives Municipales de Versailles*

> Facsimile of the birth certificate of J. Hadamard on p. 11; Courtesy of the *Archives Municipales de Versailles.*

*Archives Nationales* in Paris

> Facsimile of Hadamard's request for admission to the entrance examinations for the *École Normale* on p. 23; facsimile of the letters on p. 46 and 116; Courtesy of the *Archives Nationales.*

*Armand Colin* Publishers

> Drawings on p. 93, 110, 166; Courtesy of the *Armand Colin* Publishers.

*Bibliothèque Nationale* in Paris

> Photograph on p. 8, drawing on p. 175 (artist unknown); Courtesy of the *Bibliothèque Nationale* in Paris.

*Collection Sirot–Angel*

> Photograph on p. 81; Courtesy of the *Collection Sirot–Angel.*

Emilio Segrè Visual Archives, Center for History of Physics, American Institute of Physics

> Photographs on p. 158, 301, 483; Courtesy of the Emilio Segrè Visual Archives.

*École Normale Supérieure*

> Photographs on p. 15, 27, 31, 40-43, 177; Courtesy of the *École Normale Supérieure*.

Faculty of Mathematical and Natural Sciences, University of Leiden

> Photograph on p. 53; Courtesy of the Faculty of Mathematical and Natural Sciences, University of Leiden.

Harvard University Archives

> Photograph on p. 250; Courtesy of the Harvard University Archives.

Hebrew University Photo Archives, Jerusalem

> Photographs on p. 136, 279; Courtesy of the Hebrew University Photo Archives, Jerusalem.

Institute for Scientific Information

> Data on p. 501 from the Science Citation Index; Courtesy of the Institute for Scientific Information.

*Institut Mittag–Leffler*

> Photographs on p. 29, 33, 34, 35, 38, 51, 68, 70, 76, 90, 101, 148, 150, 373 396, 454, photograph of É. Goursat on p. 37; photograph of C. Jordan on p. 401; photographs from *Acta Mathematica* 1882-1912, *Table Générale des Tomes* 1-35, on p. 26, 73, 98, 184, 317, 319, 355, 384, 392, 399, 423, 448, 474; Courtesy of the *Institut Mittag–Leffler*.

*Det Kongelige Bibliotek*, Copenhagen

> Photograph on p. 86; Courtesy of *Det Kongelige Bibliotek*, Copenhagen.

*Kungliga Biblioteket*, Stockholm

> Photograpns on p. 79, 141, 144, 146, 400, 424; Courtesy of *Kungliga Biblioteket*, Stockholm.

Library of the *Accademia dei Lincei*

> Photographs on p. 106, 191, 206, 207; telegram on p. 119; Courtesy of the Library of the *Accademia dei Lincei*.

Massachusetts Institute of Technology Press

> Photograph on p. 171; Courtesy of the Massachusetts Institute of Technology Press.

MIT Museum

> Photographs on p. 190, 196, 243; Courtesy of the MIT Museum.

*Mathematisches Forschungsinstitut*, Oberwolfach

> Potographs on p. 47, 64, 89, 173, 302, 303, 330, 332, 333, 342, 343, 356, 366-368, 404, 414, 416, 419, 421, 433, 434, 471; photograph of P. Appell on p. 37; photograph of E. Betti on p. 401; Courtesy of the *Mathematisches Forschungsinstitut*, Oberwolfach.

*Stadtarchiv Wuppertal*

> Photograph of K. Krall on p. 143; Courtesy of *Stadtarchiv Wuppertal*.

Trinity College, Cambridge

> Photograph on p. 248; Courtesy of the Master and Fellows of Trinity College, Cambridge.

S.I. Vavilov Institute of the History of Natural Sciences and Technics, Russian Academy of Sciences

> Photographs on p. 197, 199, 211, 331, 409; portrait of A.N. Krylov on p. 215; Courtesy of the S.I. Vavilov Institute of the History of Natural Sciences and Technics, Russian Academy of Sciences.

*Uppsala universitetsbibliotek, kart- och bildavdelningen*

Print of Yale University on p. 133, post card with a view of Columbia University on p. 237; Courtesy of Uppsala universitetsbibliotek.

Woodson Research Center, Rice University

Photograps on p. 131, 132; Courtesy of the Woodson Reseach Center, Rice University.

The following figures and photographs are in the public domain

First page of the article [III.292] on p. 6; title page of the book *Mémoire sur les moyens de hater la régénération des israélites de l'Alsace* by P. Wittersheim on p. 7; illustration on p. 9; first page of A. Hadamard's lecture on p. 18; first page of the paper [I.13] on p. 50; first page of the paper [III.379] on p. 54; first page of the paper [I.263] on p. 65, first pages of the articles [I.399] and [I.401] on p. 84; title page of the book [I.67] on p. 88; title page of the book [I.192] on p. 102; title page of the book [I.223] on p. 134; title pages of the books [I.258], [I.291] on p. 138; first page of the article [III.234] on p. 142; photograph of S.N. Bernstein from the book "Istoriya otechestvennoĭ matematiki", *Naukova Dumka* Publishers, Kiev, 1970 on p. 180; title page of the book [I.253] on p. 182; beginning of the article [I.279] on p. 202; photograph of I.G. Petrovskiĭ on p. 213 from Uspekhi Matem. Nauk **26**:2 (1971); title page of the book [I.333] on p. 219; *Légion d'honneur* on p. 220; title page of the monthly *La politique de Pékin* on p. 224; title page of the journal *Renaissance* and the beginning of the paper [I.363] on p. 238; press-cutting on p. 241; title page of Hadamard's open letter [I.367] on p. 242; cover of the book [I.404] on p. 251; photograph of A.M. Liapunov on p. 72; photograph of A.N. Kolmogorov on p. 255 from Uspekhi Matem. Nauk **28**:5 (1973); photograph of P.S. Alexandrov on p. 256 from Uspekhi Matem. Nauk **21**:4 (1966), photograph of Yu.V. Linnik on p. 257 from Uspekhi Matem. Nauk **28**:2 (1973); title page of the book [I.405] on p. 268; press-cutting on p. 278; leaflet of the *Union Rationaliste* and title page of the article [I.389] on p. 288; first page of the article [I.42] on p. 336; title page of the book [I.159] on p. 377; article [I.68] on p. 397; first page of the article [I.258] on p. 402; first page of the article [I.163] on p. 403; photograph of J. Fourier on p. 420; title page of the book [I.109] on p. 425; title page of the book [III.114] on p. 426; title page of Cauchy's *Exercices de Mathématiques* and first page of the article [III.75] on p. 478; figure on p. 485 from Hadamard's article [I.310]; first page of the article [I.400] on p. 497; headpieces: coat of arms of the town Hadamar on p. 5, *Notre Dame de Paris* on p. 85; Athena, the emblem of the *Académie des Sciences* on p. 103; a goatsbeard on p. 112; Wailing Wall in Jerusalem on p. 135, a sight of Prague on p. 137; mountain Matterhorn in the Alps on p. 164; a fern on p. 167; a pagoda on p. 221; statue of Liberty on p. 235; Clock Tower on p. 245; Eiffel Tower on p. 249; Kremlin (Moscow), Peter and Paul Fortress and the Bronze Horseman (Saint Petersburg) on p. 253; Taj Mahal on p. 258.

# Part I

# Hadamard's Life

# Prologue

As a boy, Jacques Hadamard loved travel books, played the violin, excelled at languages, and had a passion for plants, but he detested having to solve problems in arithmetic. As an adult, however, he was to devote his life to mathematics, and make great discoveries, without losing any of those childhood interests.

Hadamard lived a long life, his centenary being celebrated just two years after his death. He was born when the only means of transport were horses and steam engines; by the end of his life, man had flown around the earth. His childhood was during the time when cold steel and gunpowder were used in warfare; he lived to the era of nuclear weapons tests. When a young academic, he was marked by the Dreyfus affair, in which a French army captain was falsely imprisoned for treason; in his old age he was to see a world where millions were murdered in concentration camps.

He knew great joys: he was happily married with two daughters and three sons, and he became world famous before the age of thirty. He also endured the most terrible sorrows: all three of his sons died during the two world wars, and he himself had to escape from his native France when the World War II broke out. Despite all trials and tribulations, he kept his courage and never broke.

Endowed with a brilliant memory, he thirsted for and absorbed knowledge. He adored music and was an indefatigable traveller. His proverbial absent-mindedness made him the hero of many an anecdote. Witty and easy-tempered, he had an attractive disposition; some of the most famous people in science and culture where among his friends. He was a tireless campaigner for human rights, fighting against discrimination and injustice.

He worked incessantly, writing on function theory, calculus of variations, number theory, analytical mechanics, algebra, geometry, probability theory, elasticity, hydrodynamics, partial differential equations, topology, logic, education, psychology, and the history of mathematics. Every undergraduate learns the Cauchy-Hadamard formula for the radius of convergence for a power series. Among the classical results of the theory of functions of a complex variable are Hadamard's theorems on three circles, gap series, and multiplication of singularities. One of the greatest achievements of the theory of numbers, the proof of the asymptotic law of distribution of primes, was obtained by Hadamard. Hadamard's inequality for determinants is well known. Well-posedness in the sense of Hadamard is one of the most important characteristics of problems in mathematical physics, and his counterexamples are included in all lecture courses on this subject. Hadamard's matrices, his variational formula for Green's function and the construction of solutions to the Cauchy problem for hyperbolic equations, the Legendre-Hadamard condition for the positivity of the acoustic tensor of a solid body and Hadamard's equation for water waves is far from a complete list of mathematical topics connected with the name of Jacques Hadamard.

Mathematics of the twentieth century is a kaleidoscopic conglomeration of methods and ideas and is branched so much that an encyclopedic knowledge of it has become practically impossible. Even the best mathematicians who work in different areas as a rule do not understand each other. Nevertheless Hadamard was able to know and enhance almost all areas of the mathematics of his time. Of course, the merits of his contributions to different domains are sometimes incomparable. However, it is this versatility which is such the characteristic of Hadamard's work that is so little known. Indeed, specialists in number theory usually regard Hadamard as one of the classic mathematicians in their area, unaware that he is also one of the great pillars of mathematical physics. Their ignorance is comparable with that of mathematical physicists, who do not know of Hadamard's contribution to number theory.

During the course of Hadamard's life, mathematics changed completely, and his own colossal work played a fundamental role in this. On the one hand, he contributed greatly to the solution of concrete problems which the nineteenth century had left unsolved, using classical analysis with profound virtuosity. On the other hand, he initiated a number of directions which are still a vital part of modern mathematics.

# Chapter 1

# The Beginning

## 1.1 Family and childhood

Jacques Salomon Hadamard was born on December 8, 1865, in Versailles into a Jewish family. His father's forebears came from Metz, and it is possible that the ancestors of this part of the family were among those Jews who came to Lorraine with the Romans.

*Coat of arms of the town Hadamar*

The name Hadamard appears to originate from the old German town of Hadamar (meaning a place with swamp water in old German), in Hessen, which goes back as far as the year 832. The town was one of the seven 'Tötungsanstalten' [extermination institutions] of the Third Reich in whose gas chambers the mentally ill were systematically killed from the beginning of the euthanasia programme [III.134, p. 387].

Hadamar is both a German and a Jewish surname in Germany, where the Jewish Hadamars lived as long ago as 1650 [III.242]. Its use as a surname can be traced back to the tenth century, to the German Cardinal Hadamar (died 956). The gothic poet Hadamar von Laber who lived in the first half of the fourteenth century was another bearer of the name.

The earliest mention of the Hadamards of Metz seems to be in a court case in 1715.[1] Several Hadamards are mentioned in the *État civil* of Metz

---

[1] The manuscript *Factum pour le Sieur Paul Guerre Maître de Forge* [Master of the Ironworks] *à Moyeuvre...contre Nathan Hodomard, Juif, résident à Metz...* [IV.43]. This document refers to a court case in which Nathan Hodomard and Isaac Spire Levy, a Jewish iron merchant, were involved and in which the *Maître de Forge* was the appellant. The dispute was connected with trade in iron in Lorraine, and it became so acrimonious that the *Maître de Forge* complained bitterly "...the position of banker entails a public function which the Jews are incapable of...But why should this Jew meddle in the affairs

during the eighteenth century, when the Duchy of Lorraine lost its sovereignty and became a province of France [IV.4]. Nathan Mayer Hadamard, Jacques Hadamard's great-great-grandfather, was a merchant who represented the Jews of Metz in Paris. One of his sons, David Mayer Hadamard (1752-1802) also became a merchant and married Rebecca Lambert (1760-1843), the daughter of an old Jewish family in Metz. She was such an interesting woman that her obituary was published in the Jewish monthly *Les Archives Israélites de France* of 1843 [III.292].

Being *la femme forte*, she was very active in the social life of the Jewish community of Metz, which at that time was rare for a woman. Once she

protested to the authorities about the desecration of Jewish tombs and was able to persuade them to have the perpetrators brought to justice. As the obituary says: "This devotion in Madame Hadamard was moreover hereditary. At the time of the Terror, when people gloried in repudiating all religious practices, the Israelites of Metz could not decide whether to celebrate Passover without unleavened bread, and Metz saw the repetition of scenes from Spain in the Middle Ages when the deepest mystery surrounded the practices of a religion which has no mystery. They made the unleavened bread but were afraid of denunciations which were so frequent at that time, and this made Madame Hadamard's mother go and search out the people's representative.

The first page of the obituary of Rebecca Hadamard

'What do you want, citizen?', asked the proconsul of Metz.

'To ask for permission to celebrate Passover'.

'What? Do you still keep to such nonsense when the sun of reason shines on the horizon?'

'The bread is ready and it is a custom close to our hearts as a reminder of freedom'.

'Well, since the wine has been poured, one has to drink it'."

of *Monsieur le Duc de Lorraine*...why involve two other Jews."

Widowed at the age of forty-two, Rebecca Hadamard was left to bring up nine children. "Her skill and her commercial probity became proverbial" [III.292]. She was well acquainted with books on liturgy and had an excellent knowledge of French and French literature, being able to recite long passages from Racine and Corneille. Her last years were spent in Paris with her children.

Her son Ephraïm Hadamard (1787-1854), Jacques Hadamard's grandfather, became a well-known printer in Metz.[2] Towards the end of the eighteenth century the printing of Hebrew literature in Metz had virtually ceased, and liturgical and educational books necessary for the considerable Jewish community were brought in from abroad. The young Ephraïm Hadamard conceived a plan to reestablish the printing of Hebrew books in Metz, and in 1813 he started to realize it. He began his career in the printing trade in Metz and then worked for several years in printing ateliers in France, Germany and the Netherlands in order to improve his skills and study languages. On returning to Metz he bought an old printing workshop and, with some difficulty, obtained permission to start his own printing firm. He succeeded in collecting Hebrew characters from different places, and over the years Ephraïm Hadamard's atelier flourished, publishing many books in Hebrew, French and German, as he had many pupils. In 1816 he married Fillette May (1791-1882) and they had five children together.

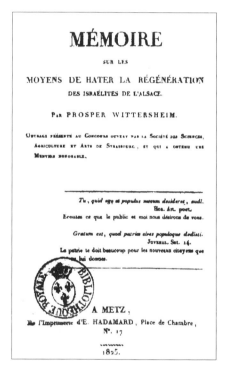

*The title page of a book printed in E. Hadamard's atelier*

One of their sons, David (1821-1849), Jacques Hadamard's uncle, became a noted arabist. He is mentioned in the *Archives Israélites de France* of 1845 [III.82, p. 170]:

"Mr. David Hadamard, son of the printer of that name, after having studied the Arabic language in Paris with success, was

---

[2]A short description of his life and work can be found in the book *Essai philologique sur les commencements de la typographie à Metz* (1828) [III.394, p. 222-231].

sent by the government to Algiers as a secretary-interpreter in
the department of finance. After having performed these duties
for several years, to the great satisfaction of his superiors, he
was cruelly afflicted by ophthalmia. Deprived of his sight at the
age of twenty-three, at the moment when a brilliant career was
opening up before him, he has returned, and wishing to provide
for himself in an honourable manner at the same time as pass
on his rare knowledge to his fellow countrymen, he has started
a Course of Arabic which he has been giving for some days.
His good method will soon put his pupils in a position to speak
Arabic.

His course takes place at 148 *rue Montmartre*, where one can
obtain the prospectus".

In 1847 David Hadamard became professor of Arabic in Oran, but two
years later he died of cholera, leaving a wife and a six-week-old daughter,
Zélie (1849-1902).

*Zélie Hadamard on stage*

Zélie Hadamard was a talented dramatic actress. At the age of sixteen
she won the first certificate of merit for tragedy and second prize for comedy
in a competition. The following year, she won the first prize for comedy and

was accepted by the *Odéon* theatre. She toured Europe and America and on her return played in the best French theatres. "Mademoiselle Hadamard possesses to a remarkable degree the skill of projecting her voice. She is one of the very few comediennes who know how to recite verse which gives her, without singing, her sonority and colour. She excells in expressing the emotions of modesty, nobility and tenderness" [III.60].

Another of Ephraïm Hadamard's sons was the painter Auguste Hadamard (1823-1886). In his youth he was obliged to earn a living at the same time as studying with the historical painter, portraitist and sculptor Paul Delaroche (1797-1856). Many of Auguste Hadamard's genre and portrait paintings were exhibited in the *Salon*. He also worked as an illustrator of books and journals.

*Auguste Hadamard: An illustration to the book Chansons nationales* [III.76]

The Hadamards were a family with cultural and liberal traditions, and Ephraïm's youngest son, Amédée (1828-1888), Jacques Hadamard's father, was no exception to this. He was sixteen when Ephraïm died, and after obtaining the *licence ès lettres* in 1849 at the age of twenty-one Amédée Hadamard was compelled to end his studies in order to earn a living as a freelance tutor in Paris. It was only ten years later, in 1859, that he obtained the *agrégation* examination which allowed him to teach in *lycées* and *collèges*. He changed schools several times at the beginning of his career, staying no more than three years in each one – Nîmes (1859-1861), Douai (1861-1864), *Collège Rollin*, Paris (1864-1865) – teaching classical languages, French grammar, history, geography and even elementary arithmetic.

During his years as a tutor in Paris he had a liaison with a woman who bore his child. While in Nîmes, fearing that she would visit him, he wrote to her, saying that he was married, and entered this in his *Notice individuelle*, the official Ministry of Public Instruction document in which the career of each teacher was carefully noted. When his Parisian mistress found another admirer, and the child died, Amédée, by then working in Douai, wrote in his documents that he was a bachelor. This discrepancy in the papers was discovered and the affair was brought to the attention of the Minister of Public Instruction [IV.24]. It was an indiscretion which Amédée Hadamard's superiors mentioned in their reports about him for many years afterwards [IV.24].

*Lycée Hoche, the former Lycée Impérial de Versailles, where Amédée Hadamard worked in 1865–1869*

On June 6, 1864, he married Claire Marie Jeanne Picard, who was fourteen years younger, and in 1865 he obtained a position at the *Lycée Impérial de Versailles*. "Hadamard, the sound grammarian, came to us from Rollin," as was written in an article about the *Lycée* at that time [III.233, p. 100].

On Monday eleventh December eighteen hundred and sixty-five, at ten o'clock in the morning, Birth Certificate of Jacques Salomon, of the masculin sex, born the eigth of this month, at ten o'clock at night in the home of his father and mother at Versailles, Boulevard du Roi 1, son of Amédée Hadamard, teacher at the Lycée of this town, aged thirty seven years, and of Claire Marie Jeanne Picard, his wife, aged twenty-three years. The witnesses are Jean Baptiste Adolphe Aderer, teacher at the said Lycée, aged thirty-three years, resident in Versailles, rue de la Nativité 75, and François Jacques Nomilly, also a teacher at the Lycée of Versailles, aged thirty-nine years, resident in Versailles, rue Royale 3. On the presentation of the child and the declaration of his father, who has signed with the witnesses and Us, Deputy Mayor executing by delegation the duties of public registrar, after reading.

The birth certificate of Jacques Hadamard

The young couple rented a flat with three rooms and a kitchen, sharing it with their house-maid, in an apartment block at *1, boulevard du Roi*, in Versailles, and it was here that Jacques, their first child, was born at the end of 1865.

The house in Versailles where Jacques Hadamard was born

Three years later Amédée Hadamard moved to the *Lycée Charlemagne*

and the family settled in Paris. The little we
know of Jacques Hadamard's early childhood
is from the reminiscences of his daughter Jacqueline Hadamard: "...my aunt told me that
when he was about four years old he was prone
to terrible tantrums, which were so severe that
my grandmother was summoned to the *commissariat* because of the neighbours' complaints
about the 'ill-treatment of a child'. A very intelligent doctor advised my grandmother to teach
him to read; the tantrums immediately stopped,
forever" [IV.1, p. I(5)].

1870 was a very difficult year for Amédée
and his wife, beginning with the death of
their five-month-old daughter Jeanne Hortense.
Then came the siege of Paris by the Prussians,
bringing disease and hunger for many Parisians.
As Edmond Goncourt noted in his diary entry
for December 8: "People talk only about what
they eat, what they can eat, and what there
is to eat" [III.153, p. 130]. Some seventy-two
thousand horses were slaughtered during the
siege, and those who could not afford to buy
horsemeat had to do with cats and dogs, which

were cheaper. It is from that time that Jacques remembered eating ele-
phant's trunk: the elephants Castor and Pollux, the pride of the *Jardin
d'Acclimatation*, were shot, and their meat was sold at an inflated price.

The suffering of the people of Paris did not come to an end with the
signing of the armistice on February 26, 1871. In April of that year, the
*Commune* was installed in the city, which resulted in a civil war, and when
government troops entered Paris the fighting continued on the barricades
erected in the streets. It was during one of the ensuing battles that the
house where the Hadamards lived was burnt down. A *député* of the National
Assembly wrote to the Minister of Public Instruction on their behalf [IV.24]:

Versailles, August 28, 1871

Mister Minister,

I have the honour of informing you of the unfortunate position
of M. Amédée Hadamard, a teacher at the *Lycée Charlemagne*,
who has lost his whole fortune, all his personal belongings in one
of the fires lit by the Insurgents. Married and father of a family,

M. Hadamard wishes to be appointed to a sufficiently important *lycée* so as to allow him to give lessons.

Permit me to remind you that this situation deserves real attention.

I am, Mister Minister, your most devoted servant

Bamberger.

This intercession produced no effect. In June 1874 another tragedy befell Jacques' parents when their three-year old daughter Suzanne Jeanne died. A year later, his sister Germaine was born. They were similar in features and despite the difference in age, remained close to each other.

Jacqueline Hadamard writes that her father never talked about his childhood. Was there any reason for this? She quotes her mother, who said that the Hadamard household was strict. Judging from the reports of the *lycée* inspectors and administrators which have been preserved, Amédée Hadamard could hardly be called an easy person. "A firm and forceful character, dedicated to his work to the point of obstinacy in spite of his poor state of health, hard on himself, he was one of those who could impose discipline on himself so as to get others to accept it", as Marguet, one of his colleagues, wrote in his obituary [III.267].

His mother seems to have been similar in character. She gave piano lessons at home and was such a stern teacher that her pupils were afraid of her. Jacqueline Hadamard describes this vividly:

"My aunt, her daughter, claimed that the pupils would cry when climbing the stairs! As a child I was surprised that mushrooms didn't grow in her piano: it was so often watered with tears" [IV.1, p. I(5)].

However, she was a good teacher. Paul Dukas, who later became a well-known composer and is the author of the symphonic scherzo *The Sorcerer's Apprentice*, based on Goethe's ballad of the same name, was one of her pupils.

Music played a very significant role in the Hadamard household, and Jacques learned the violin at an early age. Thanks to his mother he got a thorough musical education while his sister became a professional teacher of music who later taught the mathematician Laurent Schwartz the piano.

Jacques also developed an early passion for reading. "I know some things he liked to read," writes Jacqueline Hadamard, "the *Magazine d'Éducation et de Récréation*, which was very good, and published all of Jules Verne, P.J. Stahl and others, with excellent engravings which, even when they were not by Gustave Doré, were of that quality; the word 'Education' was taken seriously there. Most certainly, my father's incredible memory got a lot of knowledge from it, so one could ask him questions on anything. How many times have I heard someone saying to him: 'Jacques, you know everything, what is...?' And almost always, he knew!" [IV.1, p. I(6)]

## 1.2  The years at the *lycée*

When Hadamard was a schoolboy, secondary education was offered by both state schools (*lycées*) and municipal schools (*collèges*), the former enjoying a better reputation and being more difficult to enter. Education started at the age of six, in the *petit collège* or primary school, beginning with the eleventh class and going on to the seventh. The next stage was the first part of secondary education (the *premier cycle*), from the sixth class to the third in the *moyen collège*. Then, at the age of fifteen, one would move on to take the *second cycle* at the *grand collège* for three years, the classes being called *seconde*, *rhétorique* and *terminale*, respectively. At the end of this second part was the *baccalauréat* examination which entitled students to university entrance.

In the *terminale* students chose between *philosophie* and *mathématiques élémentaires* (*math élém*) according to whether they wished to become *bachelier ès lettres* or *bachelier ès sciences*. After the *baccalauréat* one could go on to university or prepare for entry to one of the *Grandes Écoles*, selective professional schools where the intellectual élite of France were trained. Some *lycées* had classes to prepare students for the rigorous entrance examinations to these *Grandes Écoles*.

A characteristic feature of the French educational system was the competitiveness which manifested itself particularly sharply in the *lycées*. At the end of each teaching year, the pupils of the same year were ranked according to their academic achievement in each subject and a *prix d'excellence* was given to the best overall student. In addition to this, yearly national competitions, the *Concours Généraux*, were arranged for pupils of the *moyen* and *grand collège*, and the same tests were given at all the *lycées* and *collèges* throughout France.

Jacques Hadamard attended the *Lycée Charlemagne*, where his father taught, up to the end of the fifth class. In the register of pupils who won awards in the academic year 1873-1874 we find that the eight-year old Jacques was in the seventh class, so that he was then two years younger than the rest of the class [III.255]. From 1875 onwards, Jacques Hadamard was winning prizes in the *Concours Généraux*, which must have made his father proud. His results for the first competition are as follows [III.255]:

| | |
|---|---|
| *Thème Latin* | *2ème prix* |
| *Version Latine* | *2ème prix* |
| *Exercices Grecs* | *1er prix* |
| *Grammaire Française* | *2ème prix* |
| *Récitation* | *2ème prix* |
| *Histoire et Géographie* | *5ème accessit* [honourable mention] |
| *Calcul* | *6ème accessit* |

The prize-giving ceremony was a very solemn occasion at the end of the school year: the headmaster opened the proceedings, after which one of the teachers would give a lecture on a chosen topic, and then the winners were presented with beautifully bound books as their prizes.

Jacques at first showed no mathematical ability whatsoever, solving arithmetical problems with great unwillingness: "...until the end of the fifth class I was last or almost last in arithmetic" [II.27, p. 52]. He undoubtedly had in mind the sixth or seventh classes, since it was in the fifth class that he won second prizes in arithmetic both in the *lycée* and the *Concours Général*. Once, when passing the *École Normale* with his father, they had the following conversation [II.27, p. 52]:

*The entrance to the École Normale*

"Is this where they learn mathematics?"

"Yes, at the *École Normale*, in the Department of Science."

"Then I won't come here."

It is interesting that Hadamard's future teacher, Émile Picard, showed a similar dislike for mathematics when he was young. In the *Lycée Henri IV* (then called *Lycée Napoléon*), where Picard obtained his secondary education, he was brilliant in translating poetry from Latin and Greek, as well as in history, but he loathed geometry which he "...learned by heart in order to avoid being punished!" [I.362, p. 114].

The person to awaken Hadamard's interest in mathematics was Launay, his new teacher in the fifth class. In Hadamard's words, Launay evoked in him the first childlike appreciation of the beauty of science [II.27, p. 52].

By that time, Jacques' father had already stopped teaching at the *Lycée Charlemagne*, where he was able to come to terms neither with the pupils nor with his colleagues. The deputy headmaster, in a report on Amédée Hadamard's behaviour to the Minister of Public Instruction [IV.24], wrote:

> He told young Gari, a pupil in his class from Chile, that "Chileans are all liars and thieves."
>
> To the young Villain, son of a *député* from Aisne: 'you will be called to order just like your father was yesterday at Versailles.'
>
> You will find the details concerning the pupils Gari and Villain and a more offensive case going back to October 1872: Mr. Hadamard told a child who had been recommended to him: "you cannot be the son of Mr. X. I knew him six years ago in Versailles, and he was not married."

The report ends with a request to transfer Amédée Hadamard to another *lycée* in Paris, and in 1875 he was assigned to the *Lycée Louis-le-Grand*. The oldest *lycée* in France, its history goes back to 1563 when the Jesuits bought a large building in the *rue Saint-Jacques* and founded a *collège* there. After their expulsion from France in 1762, the *collège* continued to grow, absorbing a number of small Parisian *collèges*. Its name changed many times, following the twists and turns in French history, finally becoming the *Lycée Louis-le-Grand* in 1873. Among its former pupils are many eminent names in French life: Baudelaire, Becquerel, Borel, Cyrano de Bergerac, Degas, Delacroix, Durkheim, Galois, Hadamard, Hermite, Hugo, Jaurès, La Fayette, Lebesgue, Molière, Painlevé, Raymond Poincaré, Poinsot, the Duc de Richelieu, Robespierre, the Marquis de Sade, and Voltaire.

*The court of the Lycée Louis-le-Grand*

It seems to have been a fortunate move for Amédée Hadamard, as the following extract of a letter written by the headmaster shows [IV.24]:

> ...Mr. Hadamard has been accredited with the reputation of being a bad teacher. This is an exaggeration. He is not a man with a superior air; he is a very acceptable teacher. He is merely of a somewhat sad disposition and often ill, he gives exercises which are perhaps a little too hard for the beginners in the sixth class, but the pupils are knowledgeable, they answer well and leave his care as well-prepared as those in other classes. He asks to be promoted to the fifth [class] to replace Mr. Béchet. I proposed him for this position and I renew my proposal in his favour.

After having worked at *Louis-le-Grand* for one year, Amédée Hadamard brought over his ten-year-old son, who was now in the fourth class. In that school year, Jacques won eight prizes at the *lycée* in almost all subjects, but his sole achievement at the *Concours Général* was *4ème accessit* in German [III.256]. At the end of the third class he was awarded the *prix d'excellence* together with six other prizes at school. However, he was only able to receive *7ème accessit* in mathematics at the *Concours Général* [III.257].

It was then that Jacques' parents took the unexpected step of making him repeat the third class. This proved to be an inspired decision: Jacques won first prize in Latin translation, second prize in Greek translation and in mathematics at the *Concours Général* [III.258]. (During all his life he felt very keenly that this second place in mathematics was a failure, as witnessed independently by S. Mandelbrojt: "It is true that Hadamard also regretted not having come first in a *Concours Général*" [III.265, p.30] and Kolmogorov: "I came out second, – he said, – but the one who was first also became a mathematician, a much weaker one – he was always weaker". And it was evident that even then Hadamard took his "defeat" at the *Concours General* painfully" [III.363, p. 170].)

During the prize-giving ceremony for 1878 his father gave the address, choosing as his subject the famous philologist and pedagogue Lhomond (1727-1794). Condemned to be guillotined during the period of the Terror, Lhomond was saved by the intervention of Tallien, a former student of his and one of the leaders of the Revolution. The lecture ends with the following words addressed to the pupils [III.107, p. 208]:

LYCÉE
DE
**LOUIS - LE - GRAND**
—
DISTRIBUTION SOLENNELLE DES PRIX
FAITE AUX ÉLÈVES DU PETIT COLLÉGE
**Le 2 août 1878**
SOUS LA PRÉSIDENCE DE M. LE PROVISEUR.
—
AN 1878
—

La séance a été ouverte à 2 heures par le discours suivant de **M**. HADAMARD, professeur de septième :

CHERS ÉLÈVES,

Voulez-vous me permettre d'introduire dans cette fête de nos études un personnage bien connu de vous ? C'est un hôte familier de la maison, du moins de cette partie de la maison que nous habitons ensemble. Il a été le guide assidu de nos laborieux exercices pen-

*The first page of Amédée Hadamard's lecture at the prize-giving ceremony at the Lycée Louis-le-Grand, August 2, 1878*

These first years of study, to which we have opened the way for you, will be followed by much higher studies. Success in these latter can only be assured by hard work from the start. Should your efforts match ours, then we shall be allowed to hope that within different walls and in a vaster arena, your names, proclaimed with honour, will be written into the annals of our competitions and continue the ancient fame of the *Lycée Louis-le-Grand*.

Jacques continued his successes at the *Concours Généraux*, obtaining first prize in mathematics in the second class and second prize in geometry and cosmography when in *rhétorique* [III.259]. His teachers heaped praise on him and predicted brilliant futures as a historian, geographer or linguist, according to the subject the teachers taught.

After *rhétorique* Jacques entered *philosophie* and in 1882 became *bachelier ès lettres*. He spent the academic year 1882-1883 in *math élém* and took his *baccalauréat ès sciences*, with outstanding results: first prize in algebra and mechanics at the *Concours Général*, the *prix d'excellence* together with first prizes in algebra, geometry and mechanics, physics and chemistry, in the *lycée* [III.259]. A possible explanation for Hadamard taking the double baccalaureate is that it gave an advantage to those applying for admission to the *École Polytechnique* (twenty-five extra marks were added to the results of the entrance examinations).

His solutions to two geometrical problems set at the *Concours Généraux* were published in the *Journal de Mathématiques Élémentaires* [I.1], [I.2]. The following question from the *Concours Général* for the class of *philosophie* in 1881 can give an idea of the level of difficulty:

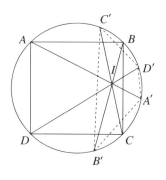

*Given an inscribed square ABCD and a point I in the plane, I is joined to the four vertices. The straight lines obtained cut the circle in four new points A',B',C', D'.*

*1. Show that*

$$A'B' \cdot C'D' = A'D' \cdot B'C'.$$

*2. Given the quadrilateral A'B'C'D' satisfying this condition, place the point I in such a way as to recover the square ABCD.*

The subtlety here consists in the fact that the point *I* does not necessarily lie inside the circle as depicted in the above figure. Hadamard first gives a simple, naive solution, explains its inadequacy and then proceeds to present a detailed analysis of the general case, taking up six pages of the journal. This seems to be his first mathematical publication.

Eighteen eighty-three was the end of Hadamard's obligatory secondary education. A student intending to enter a *Grande École* and interested in natural sciences could spend two years in the classes *mathématiques supérieures* and *mathématiques spéciales* (*math spé*), which were called *hypotaupe* and *taupe* (mole) in student jargon. Hadamard decided to take only *math spé.* and became a *taupin*. He received first *accessit* in mathematics at the *Concours Général*, and in the *lycée* he was awarded the first *prix d'excellence* and the *prix d'honneur* in mathematics together with first prizes in physics and German [III.259]. Moreover, two difficult geometric problems which he solved during *math spé* became the subject of his mathematical papers: *Sur l'hypocycloïde à trois rebroussements*, which appeared in the *Journal de Mathématiques Spéciales* [I.4], [I.6]. In this and similar journals, several problems proposed by Hadamard for school students were published.

*Hadamard's awards at the Lycée Louis-le-Grand and at the Concours Généraux*

Here is, for example, his problem which appeared in *Journal de Mathématiques Élémentaires*, Sér. 2, 1885, p. 22-23:

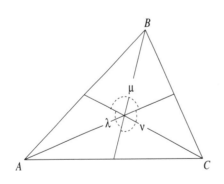

Let $A, B, C$ be the angles in a triangle, then the corresponding angles $\lambda, \mu, \nu$ between the medians are given by formulas

$$3 \cot \lambda = \cot A - 2 \cot B - 2 \cot C$$

$$3 \cot \mu = \cot B - 2 \cot C - 2 \cot A$$

$$3 \cot \nu = \cot C - 2 \cot A - 2 \cot B$$

A medal which Hadamard won at the *lycée* was returned to him after the Second World War by a French army general who was a former student of the *École Polytechnique*. The Hadamards' apartment was plundered by the Germans in 1942 when the family was in exile, and the medal was lost together with other belongings. The general came across it in a Paris antique shop and returned it to its rightful owner, his teacher at the *École Polytechnique.* However, the family's papers were never recovered, and with them disappeared a great deal of valuable information about Hadamard. It is a loss compounded by the dearth of material which Hadamard kept about himself.

One sometimes says that "the child is the father of the man", and Hadamard was no exception in this: what made him great can be traced back to the years in the *lycée.* This pertains not only to his phenomenal mathematical activity and broad humanitarian culture, but also to his general political leanings. While still at the *lycée* he wrote an essay entitled *La Patrie* which gave him the reputation of being a revolutionary [IV.25]. His passion for botany also stems from his youth, as does the extraordinary absent-mindedness for which he became renowned. In-

deed, on one occasion, in 1882, when the Hadamards spent their summer vacation in the Alps, Jacques went to the *glacier des Bossons* together with his seven-year old sister Germaine, to collect specimens for his herbarium. Leaving her at the edge of the glacier, he went off to collect plants. When he returned home, his mother asked him about Germaine. He confessed that he had forgotten her at the glacier and hurried back to find her [IV.1, p. I(7)].

## 1.3 The *taupin* becomes a *gnouf*

In 1884 the eighteen-year old Hadamard took the entrance examinations for both the *École Polytechnique* and the Department of Science of the *École Normale Supérieure.* This was a common practice: Gaston Darboux, Henri Poincaré and Émile Borel all did the same at their time. Those accepted to these *Grandes Écoles* became civil servants, unlike university students, and they received a small salary. They were then obliged to work for the state during a period of ten years which included the years of study.

The *École Polytechnique* was founded in September 1794 in order to train military engineers and army officers. Its symbol is the letter X, which is said to be a simplified version of two crossed gun barrels. In Hadamard's time, and much later, the *École* was situated on the *Mon-*

*tagne Sainte-Geneviève* in the centre of Paris, but it is today located in large grounds at Palaiseau, about 20 kilometers South-West of the city.

Soon after its founding, the *École Polytechnique* became a model for establishments of higher education because of its brilliant professors, rigid discipline, demanding examinations and excellent textbooks. For a long time it was the most prestigious educational institute in France and had the reputation of being the best mathematical centre in the world. Ampère, Cauchy, Fourier, Lagrange, Laplace, Legendre, Monge, Poisson, and Poncelet, all taught there. However, in the second half of the nineteenth century, the level of the mathematical education fell somewhat and the prestige of the *École Normale Supérieure* grew.

*The old building of the École Polytechnique*

The *École Normale Supérieure* was founded one month after the *École Polytechnique* and was originally called simply the *École Normale* with the aim of training teachers. Its first courses began on January 19, 1795, in the *Muséum d'Histoire Naturelle.* In spite of having such illustrious professors of mathematics as Lagrange, Laplace and Monge, the quality of education was criticized and within half a year it was closed down. Napoléon reopened it in 1810 in the building of the *Collège du Plessis.* Closing again in 1822, it reopened a second time in 1826 at a new address and with a new name, the *École Préparatoire.* Finally, in 1847, it moved to its present place in the *rue d'Ulm,* having by then changed its name to *École Normale Supérieure.* The name of the street gradually became inextricably associated with the *École,* and the abbreviation *ÉNS Ulm* is universally recognized.

Hadamard's request for admission to the entrance examinations for the *École Normale*

These two *Écoles* were extremely difficult to enter. In the 1880s, for instance, about one thousand candidates applied for forty-five places in the *École Normale Supérieure*, roughly twenty places for the science section and the remainder for the humanities section. The examinations took place in the candidates' own colleges and *lycées* on the same two days all over France: first there were six hours of written examinations in mathematics, physics and philosophy, and on the second day the candidates had four hours of Latin translation. It was on the basis of these examinations that the candidates were then admitted to the final oral and written examinations, which were held at the *École Normale Supérieure*. The oral consisted of questions from the *math spé* programme and lasted at least one hour. The candidates also had to make a drawing in descriptive geometry. A list of those admitted, ranked according to their marks, was then published.

ÉCOLE POLYTECHNIQUE.

CONCOURS D'ADMISSION EN 1884. *Ville d'examen.*

*Paris*

Nº du Bulletin d'admissibilité: *203*

M. *Hadamard* Jacques, Salomon.

| Désignation des Épreuves | | | M. | m. | P | N | n. | p. | S. | Observations |
|---|---|---|---|---|---|---|---|---|---|---|
| Examens oraux. | Mathématiques | (T.) | 20 | 29 | 580 | | | | | |
| | | (L.) | 20 | 29 | 580 | | | | | |
| | Physique et Chimie | (L.) | 16 | 15 | 240 | | | | | |
| | | | | | 1400 | | | | 1400 | |
| Compositions écrites | Composition mathématique | | | | | 19 | 4 | 76 | | |
| | Composition de Géométrie descriptive | | | | | 17 | 4 | 68 | | |
| | Résolution de triangle. Calcul logarithmique | | | | | 12 | 1 | 12 | | |
| | Composition française | | | | | 13 | 7 | 91 | | |
| | Dessin | | | | | 14 | 5 | 70 | | |
| | Lavis | | | | | 12 | 1 | 12 | | |
| | Allemand (composition et examen) | | | | | 18 | 5 | 90 | | |
| | | | | | | | | 419 | 419 | |
| Immunité de 15 points attribuée aux Bacheliers ès-lettres. | | | | | | | | | 15 | |
| | | | | | | | | | 1834 | |

Hadamard's results for the entrance examinations for the École Polytechnique

In 1884 Hadamard was placed first in both *Écoles*, as Darboux in 1859 and Picard in 1874 had been before him and Borel was to be after him in 1889. In the examination for the *École Polytechnique* he beat all previous records by scoring 1834 marks out of a total 2000 possible. "You can imagine the prestige among the *taupins* which Hadamard's name enjoyed for a long time. When I was one of them in 1902, Hadamard was a fabulous being for me and my companions," wrote Denjoy [II.8, p. 33].[3]

Later on, there were many stories told about Hadamard, but they are sometimes not to be trusted. Here is an amusing one about the entrance examinations, recounted by Jacqueline Hadamard : "As for the *École Normale*, there was keen competition between him and Émile Borel (who later became his colleague) for first place. Émile Borel's sister (Germaine Duclaux) told my cousin that her brother had returned home delighted after the examination: 'Hadamard has a terrible migraine; I have a chance to be first!' Needless to say, the migraine was not enough" [IV.1, p. I(7)]. Unfortunately this story does not fit the facts, since Hadamard left the *École Normale* in 1888 and Borel entered in 1889, also achieving first place in both *Écoles*. There are other stories, but they are more difficult to confirm or refute.

We do not know whether Jacques Hadamard was unsure about which of the two *Écoles* to choose, but he wrote the following about Émile Picard who had been in a similar position ten years earlier (this is taken from Hadamard's obituary of Picard, published two years after Picard's death): "As any young Frenchman of our time who was gifted in science, he was obliged to choose between the *École Polytechnique* which, in principle, prepared one to be an engineer, and the *École Normale*, with its pure scientific orientation. He was ranked first and chose the latter. It is said that he made this decision after an exciting visit to Pasteur, during which the father of bacteriology spoke about pure science in such lofty terms that the young man was completely persuaded" [I.362, p. 114]. Here one should also mention that Pasteur could hardly be objective as he was a former student of the *École Normale*, graduating in 1843, later becoming a professor there, and served as its *directeur des études scientifiques* during ten years, from 1857 to 1867.

Hadamard chose the *École Normale* just like Picard, and moved to the *rue d'Ulm*. He had become a *normalien*, officially, at least, but not as far as student tradition would have it. In the jargon of the *École Normale*, he was merely a *gnouf*, a term originating from the word *pignouf*, meaning peasant

---

[3]A. Denjoy (1884-1973) became a member of the *Académie des Sciences* in 1942, professor of the *Sorbonne* in 1922-1955. His principal works are in the theory of functions of a real variable (there exists a Denjoy integral), in the theory of differential equations, and measure theory. He worked also in the complex function theory and topology.

or boor.  Before becoming a *conscrit* (conscript), *gnoufs* had to undergo
the *canular,* a series of initiation ceremonies organized by the *carrés* and
*cubes* (second and third year students).  Short-lived by their very nature,
the *gnoufs* were treated as transient beings: "Indefinite, spectral forms, like
dreams of the night, the *gnoufs* pass in pale processions, trailing their vain
worries" (from the *Ballade des gnoufs* by Gaston Rageot [III.311, p. 112]).

*"Your entry to the school was preceded
by your reputation as an exceptionally
talented student, a celebrity,"* recalled
Hadamard's classmate Ernest Vessiot.
*"In you, our class found an excellent
comrade who never lost his amiability
nor his simplicity"* [II.27].

E.P.J. Vessiot (1865-1952) was *directeur des études* (1920-1927) and *directeur* (1927-1935) of the *ÉNS*, a member of the *Académie des Sciences* from
1943. He worked on various problems
of group theory, differential equations,
mechanics and general relativity.

The *canular* varied from year to year, including, for instance, the public
mocking of the new arrivals: the *gnoufs* would be ordered to get up, one
by one, onto a large stove in a lecture room, and then subjected to booing,
hissing and jokes about their names, appearance, habits and idiosyncracies
(real and imagined).  The whole event would end with a wild dance, the
*sarabande des conscrits.* Romain Rolland, who was a *gnouf* two years after
Hadamard, recalls the initiation rites in his diary:

> Having gone to bed, we had our mattresses turned, our noses
> against the springs. The next day, a rag procession organized by
> the *carrés.* Our mahouts lead us, from one end of the *École* to the
> other, visiting the dirty places, making us kneel in front of the
> skeleton of the fossil elephant (the *Méga*) and respectfully kiss
> the end of its tail, snaking around in the courtyards, round the
> fountain, on the well-top, each of us going on all fours under the
> legs of the other twenty-three. The *cubes* make us cut [the pages

of] their books or copy their notes. The *carrés* give us a series of vile tasks and make us pass an examination "in morality" which is the most disgusting thing one can imagine. I was lucky to be forgotten in the roll-call of the *gnoufs* who had to go in front of the learned assembly... the stories my fellow students tell me about the cross-examination make me appreciate my good fortune [III.311, p. 113].

*A visit to Méga, 1895*

The *Méga* mentioned here is the megatherium, bequeathed to the *École* by the famous zoologist and paleontologist George Cuvier (1769-1832), and which stood in the library. The *canular* culminated with the ceremony of admission to the ranks of the *conscrits,* named after the *Méga.* It ended with the incantation "There have never been any *gnoufs*, there has never been a *canular;* there are only *cubes, carrés* and *conscrits*!"

## 1.4   Pupils

It is difficult to imagine any other institution in France producing such intellectual ferment in cultural life. The young men who lived at the *École* became philosophers, priests, politicians, historians, writers, philologists, chemists, physicists, and mathematicians. As Romain Rolland stated: "... the house in the *rue d'Ulm* had the jealous pride of being sufficient in itself. It was like a noble intellectual cloister.... We were about a hundred and thirty or forty young intellectuals – letters and sciences – enjoying exceptional privileges... Three years of austere and intoxicating mental games, breaking free from tutelage and launching out to discovery..." [III.311, p. 145].

Indeed, with the *École Polytechnique*, the *École Normale Supérieure* was a cradle of the French scientific and intellectual elite, a place where the country's brightest and most gifted youth could spend three highly formative years studying, debating and socialising together in a closely knit community of scholars. One who deserves special mention is the physicist Pierre Duhem, a *cube* at the time Hadamard arrived at the *École*. Many years later, Hadamard recalled how Duhem "had instantly adopted the young *conscrit* who had just disembarked at the *École*" [II.27, p. 53]. In Vessiot's words Duhem was Hadamard's friend "more than anyone else" [II.27, p. 25], in spite of Duhem being almost five years Hadamard's senior. This friendship was later to play a crucial role in forming Hadamard's scientific interests. In his *L'œuvre de Duhem sous son aspect mathématique* [I.263, p.638]. Hadamard gratefully acknowledges his debt:

> In these long and precious conversations during which, from the moment of my entry to the *École,* our friendship grew, how I felt him being thrilled by the genius of Hermite, or that of Poincaré, whose works he followed better than most of us could (I mean the most specialized in mathematics)! But in a general way, all the great mathematical ideas, all the ones which were truly fruitful, were familiar to him. From this time I owe him revelations, insights (how broad, how disdainful of details to the profit of that which was really essential!) which for me, effortlessly and as if unconsciously, replaced long months of study.

A year earlier than Hadamard, another brilliant student, Paul Painlevé, had entered the *École*. He was also a former pupil of *Louis-le-Grand* and became Hadamard's lifelong friend. Painlevé had an outstanding career as a mathematician and an equally eminent political career, serving several times as a minister, and even as prime minister in 1917 and again in 1925.

One can gain an impression of how he was as a young man by reading the answers he gave to a questionnaire when still a student in *philosophie:*

*Paul Painlevé (1863-1933)*

The quality I want from a man? *Mine.* My main fault: *Is it pleasing?* My preferred occupation? *Discussion.* My dream of contentment? *To be famous.* What do I want to be? *Everything.* The country I wish to live in? *In the one where I would have genius.* My favourite authors? *Tacitus, Pascal, Taine.* My favourite poets? *Shakespeare, Heine, Musset.* My favourite composers? *Berlioz, and again Berlioz.* My heroes in fiction? *Achilles, Faust, Hamlet.* My favourite names? *The famous names.* The talents I would like to have? *The ones which*

*will come.* How I would like to die? *For truth.* The present state
of my mind? *Ambition.* My motto? *Parcere subjectis, debellare
superbos* [III.77, p. 205-206].[4]

Among those doing the mocking at the *canular* in 1884 were the *carrés*
Joseph Bédier, Paul Janet, and Eugène Cosserat. Bédier (1864-1938) became a distinguished philologist and is especially known for a beautiful ren-
dering of the medieval story *Tristan et Iseult* (1900) as well as the author
of the *Légendes Épiques* (1908-1913). Janet became a physicist and, in
1894, the first director of the *École Supérieure d'Électricité*. Cosserat (1866-
1931) was to make fundamental contributions to projective geometry and
continuum mechanics. Vessiot, Painlevé, Bédier, Janet, Cosserat, Hadamard and ten other students who entered the *École* during 1883-1885, were
to become members of the prestigious *Institut de France* which consists of
five *Académies: Académie Française, Académie des Inscriptions et Belles-
Lettres, Académie des Beaux-Arts, Académie des Sciences* and *Académie
des Sciences Morales et Politiques.* Few years after Hadamard the future
mathematicians É. Cartan (1888), Borel (1889), Baire (1892), Lebesgue and
Montel (1894) appeared within the precincts of the *École.*

## 1.5   Teachers

By the time Hadamard entered the *École Normale* the most influential
French mathematicians of the older generation were Bertrand, Bonnet, Hermite, Jordan, and M. Lévy. They were members of the *Académie des Sciences,* taught at the *Collège de France, Faculté des Sciences de Paris* and
at the *Grandes Écoles.* In that they were followed by the younger stars
Darboux, Poincaré, Appell, Picard, and Goursat.[5]

A special role in the education of the new generation of French mathematicians was played by Jules Tannery (1842-1910) who was the *Directeur des Études Scientifiques* at the *École Normale* during the years
1884-1910. He was a man of rare charm and erudition, as well as an excellent teacher. Tannery's lectures were the basis of his book *Introduction à la théorie des fonctions d'une variable* (1886). Later, this book was
extended, updated and enhanced by set-theoretical ideas, and was published in two volumes (1904-1910). Tannery was interested in the phi-

---

[4]To spare the vanquished and put down the proud.

[5]On the placing of actors in the Paris mathematical scene of that time see the paper
by M. Zerner [III.428].

losophy of science and in the methodology of mathematical education.   He
published Galois' manuscripts and the
correspondence between Liouville and
Dirichlet.    In the *Dictionary of scientific biography* P. Speziali writes: "Tannery possessed considerable gifts as a
writer.   The pure and elegant style of
the poems he composed in his free hours
clearly bears the stamp of a classic sen-
sibility.    His vast culture, nobility of
character, and innate sense of a ratio-
nally grounded morality are reflected in
each of his *Pensées*, a collection of his
thoughts on friendship, the arts, and
beauty.   Often they exhibit a very re-
fined sense of humor" [III.373, p. 250].

On the day of the celebration of his
seventieth birthday by the French scientific community, Hadamard, in replying
to the addresses, spoke of Tannery in
the most poignant terms, saying:

*Jules Tannery (1848-1910)*

> The young people of today are unable to imagine what the
> luminous figure of Jules Tannery was for our generation. They
> cannot imagine this because there is no personal scientific pro-
> duction remaining after him. But for us, he was the scientific,
> intellectual and moral guide. As for myself, I shall never forget
> the meeting when, from the very first words exchanged, I dis-
> covered the serene and, at the same time humane, superiority
> of the man I was to admire and cherish all my life. Everything
> that each and every one of us has been able to do is in some way
> his work, because he has left something of himself and his spirit
> in the personality and spirit of each one of us. You know this
> well, Vessiot, you who have, like me, kept his memory, and re-
> served him first place just now, when you recalled our wonderful
> memories of the *École* [II.27, p. 52].

At the same time, Hadamard expressed his deep regard for his other
teachers. While mentioning that a significant part of life of the students of
the *École Normale* was spent at the *Sorbonne,* he recalled the lectures of
Hermite, which he attended there:

I do not think that those who never listened to him can realize how magnificent Hermite's teaching was, overflowing with
enthusiasm for science, which seemed to come to life in his voice
and whose beauty he never failed to communicate to us, since
he felt it so much himself to the very depth of his being [II.27,
p. 53].

*The Sorbonne*

After the death of Cauchy, Hermite was considered to be the leading
mathematician in France although at the age of twenty he had been placed
only sixty-eighth at the entrance examinations to the *École Polytechnique*.
He worked there in 1848-1876: first, as a *répétiteur* and admissions examiner,
in 1862 he became a *maître de conférence* and in 1869 was appointed to a
professorship in Analysis. The same year 1869 Hermite was appointed to
the Chair at the *Sorbonne* which he held until retirement in 1897. In 1856
he was elected to the *Académie des Sciences*. Hermite's principal works are
in the theory of elliptic functions[6] and applications, algebra, analysis and

---

[6]Elliptic functions are doubly periodic analytic functions of a complex variable.

number theory. He was a legendary figure due to his solution of fifth-order algebraic equations by means of elliptic functions (1858) and the proof of the transcendentality[7] of the number $e$ (1873). Engineers and physicists know Hermite for his polynomials and quadratic forms. Among his numerous pupils were such brilliant names as Darboux, Picard, Appell, and Poincaré. Hermite's prestige was enormous, which his pupils, friends and even family ties also contributed to.[8]

*Charles Hermite (1822-1901)*

Under Cauchy's influence Hermite became a devoted Catholic after his illness with smallpox in 1856. His mathematical philosophy was a kind of Platonic idealism. He believed that mathematicians by no means invent anything but they are sometimes allowed to discover the harmony of the mathematical world which exists independently of the human mind.

In 1924, Hadamard recounted his meetings with Hermite:

---

[7] A real number is called transcendental if it is not a root of a polynomial with integer coefficients.

[8] He was married to Louise, the sister of Joseph Bertrand, his younger daughter became the wife of Émile Picard. Picard's classmate and friend Paul Appell also became Hermite's relative through his in-laws.

When I was a young student, a happy circumstance enabled
me to visit the master regularly for a few minutes. At the time,
he was making a deep impression on us, not only with his methods and those of Weierstrass, but also with his enthusiasm and
love of science; in our brief but fruitful conversations, Hermite
loved to direct to me remarks such as: "He who strays from the
paths traced by Providence crashes." These were the words of
a profoundly religious man, but an atheist like me understood
them very well, especially when he added at other times: "In
mathematics, our role is more that of servant than master." It
goes without saying that gradually, as years and my scientific
work unfolded, I came to understand more and more deeply the
aptness and scope of his words [I.239, p. 66].

*Gaston Darboux (1842-1917)*

In 1880, Darboux was appointed to the Chair of Geometry at the *Sor-*
*bonne* which he held until his death. His principal works were devoted to
differential geometry, but he also contributed to differential equations (both
ordinary and partial), analytical mechanics, and function theory. The four
volumes of his fundamental treatise on surfaces, the *Leçons sur la théorie*

*générale des surfaces et les applications géométriques du calcul infinitésimal* (1888, 1894, 1896) are famous, as well as his *Leçons sur les systèmes orthogonaux et les coordonnées curvilignes* (1898, 1910) and *Les principes de géométrie analytique* (1917). Darboux stressed the great importance of the history of science and he wrote the biographies of some French scientists. "It was also a singularly good fortune in these lectures, in which Darboux, with his voice, whose musical gentleness has always remained in my mind, introduced us to infinitesimal geometry, of which he was one of the founders," Hadamard recalled at his jubilee in 1936 [II.27, p. 63].

*Émile Picard (1856-1941)*

After paying tribute to Hermite and Darboux, Hadamard continued: "I have mentioned two of the great masters to whom I owe my scientific education. Another is sitting in front of me." This was Émile Picard who in 1881-1886 served as *maître de conférence* in mechanics and astronomy at the *École Normale*. In 1885 Picard became his own *suppléant* to the chair of differential calculus at the *Sorbonne*. He was appointed to the chair when it fell vacant after the death of Bouquet,[9] but was not allowed to become

---

[9]At the very beginning, in 1884, Hadamard's class attended the last lectures of Bouquet, who held a professorship at the *Sorbonne* from 1870. A former student of the *École Normale* and an outstanding pupil of Cauchy, Bouquet was a well-known specialist in

professor until he had reached the age of thirty (the minimum age for a professorship).

Picard's principal works concern function theory, differential equations, and algebraic geometry. Every student of mathematics studies his two theorems in complex function theory and his theorem on the unique solvability of the Cauchy problem for non-linear ordinary differential equations. In order to construct solutions of these equations (and for other purposes) he applied the method of successive approximations, the first of numerous iteration procedures used nowadays in numerical analysis.

Congratulating Hadamard at his jubilee, Picard remarked that Hadamard probably did not remember his lessons "on modest problems in rational mechanics." Hadamard replied:

> In addressing myself to Monsieur Picard, I must first of all protest. No, I have never forgotten the lessons you gave during the second year of the *École,* and I am even able to add to your recollections. It is perfectly true that you had assumed the task (should I say burden?) of training us in this artificial and lamentably monotonous exercise, which Mechanics is in the degree programme. You were able to make it almost interesting; I have always wondered how you went about this, because I was never able to do it when it was my turn. But you also escaped, you introduced us not only to hydrodynamics and turbulence, but to many other theories of mathematical physics and even of infinitesimal geometry: all this in lectures, the most masterly I heard, in my opinion, where there was not one word too many nor one word too little, and where the essence of the problem and the means used to overcome it appeared crystal clear, with all secondary details being treated thoroughly and at the same time consigned to their proper place.
>
> All mathematicians know, on the other hand, what a marvellous stimulus for research your mysterious and disconcerting theorem on entire functions was, and still is, because the subject has lost nothing of its topicality. I can say that I owe to it a great part of the inspiration of my first years of work [II.27, p. 54].

Much later, in Picard's obituary, he wrote: "A striking feature of Picard's scientific personality was the perfection of his teaching, one of the most marvellous, if not the most marvellous that I have known" [I.362, p. 128].

---

mathematical analysis and mechanics. According to Vessiot, Bouquet was "a rigorous apostle of precision, but whose courses opened few new horizons. M. Émile Picard, a young and already famous maître, succeeded him" [II.27, p. 25].

At his jubilee, Hadamard also mentioned the lectures of Paul Appell "of a clarity which has remained proverbial" and those of Édouard Goursat "so perfect even in their simplicity" [II.27, p. 54].

Appell acceded to the chair of mechanics of the *Sorbonne* in 1885. He had already made remarkable discoveries in analysis and geometry, in particular on elliptic and hypergeometric functions with two or more variables. He introduced polynomials, which were named after him, and obtained the most general ordinary differential equations of the motion of mechanical systems. In the years 1893-1896 Appell published the five volumes of his fundamental *Traité de mécanique rationnelle*.

*Paul Appell (1855-1930)*        *Édouard Goursat (1858-1936)*

Another outstanding analyst, Goursat was appointed as *maître de conférence* at the *ÉNS* in 1885. Among his results was a proof of the Cauchy integral theorem without the a priori assumption of continuity of the first derivative: the integral of a function along a closed contour is zero if the function is analytic inside and on the contour. For second-order hyperbolic equations, he stated and solved a boundary value problem with data on the characteristic curve, which is called the Goursat problem. Goursat was a brilliant teacher, and his *Cours d'analyse* became a classic textbook.

Starting from 1881, Henri Poincaré also taught at the *Sorbonne*. Before the age of thirty, that is before the time when Hadamard became a *normalien*, Poincaré had already developed an asymptotic theory of ordinary differential equations with a small parameter, studied curves determined by differential equations, discovered the so-called automorphic functions of a complex variable, which generalize the elliptic functions and also made contributions to number theory, algebra, and mathematical astronomy.

*Henri Poincaré (1854-1912)*

Hadamard wrote about Poincaré's early mathematical work that one of its most remarkable features was "the number and importance of the results which appeared almost simultaneously in the short period between the years 1879 and 1884-1885, changing entirely the state of mathematical science and opening new paths for it in all its directions. So rapid was the outburst of his Memoirs devoted to the most varied branches of mathematics that volume XI of the *Bulletin de la Société Mathématique de France*, for instance, contains three which are almost immediately consecutive (two of them having been presented at the same fortnightly meeting of the Society) and each of them constituting an entirely new chapter in the Theory of Functions" [I.220, p. 111].

However, as Hadamard stated at his jubilee in 1936, Poincaré did not have the immediate and direct influence on his students that one might expect. His works were admired, but no one "dared touch them" [II.27, p. 54].

## 1.6 The *École* at the *rue d'Ulm*

What kind of institution was the *École Normale Supérieure*, with its exclusiveness and the talent of the students who lived under its roof? The *École* occupied a quadrangular, three-storey building, situated between *rue d'Ulm* and *rue Rataud.* In the middle of the large inner garden is a fountain whose water is caught by a small, round cement basin with red fish swimming

*A courtyard in the ÉNS Ulm with the bassin d'Ernests in the foreground*

in it. Each fish is called Ernest, in memory of Ernest Bersot, a director of the *École* (1871-1880) who had the basin built. It is part of *Normalien* "mythology" that the Ernests are divine, and the *gnoufs* were required to worship them during the *Méga.*

The eastern side of the building, which faces onto the *rue d'Ulm,* was given over to offices and apartments of the administration. The student dormitories were on the first and second floors in other parts of the building, with the students sleeping in *piaules,* narrow, single cells furnished with just an iron bed and a wardrobe. A jug, a bucket and a basin were provided for washing. Instead of a door there was a curtain to facilitate surveillance. A. Weil, who entered the *École Normale* almost forty years later, wrote in his book *The apprenticeship of a mathematician*: "They slept in large dormitory rooms, their beds separated from one another by thin screens which afforded the merest semblance of privacy: most horses are better off in their stables" [III.418, p. 45].

The students were grouped in *thurnes* (from the Alsatian *türn*, meaning prison), or studies. It was much more comfortable in a *thurne* than in a *piaule* because it had a stove as well as tables and chairs. The walls were decorated with pictures and prints, according to the tastes of the inhabitants.

*A thurne in 1895*

During Pasteur's directorship the regulations were somewhat reminiscent of a military unit. Reveille was at five o'clock in the morning, except for Mondays, Wednesdays and some special days, when one could get up at six o'clock. Leave of absence from the *École* was strictly controlled, and students were not allowed to leave the grounds without their uniform. Smoking and playing cards were forbidden, as were reading newspapers, illustrated magazines and novels, all being considered as distractions from studies. Making a fetish of work and order, Pasteur was also obsessed with time-wasting: "The students of the first and second years take at least twenty minutes to dress properly and to make the journey from the *École* to the *Faculté* [the *Sorbonne*]. The same to return. That makes at least forty minutes lost outside for each lesson." Since students of the science section had at least five lectures at the *Sorbonne* each week, it made "a loss of time of five times forty minutes, or three and a half hours each week. It is in reality more than that. I have taken the minimum" [III.195, p. 83].

However, the regulations became less rigid with time, and when Hadamard entered the *École* early rising at five o'clock and wearing uniforms were not required.  Moreover, students were allowed to read any newspapers, journals, or books they wished.

Hadamard's day was usually as follows: the bell would ring in the dormitories at six o'clock and twenty minutes later, at the second ringing, everyone rushed out to be at his *thurne* by six-thirty.  Here, they would work until the next bell called them to the *pot*, the large student dining-hall in the north side of the building "somber and cold with no furniture other than benches and long tables where the cutlery, the carafes of water, and the 'quarters' of red wine were placed on the marble" [III.195, p. 173].  The *pot* was the constant butt of jokes about "nourishment or poisoning."

After breakfast, the *maîtres de conférences* (tutors) would arrive and the *Normaliens* separated according to their sections and took part in *conférences* (lessons).  As far as mathematics was concerned, the students would solve problems and discuss theoretical material.  Lunch followed, with a long break during which one could participate in sports, play whist or billiards, walk or even doze in the garden, if the weather was fine.  When it rained, the time was spent in talking and debating in the *thurnes* and corridors.  Politics and metaphysics were constantly the subjects of heated debates.  Then there were more *conférences* in the afternoon, with three hours of individual work in the *thurnes* to finish off the day.

*The library of the ÉNS (engraving from 1895)*

A popular place in the *École* was the library, whose two main rooms, for humanities and sciences, were on the first floor of the eastern side of the building. The library had tens of thousands of books kept in double rows in high oak bookcases.

In the evenings, the students provided their own entertainment, with small parties where they would sing, read poetry and dance with each other. If plays were performed, then any women roles were played by the students (women were first admitted to the *École* in 1917). During winter, the director gave receptions on special occasions, all the University celebrities and the Minister of Public Instruction, together with their families, were invited, and the *Normaliens* became *hommes du monde* for a few hours. In addition to these amusements there were visits to the theatres, the opera, and even city balls were allowed. There were also the traditional, if somewhat less innocent, walks on the roofs of the houses in the *rue d'Ulm,* causing complaints from residents and anger in the administration.

*A walk on the roofs, 1895*

The *Normaliens* attended lecture courses outside of the *École,* usually at the *Sorbonne* or at the *Collège de France,* one of the oldest institutions of learning in Paris offering public lectures at a high academic level. The lessons at the *École* were designed to complement these courses.

The subjects studied by the *conscrits* and *carrées* in the science section included organic and inorganic chemistry, infinitesimal calculus, mineralogy, physics, rational mechanics, astronomy, zoology, geology, as well as drawing.

Experimental work, both for their studies and research, was done in the laboratories of physics, chemistry and natural sciences. There was even a small botanical garden to the east of the *École*, in the *rue Rataud*. The organic chemistry laboratory became especially famous, being founded by Pasteur, who did his celebrated work on fermentation and immunology there. Indeed, it was at the end of Hadamard's first year at the *École* that Pasteur developed his vaccine against rabies.

*The class (sciences) of 1884 at the ÉNS. Hadamard is seated second from the right. Vessiot is in the back row, second from the left.*

Twice a week for two hours, the students had military training, and the noise of shouted orders, rattling of sabres and tramping of boots could be heard within the courtyard of the *École* and in the neighbouring streets.

At the end of his first academic year, Hadamard had to take two *demi-licences ès sciences physiques et ès sciences mathématiques*. A year later, he passed the remaining part of the *licence* (degree) examinations. He devoted the third year to preparing for the *agrégation* examination, which gave the right to teach in *lycées*. After that, he was allowed to stay for one more year (1887-1888) for free research. Vessiot recalled: "The years you spent at the *École* were untroubled and filled with work, which allowed your talent to mature, and you were able to choose your way unhurriedly. With your usual active curiosity, which no knowledge or experience was able to elude, you benefitted from talking with writers as well as with scientists; with physicists as well as with mathematicians" [II.27, p. 25].

It was during these years that Hadamard began to show his creativity, obtaining his first theorems which were published in the *Comptes Rendus de l'Académie des Sciences (séances du 23 janvier 1888 et du 8 avril 1889)* and later included in his doctoral thesis. As part of these investigations he was trying to find an estimate for a determinant generated by the coefficients of a power series. It was a difficult problem, with which Hadamard wrestled for some time until, as he wrote, "On being very abruptly awakened by an external noise, a solution long searched for appeared to me at once without the slightest instant of reflection on my part – the fact remarkable enough to have struck me unforgettably". These lines are taken from his book *The Psychology of Invention in the Mathematical Field* [I.372, p. 8]. Hadamard adds that he knows from personal experience that powerful emotions may favour mathematical creation and comments: "The above mentioned finding of a solution on a sudden awakening occurred during such a period of emotion" [I.372, p. 10].

We do not know what were these feelings which he mentions, but around that time, the health of Jacques' father, never having been strong, became worse, and in the autumn of 1888, when sixty years old, he asked for three months' leave from the *lycée*. Soon after that, he applied for a pension because of his infirmity. The authorisation for this came the day after his sudden death on November 28. Hadamard's mother never remarried, living on the money she earned from her piano lessons, and died in 1926.

## 1.7   Failures at the *Lycée Buffon*

On graduating from the *École Normale* on October 30, 1888, Hadamard went on to complete his education in 1888-1889, as a *boursier* of the City of Paris and the *Collège de France*. (Formally, from November 1, 1888, till September 30, 1889, he was a teacher at the *Lycée de Caen*, free of duties.)

From June 1, 1889 until September 4, 1890, Hadamard was a temporary *suppléant* [supply teacher] at the *Lycée Saint-Louis*, and then he was appointed as a mathematics teacher at the *Lycée Buffon*. Here he worked from September 5, 1890, until September 30, 1893.

At the beginning, he had many problems with teaching as can be understood from the following report, written by the headmaster of the *lycée* to the Ministry of Education:

> Despite all his distinctions and his real value, Mr. Hadamard achieves only mediocrity in the course which has been given to him. He does not have an exact feeling of what he should teach the pupils...
>
> In the humanities class, where there are more pupils, he has the same difficulties with the discipline and he punishes indiscriminately.
>
> I have been obliged to advise Mr. Hadamard to take his teaching a little more seriously, to make more of an effort to put himself at the level of the pupils, and I warned him that, if he does not obtain better results, the question whether he should continue in a Paris *lycée* will have to be considered [IV.26].

There is another letter of the same kind, written to the Minister of Education by the vice-rector of the *Académie de Paris* (one of the administrative districts, or *académies*, into which Napoléon divided France as part of an educational reform):

> Mister Minister,
>
> The last bi-monthly report from the headmaster of the *Lycée Buffon* contains the following note concerning Mr. Hadamard:
>
> "Mr. Hadamard's classes leave more and more to be desired. No concern for the moral interests of the pupils, both young and old. No authority over them. Inadequate and arbitrary discipline. Continual complaints and requests for punishment which are easy to avoid with a little firmness and kindness. No practical preparation for the classes. Mr. Hadamard believes himself exempted from everything because of his remarkable mathematical abilities. The longer we continue, the longer we sacrifice the public good in favour of the personal convenience of this young scientist."
>
> I have invited the *Inspecteur de l'académie*, Mr. Piéron, to see Mr. Hadamard's class.
>
> I shall write to inform you [IV.26].

Were his early failures in teaching due to his lack of care and diligence, or was he unable to appreciate the level of his pupils? It seems that the latter is more probable. In fairness, it should be noted that by the end of Hadamard's time as a teacher, official reports about him became less critical. Indeed, he seemed to have developed a liking for teaching: "It is most certain that it was during those years, when he taught in the *lycée,* that his interest in teaching and in the curriculum grew; probably it was during that period that he prepared the plan for his *Leçons de géométrie élémentaire,* which he wrote and published ten years later, and which had such an influence on the students of my generation," wrote Paul Lévy [II.37, p.2].

## 1.8   Hadamard's first pupil

While teaching at the *Lycée Buffon*, Hadamard's attention was caught by a gifted young man, Maurice Fréchet, who was more interested in mathematics than his fellow students. Hadamard gave him extra lessons and encouraged in many other ways his enthusiasm for mathematics. They wrote to each other during summer vacations, and continued writing when Hadamard moved to Bordeaux and Fréchet entered the *École Normale*.

*Maurice Fréchet (1878-1973) in 1954*

Many years later, Fréchet, by then one of the masters of functional analysis, said that he would never forget the attention paid to him by Hadamard: "You played a decisive role in my becoming a mathematician" [II.27, p. 34]. On the occasion of Hadamard's ninetieth birthday, the newspaper *Nouvelles Littéraires* asked several mathematicians "What do you think about the works of Hadamard?" Again, Fréchet recalls how Hadamard noticed his abilities, talked about this with his parents, and directed and helped him

[II.18]. In the obituary notice of Jacques Hadamard, written by the eighty-five year old Fréchet, one reads the following: "I must confess that, as well as being happy at meeting my benefactor, I was also haunted by the fear of not being able to answer his questions" [II.13, p. 15].

A.E. Taylor comments on the letters from Hadamard to Fréchet which have survived:

> Fréchet saved a group of more than twenty letters written to him by Hadamard during a period (approximately 1890-1899) prior to his entry into the *École Normale Supérieure*. They are preserved in the possession of his daughter Mme. Hélène Lederer, who allowed me to examine them. A few are dated but most are not (which was characteristic of Hadamard). Some have been slightly annotated (by Fréchet or someone else, many years later, I think) with surmises of the year or other brief remarks. The letters are mostly part of the tutorial efforts Hadamard was expending on Fréchet, but there are also some expressions of personal concern and advice. The communications of Fréchet to Hadamard have not survived.
>
> In an early letter (probably of 1890) Hadamard asks Fréchet to come to see him (in the *Avenue de Wagram* in Paris) at a specified time. He writes that he is happy to know that the boy has taken a first place in physics. *'Vous savez que je tiens beaucoup à cela'*,[10] he wrote.
>
> In 1893 Hadamard had Fréchet writing out and sending him solutions to problems in geometry and algebra. One problem: If $p(x)$ is a polynomial with $p(a) = m$, $p(b) = n$, find the remainder when $p(x)$ is divided by $(x - a)(x - b)$. In one letter Hadamard asks Fréchet to prove Pascal's theorem (the one about a hexagon inscribed in an ellipse, I suppose). In a later letter one can see that Fréchet has had difficulty with this; Hadamard responds with corrections and hints and tells Fréchet he must always make certain that he has used all of the hypotheses. With one letter there is a torn sheet on which Hadamard advises Fréchet: *'travaillez l'allemand; une insuffisance en cette langue vous sera une grande gêne plus tard'*.
>
> On a postcard of 1895 Hadamard writes to Fréchet *"Bravo pour le Concours Général"*. In a letter of 1896 Hadamard assigns his young friend some algebra problems from a book by

---

[10]'You know that I value it greatly'.

Laurent, and asks if he has studied poles and polars. In other letters of about this period we find Hadamard correcting Fréchet's mistakes on determinants and on division of one polynomial by another, also telling him some things abour voltaic cells, about electrostatics, about Newtonian potentials of a system of mass particles, and about friction. In a letter, on which the annotation suggests that it was written during the 1897 Easter vacation, there is reference to a book on conics by Salmon. Hadamard writes that Fréchet should not suppose that the teaching of mathematics in England is more rigorous than in France. He says it is the other way around – that English texts are rarely rigorous. Hadamard says that to obtain a definitive form of a theory it is always necessary to go to a French or German book. Then he talks about Euclid and points out some deficiencies in rigor in what Fréchet has sent him. In this same letter Hadamard writes about the method of successive approximations, saying that the rule for extracting square roots is an application of the method. Proof about this, he writes, lead into the domain of differential calculus.

On occasion Hadamard admonished Fréchet firmly when he was displeased with what his pupil had written. Evidently Fréchet had, in one case, asserted that a certain proposition was easy to prove and then made use of the alleged results. Hadamard told him that the proposition was, in fact, false, and that he should never make such an assertion unless he was actually able to give the proof. In another case Hadamard was severe with Fréchet about a mistake he had made in a geometry problem. One can only conjecture as to the exact nature of Fréchet's fault, which elicited the following response: *'mais vous ne devez <u>en aucun cas</u> faire de déduction fausse, <u>jamais</u>* [with double underline!] *sous <u>quelque prétexte et en quelque circonstance que ce soit</u>'.*[11] (The underlining is that of Hadamard) [III.391, p. 238-239].

## 1.9  Doctoral thesis

In spite of his teaching duties at the lycée (he had twelve hours of classes per week), Hadamard finished his doctoral thesis *Essai sur l'étude des fonctions*

---

[11]'but you must <u>in no case</u> make a false deduction, <u>never</u> [with double underline!] under any pretext or in any circumstance whatsoever'.

*données par leur développment de Taylor*. It was devoted to complex function theory, i.e., the theory of functions of a complex variable $z$ that in a neighbourhood of an arbitrary point $z_0$ of a plane domain can be represented by a convergent power series

$$\sum_{n \geq 0} a_n (z - z_0)^n.$$

Such functions are called analytic functions.

*Essai sur l'étude des fonctions données par leur développement de Taylor;*

**Par M. J. HADAMARD,**

Ancien élève de l'École Normale supérieure.

———

**INTRODUCTION.**

Le développement de Taylor rend d'importants services aux mathématiciens, en raison de sa grande généralité. Lui seul, en effet, permet de représenter une fonction analytique quelconque, à certains cas singuliers près.

Depuis les travaux d'Abel et de Cauchy, on sait qu'à toute fonction régulière dans un certain cercle correspond un développement de Taylor, et réciproquement. C'est même ce développement que M. Weierstrass, et, en France, M. Méray emploient pour définir la fonction.

Un point $a$ étant donné au hasard, on pourra, en général, former une série ordonnée suivant les puissances entières et positives de $x - a$ et qui représentera notre fonction dans le voisinage du point $a$. Il pourra y avoir exception pour certaines positions particulières du point $a$. C'est à ces points particuliers que l'on donne le nom de *points singuliers*.

On peut donc dire que se donner une fonction analytique non singulière au point $x = o$, c'est se donner une suite de coefficients $a_{,,}$

*Hadamard's thesis*

The complex function theory goes back to the eighteenth century, but the foundations of this discipline have been laid by Cauchy in the first half of the nineteenth century. The principal contributors to the subsequent development of the theory in the middle of the century were Riemann and Weierstrass. In France Cauchy's studies were carried on by Liouville, Laurent, Puiseux, Hermite, Briot, Bouquet *et al.*[12] In the second half of the nineteenth century the French school was reinforced by Laguerre, Darboux, Poincaré, Picard and others. At that time analytic functions were considered as a universal and effective means of solving problems of mathematical analysis and applications. Hence complex function theory attracted general attention and the best mathematicians worked in the field.

Tremendous work has been done on particular classes of analytic functions (hypergeometric series, elliptic and elliptic modular functions, automorphic functions by Klein and Poincaré).[13] On the other hand, there were just a few publications on their general properties including Hadamard's thesis. He studied the general series

$$\sum_{n \geq 0} c_n z^n \qquad (1.1)$$

---

[12] For the history of complex function theory see [III.53], [III.54], [III.55] [III.207].

[13] A historical analysis of this work is given in J.J. Gray [III.157].

and its analytic extension. (To get an idea of the topic, take the geometric series $1+z+z^2+\dots$ which converges for $|z| < 1$ and which can be analytically extended to the function $(1 - z)^{-1}$ defined on the whole complex plane excluding $z = 1$.) Hadamard started with the question: given the series (1.1), determine its radius of convergence, that is the maximum value $R$ of the radius of the disc where the series converges, and he gave the answer

$$R = (\limsup |c_n|^{1/n})^{-1}. \tag{1.2}$$

In fact, this formula has been discovered much earlier by Cauchy, but Hadamard was unaware of this and he was the first who gave an accurate proof. What is more important, he further deduced a number of deep results on singularities of power series, that is on points which are obstacles for an analytic extension of the series. The simplest singularities are called poles. One says that the point $z_0$ is a pole of the function $\varphi(z)$ if there exists a non-zero limit of $(z - z_0)^n \varphi(z)$ as $z \to z_0$, where $n$ is a positive integer called the order of the pole. Meromorphic functions are those with only polar singularities, the term being introduced by Briot and Bouquet.

*Léon François Alfred Lecornu (1854-1940) was a mathematician and engineer, a member of the Académie des Sciences from 1910*

Cauchy-Hadamard's formula (1.2) immediately provides information on the existence of singularities stated in terms of the behaviour of the coefficients $c_n$ as $n \to \infty$, but it says nothing about their character and position on the circle of convergence $|z| = R$. Hence, Hadamard went further. He tried to investigate how the properties of the singularities can be deduced from those of the coefficients. His starting point was the article by Lecornu in the *Comptes Rendus* (1887). Lecornu claimed to have proved the converse of the following assertion, implicitly contained in a memoir by Darboux of 1878: if $z_0$ is a pole of order one and it is the only singularity of the series (1.1) on the circle $|z| = R$, then $c_n/c_{n+1} \to z_0$.[14] Although Lecornu's argument was incomplete and the result itself proved to be erroneous, it inspired Hadamard's research – a good example of the usefulness of mistakes in the progress of mathematics.

In his thesis Hadamard developed the first general theory of singularities. In particular, he found a necessary and sufficient condition for a function to have at most a fixed number of poles and no other singularities on the boundary of the disc of convergence stated in terms of a certain determinant with coefficients of the series as entries.

The public presentation of the thesis took place on May 18, 1892. Hermite, Picard, and Joubert were in the Committee, Picard being the *rapporteur*. He found Hadamard's thesis too abstract. In his report of January 9, 1892, he wrote on the Cauchy-Hadamard formula (1.2) for the radius of convergence: "This result is, no doubt, more theoretical than practical" and repeated the same about Hadamard's criterion of existence of $p$ poles: "One cannot expect, remaining at the level of generality which the author essentially has tried to confine himself, to obtain formulas of practical use". However, the critic of the results never goes without praise of the author: "... but the penetration which M. Hadamard shows in these very delicate questions seems remarkable to me", "...the courage with which he dared to delve into a difficult question", and at last "It seems to me that the talent deployed is, no doubt, superior to the results obtained but the fault may be imputed to the question itself rather than to the author of the memoir" [III.149, p. 352].

In Hadamard's *Notice* of 1901 [I.87, p. 3] one finds an implicit objection to Picard's opinion:

> It seemed, at that time, that this kind of research should not have been approached: that the attempts made in this way could only end, at the cost of much effort, in insipid results of an extremely complicated form. I did not hide from myself the value of these objections; but I thought (and the events have justified this attitude) that to stop there would be to ignore the importance of Taylor series from the point of view which was

---

[14]The same statement was made by G. Kőnig in 1876 (see B. Szénássy [III.386, p. 242]).

explained above: an importance so fundamental that no result obtained in this direction, however limited or complicated as it might seem, may be neglected.

Time showed that Hadamard was right, since in fact both the results and methods of his thesis laid the foundation of a new and vast area in the complex function theory.

## 1.10 The first mathematical triumph

In the same year 1892, Hadamard was awarded his first prize by the *Académie des Sciences*, surprising everybody. The story behind his unexpected win shows how even the best laid plans can go awry.

It had been known for many years that some properties of the Riemann zeta function

$$\zeta(z) = 1 + 2^{-z} + 3^{-z} + \cdots$$

are closely connected with those of the integers, in particular, that the solution of a long standing classic problem of distribution of prime numbers is a corollary of Riemann's conjecture that all the zeros of the zeta function lie on the line $\Re z = 1/2$.

At the time when Hadamard studied complex function theory at the *École Normale*, it was believed that Riemann's hypothesis had been verified by the Dutch mathematician Thomas Joannes Stieltjes. His way to mathematics was unusual. Nine years older than Hadamard, he entered the Polytechnic School in Delft in 1873. He was so fond of mathematics that he neglected obligatory courses, reading instead the works of Gauss and Jacobi. As a result, he got no degree. After leaving the School in 1877 he was appointed "an

*Thomas Stieltjes (1856-1894)*

assistant for astronomical calculations" at the Leiden Observatory and he also managed to find time for his mathematical hobby. In 1883 Stieltjes resigned from the Observatory to become a mathematician. He applied for a position at the University of Groningen but failed and then he settled down in Paris with his family. Here in 1886 he succesfully presented his doctoral thesis on divergent series $\sum_{n \geq 0} a_n x^{-n}$. Stieltjes was becoming well-known by that time, having been awarded an honorary doctorate by the University of Leiden in 1884 and elected to the Royal Academy of Sciences of Amsterdam in 1885. Stieltjes obtained deep results in different areas of analysis and, in particular, he was the founder of the theory of analytic continued fractions

$$\cfrac{1}{z - \cfrac{a_1}{z - \cfrac{a_2}{z - \cfrac{a_3}{z - \ldots}}}} \, .$$

While studying these fractions he introduced a new concept of integral which became named after him.

*The first page of Stieltjes' article*

In 1885 Stieltjes published a short article [III.379] claiming that he had proved Riemann's hypothesis. He wrote:

> Riemann has announced as very probable that all the imaginary roots are of the form $\frac{1}{2} + ai$, $a$ being real.
>
> I have managed to put this proposition beyond doubt by means of a rigorous proof. I shall show the way which led me to this result.

He gave no proof, restricting himself to a few words about the argument, but nobody doubted that the problem had been solved.

Stieltjes was one of Hermite's closest friends. During twelve years they exchanged letters, 432 of which constituted two volumes published in 1905 [III.89]. In 1886 Hermite helped him to get a position at the University of Toulouse, where Stieltjes worked untill his early death from tuberculosis in 1894.

In the letter dated December 3, 1890, Hermite, who was the President of the *Académie des Sciences*, suggested to Stieltjes that he take part in the annual contest for a *Grand Prix*:

> For many years, you have produced a great number of excellent works which have given you in a high rank in Science and have merited the highest esteem amongst all geometers. The moment has come when this esteem should emerge from the small circle of friends in Analysis and receive an official consecration which gives you in a rank corresponding of your great talent and the services you have rendered to Science [III.89, p. 112].

On December 29, 1890, on Hermite's initiative, the *Académie des Sciences* announced the subject of a *Grand Prix*: "Determination of the number of prime numbers less than a given quantity." In the explanation, the desirability of filling in the gaps in Riemann's work on the zeta function was stressed.

On March 4, 1891, in a letter to Hermite, Stieltjes acknowleged that his proof was incomplete [III.89, p. 154-155] and a year later he still had not sent his work to the jury. On November 30, 1891, Hermite wrote to him:

> I hasten to inform you that the statutory term for the closure of the competition for the prize of the *Académie des Sciences* of Sciences is the month of June next year, but that you may, after having sent a memoir, perfectly well send an addition provided

that it reaches the office of the secretary of the Institute before
the month of November. You have, therefore, some time in front
of you and, if the *Académie* were not to receive anything, or, at
least, no work of sufficient value, I believe I can assure you that
the prospect of receiving a memoir from you would decide the
Commission to put the same question for competition a second
time. [III.89, p. 188].

But events unfolded in a different way, as Hadamard recalled more than
fifty years later:

> When I presented my doctor's thesis for examination, Her-
> mite observed that it would be most useful to find applications.
> At that time, I had none available. Now, between the time my
> manuscript was handed in and the day when the thesis was de-
> fended, I became aware of an important question which had been
> proposed by the *Académie des Sciences* as a prize subject; and
> precisely the results in my thesis gave the solution of that ques-
> tion. I had been led solely by my feeling of the interest of the
> problem and it led me in the right way [I.372, p. 127-128].

The question Hadamard mentions in the above abstract concerns a certain gap in Riemann's paper that he was able to fill in. The problem was
connected with the function $\xi(z)$ closely related to $\zeta(z)$. Riemann's argument was essentially based on a certain representation of $\xi(z)$ similar to the
Euler product formula for the sine function

$$\sin z = z\Pi_{n=1}^{\infty}(1 - (\frac{z}{\pi n})^2).$$

His reason for the representation of $\xi(z)$ had nothing to do with the real
proof which was supplied by Hadamard thirty-four years later. In fact,
Hadamard's achievement was a by-product of his results on entire functions,
i.e., functions represented by a power series convergent in the whole plane.

In 1869 Weierstrass had showed that these functions are in some sense
"polynomials of infinite degree" by deriving the following factorization formula for an arbitrary entire function

$$f(z) = e^{Q(z)}z^m\Pi_{n=1}^{\infty}(1 - \frac{z}{z_n})e^{P_n(z/z_n)} \tag{1.3}$$

where $Q$ is another entire function, $P_n$ are polynomials and $z_n$ is the sequence of non-zero roots of the equation $f(z) = 0$. This result leads, in
particular, to the fundamental theorem on meromorphic functions due to

Mittag-Leffler (1877) which gives an analogue of the decomposition of a rational function into partial fractions. However, it was a long way from the Weierstrass theorem to the product formula for the entire function $\xi(z)$, which contained no exponential factors. A subsequent developement was due to Laguerre (1872) and Poincaré (1882-1883) whose work essentially influenced Hadamard's study. The method Hadamard applied in his thesis for power series with finite radius of convergence also proved to be very helpful. He found the relationship between the behaviour of the coefficients $c_n$ and the distribution of zeros of the entire function and made the Weierstrass formula more explicit for functions growing slower than $\exp(|z|^\lambda)$. Then it became an easy matter to check that exponential factors were absent in the case of the function $\xi(z)$.

Hadamard submitted his results on entire functions and the zeta functions in a large paper [I.15] to the jury of the contest of the *Académie des Sciences*. The jury received two memoirs. One of them was rejected as being too elementary and violating the formal requirement: it was signed instead of bearing a motto. Hadamard's paper had the words of Pascal as a motto: "The art of demonstrating truths already discovered and of illuminating them so that their proof be incontestable, is the only one I wish to give". Stieltjes did not submit his paper and on December 19, 1892, the *Grand Prix* was given to Hadamard. Riemann's hypothesis remains unproved until present.

# Chapter 2

# The Turn of the Century

## 2.1 Marriage

On June 28, 1892, in the city hall of the first *arrondissement* of Paris, Jacques Hadamard married Louise-Anna Trénel, his only love and lifelong companion. He knew her from childhood, her mother Cécile being a friend of Claire Hadamard. Her father was Isaac Trénel, the director of the *Seminaire Israelite de France*, and the descendant of an old Jewish family with roots in Metz, just like the Hadamards.

The Trénels had six children: Jacob, Marianne, Régine, Bella, Marc and Louise. Louise was the youngest of them, and three years younger than Jacques. Jacob, the eldest son, was a Latin teacher "very far ahead of his time since he astonished all the inspectors because Latin was spoken in his class" [IV.1, p. I(4)]. Marc Trénel, Hadamard's contemporary, later became a doctor.

Jacques would visit the happy and hospitable Trénel household

*Louise-Anna Trénel. A drawing from an old photograph, by N. Singer*

frequently. Jacqueline Hadamard records: "My mother often told me about the great croquet games played in the garden of the *École Rabbinique* and particularly those which took place in 1884, at the time when my father was taking the entrance examinations to the *École Polytechnique* and *École Normale*. The older sisters had been charged with telling him the time. He left, asking them not to touch the balls. When he returned, no one, it seems, was interested how the examinations had gone: 'Jacques, it's your turn, you are late' " [IV.1, p. I(7)].

*The Seminaire Israelite de France, 9 rue Vaquelin, Paris 5ᵉ, sometimes called École Rabbinique*

The circumstances surrounding the marriage are described by Jacqueline Hadamard:

> Of course he thought about Louise Trénel, his childhood playmate. However, my grandmother had made him believe that he wouldn't gain much by marrying just now. So he remained silent on the matter. My mother's parents, seeing that he didn't declare himself, put pressure on their daughter who, disappointed, gave in and agreed to become engaged to a young man they had presented to her.

During a *soirée* a cousin told my father: "You know that
Louise Trénel is engaged". Without listening to another word,
my father left her there, in the middle of a waltz and rushed to
the Trénels. He said that it was impossible, that he was only
waiting to declare himself when he had obtained a better position! Finally, Robert Debré's [1] father intervened, brought pressure on my grandmother Hadamard and was given the thankless
task of presenting excuses to the abandoned fiancé... and all
ended well [IV.1, p. I(9)].

On the day of the engagement, Jacques once again demonstrated his
absentmindedness. His mother had bought an engagement ring and Jacques
was to give it to his fiancée. In the evening, she asked him if Louise liked
it. "But it's still in my pocket, I forgot to give it to her," he replied [IV.1,
p. I(9)].

## 2.2   Bordeaux. Duhem and Durkheim

In 1893 Hadamard and his wife moved to Bordeaux, the wine capital of
France, with over 250,000 inhabitants and a university founded in the 15th

*The old building of the University of Bordeaux, now the Museum of Aquitania*

century. They settled in a small house, at 210 *boulevard de Talence*. Hadamard began life in Bordeaux as a *chargé de cours* [lecturer] in the *Faculté*

---

[1]Robert Debré, a famous doctor, was the son of the Chief Rabbi Simon Debré and
Marianne Trénel, the sister of Louise Trénel.

*des Sciences* at the University, where he gave a course in astronomy and mechanics. In contrast to his experiences at the *Lycée Buffon*, Hadamard's relations with students and the university administration left very little to be desired, as can be seen from the following report by the Dean of the Faculty of Sciences, written in 1895:

"Mr. Hadamard has acquired real authority with the students, especially the better ones. He is a fine and sharp mind who thinks admirably clearly. The Faculty is happy to have him and hopes to give him tenure at the end of the year" [IV.25]. On February 1, 1896, he became *Professeur d'astronomie et mécanique rationnelle.* A full professorship only two years after beginning his University career was an extremely fast promotion.

*The house in Bordeaux where Jacques and Louise Hadamard lived from 1893 to 1897 (now 52 boulevard President Franklin Roosevelt)*

It was here in Bordeaux that their first son Pierre was born on October 5, 1894. Jacqueline Hadamard wrote about that period as follows:

> My parents always told me that the three years they spent there were the happiest of their life because, as they explained, one really had the time to work and to live in the province. The relations between my parents and the Bordeaux society (the wines, of course) were a little different from those of other professors. The wine aristocracy had hardly any respect for 'bureaucrats', and one of the wives, receiving the wife of the prefect, explained while showing her to the door: "You will excuse me, won't you? We never return the visits of civil servants." But my parents, being very musical, were freely admitted into this society, to the surprise of everybody [IV.1, p. I(10)].

*The Hadamards as a young couple*

Once, at the beginning of their life in Bordeaux, the young couple was invited to a party in their honour. They were sitting with other guests. The party lasted into the late hours but nobody left. The clock struck midnight but still nobody left. Only in the small hours did it suddenly occur to Louise that as guests of honour they have to take their leave first. We were told about this episode by S. Agmon who heard it from Mme Hadamard in 1948. At that time young Agmon was taken by his teacher S. Mandelbrojt to Hadamard's home in Paris. There were some other people and Hadamard's wife, "a very vivacious and talkative woman" Agmon said. He added: "when I heard the story I thought it was a time we should leave as well".

*Jacques, Louise and Pierre in Bordeaux*

Hadamard again met up with Duhem, who lectured in physics at the university. Duhem was such a significant figure, that it is worthwhile giving more details about him. Miller [III.276, p. 225], wrote of him that:

> Duhem was that rare, not to say unique, scientist whose contributions to the philosophy of science, the historiography of science, and science itself (in thermodynamics, hydrodynamics, elasticity, and physical chemistry) were of profound importance on a fully professional level in all three disciplines. Much of the purely scientific work was forgotten until recently. His apparent versatility was animated by a singlemindedness about the nature of scientific theories that was compatible with a rigidly ultra-Catholic point of view...

*Pierre Duhem (1861-1916)*

While still a student at the *École Normale*, he presented a thesis in which he developed the concept of thermodynamic potential in chemistry and physics, and criticized the work done twenty years earlier by Marcelin Berthelot. Duhem was correct, but Berthelot enjoyed a great reputation and the thesis was rejected. Four years later, in 1888, Duhem presented another thesis on the theory of magnetism which was very mathematical. His first dissertation was published as a book *Le potentiel thermodynamique* in 1886, but it took more than ten years for his point of view to become more or less accepted. In Jaki's comprehensive monograph on Duhem one reads:

> In spite of having grown aware of Duhem's scientific triumph over him, Berthelot could not bring himself to acknowledge this to the extent of letting him obtain a chair in Paris. At stake was the renown of the theoretical interpretation which Berthelot gave to his vast and most valuable experimental researches. It was all too human of Berthelot to protect that interpretation from Duhem's devastating criticism which, if delivered from a chair in Paris, would have forced Berthelot into the open. Herein lies the clue to the slighting which affected Duhem for thirty years, from his first doctoral dissertation to his very death, that is, his whole academic career. Without a careful look at it a presentation of

Duhem's life would not appear that poignant drama which it actually was [III.194, p. 162].

Once, Duhem was offered a position as professor in the history of science at the *Collège de France*, but he refused, saying that he was a physicist and did not wish to enter Paris through the back door. In 1913, he became a (non-resident) member of the *Institut de France*. Three years later, he died of a heart attack at only fifty-five.

Duhem published about four hundred papers and twenty-two books in forty-five volumes. His work concerned thermodynamics, electromagnetic theory, hydrodynamics, elasticity theory, philosophy, and the history of science. Among his books are *L'évolution de la mécanique* (1903), *La théorie physique, son objet et sa structure* (1906), *Études sur Leonard de Vinci*, in three volumes (1906-1913), and *Le système du monde. Histoire des doctrines cosmologiques de Platon à Copernic*, in ten volumes, of which the first volume appeared in 1913 and the last in 1959, posthumously. In addition, he was an excellent painter and an expert on the theory of art. His political orientation was to the right: he was anti-republican and a pro-clerical catholic. There were great differences in Duhem's and Hadamard's political views, but this did not prevent them from being friends.

Hadamard wrote: "How much, in our conversations in Bordeaux, which continued our conversations at the *École Normale*, as if taken up at the same point we had left them, how perpetually intuitive this logician appeared to me!" [I.263, p. 639]. Their discussions on the theory of elasticity and the propagation of waves, as well as on variational principles in mechanics helped Hadamard to define the direction which dominated in his work during the twentieth century. Hadamard himself confirms this:

L'ŒUVRE DE DUHEM
DANS SON ASPECT MATHÉMATIQUE (1)

Nous ne saurions parler de l' « œuvre mathématique » de Duhem, car lui-même ne l'aurait pas accepté. Physicien il était, et physicien il entendait rester, et l'on sait que cette vocation n'avait pas attendu l'École normale pour s'affirmer : elle était chose définitive en lui, et non sans se manifester par des idées personnelles, dès les bancs du collège.

Pour nous, ses camarades de l'École normale, cette précocité n'était pas notre seul sujet d'étonnement. Le goût de la physique était rare à cette époque, où, il faut bien le dire, nous sentions autour de nous, en ce qui concerne cette science, quelque peu de stagnation. Combien merveilleux, au contraire, était chez nou , en face d'un Hermite ou d'un Poincaré, d'un Darboux, pour ne parler que des morts, et aussi dans la sereine et vivifiante intimité de Tannery, l'enthousiasme mathématique !

Cet enthousiasme, nul ne le sentit plus complètement, plus profondément que Duhem, dont l'intelligence, on le sait, était vraiment universelle, et qui, aussi bien qu'un physicien ou qu'un mathématicien, aurait fait aisément un naturaliste, — il avait à cet égard une érudition étendue et il lui aurait fallu peu de chose pour réunir

(1) Je dois à des physiciens tels que MM. Jouguet et Maurain des indications précieuses pour lesquelles je tiens tout particulièrement à les remercier ici.

*Hadamard's paper on the mathematical aspect of Duhem's work, 1927*

For my part, our meeting again at the *Faculté des Sciences* of Bordeaux gave me the good fortune of supplementing my reading with invaluable and constant exchanges of views. It is to this

reading, to these exchanges of views that I owe the greater part
of my later works, almost all of which deal with the calculus of
variations, the theory of Hugoniot, hyperbolic partial differential
equations, Huygens' principle.

Duhem himself returned to almost all these questions in the
continuation of his immense work, and most of the theories which
he had so happily and so clearly explained, suggested to him
sometimes some observations on details, sometimes some additions of fundamental importance [I.263, p. 644-645].

In Bordeaux Hadamard met another prominent former student of the
*École Normale*, Émile Durkheim, one of the creators of the French school
of sociology. In 1887 he was appointed as a lecturer at the University of
Bordeaux and started the first course in sociology in France, becoming professor in 1896, and taught there until 1902. He is known for his work on
developing proper scientific methods for the study of society, and attempts
to establish an understanding of the basis of social stability. Later, in his
book on the psychology of mathematical research, Hadamard mentions one
of the topics of their discussions, the difference in views on the origins of
morality:

*Émile Durkheim (1858-1917)*

It is this I had begun to become aware of in the conversations I had in Bordeaux with the great philosopher Émile Durkheim: he thought that morality could and should have a scientific basis. For me, such a basis could not, on its own, suffice to constitute morality, an opinion which, by the way, was met by Durkheim with a phrase something like the following: 'You'll see that he will still say stupid things' [I.263, p. 644-645].

Hadamard did not change his opinion on this question during his life, and his position coincided with that of Poincaré [I.404, Appendix II]. On many occasions, the history of the first half of the 20th century gave him rich material for his reflections on the problem of the interaction of morality and science (see, for instance, his papers [I.363], [I.389].

## 2.3 *Nulla dies sine linea*

For Hadamard, the time immediately preceeding his move to Bordeaux was the beginning of a long and surprisingly intensive period of his scientific activity. In 1893 he published three papers, in 1894 – five, in 1895 – eight, in 1896, as well as in 1897 – thirteen papers. But it is not the quantity which matters (the history of science knows of cases of great mathematicians who produced only a few articles). What is surprising is the ease with which Hadamard was able to switch from one area to another, creating big and small *chefs d'œuvre*. We mention some of them below starting with a smaller one, which nevertheless became well known.

The inequality

$$|\det(a_{ij})| \leq \prod_j (\sum_i |a_{ij}|^2)^{1/2}, \tag{2.1}$$

"the most famous of all determinantal inequalities" [III.22, p. 64], was proved in Hadamard's article [I.16] of 1893. He was especially interested in the case of equality and showed that for $n \times n$ matrices with entries $+1$ or $-1$ and $n > 2$, the equality holds only if $n$ is divisible by four. He gave examples of such matrices for $n = 12$ and $n = 20$, and also for $n$ equal to an integer power of 2. Later Hadamard's matrices proved to be useful in coding theory, statistics, communication engineering, and optics.

Hadamard did not suspect that inequality (2.1) would play a crucial role in the Fredholm theory of integral equations (1900, 1903). These are equations of the form

$$g(t) - \int_a^b K(t,\tau)g(\tau)d\tau = f(t) \tag{2.2}$$

where "the kernel" $K$ and the right-hand side $f$ are given and $g$ is unknown. Such equations appear in geometric and physical problems not to mention numerous applications to ordinary and partial differential equations. Abel and Liouville had already solved special integral equations with one variable limit of integration but a general theory for this case was first developed by Voterra by the method of successive approximations (1896). In the case of constant limits of integration, which proved to be more involved, Fredholm considered equation (2.2) as a limit of a sequence of algebraic systems of infinitely growing order. Here he needed Hadamard's inequality to estimate the determinants appearing in the solution of these systems. Actually, Fredholm had arrived at (2.1) unaware of Hadamard's work of 1893, but he never mentioned this fact in his papers.

*Ivar Fredholm (1866-1927)*

In the article [I.40] published in 1896, Hadamard stated his famous three circles theorem. Let $f$ be analytic in the annulus $r_1 \leq |z| \leq r_3$, let $r_1 < r_2 < r_3$, and let $M_1, M_2, M_3$ denote the maxima of $|f(z)|$ on the circles $|z| = r_1, r_2, r_3$. Then

$$M_2^{\log(r_3/r_1)} \leq M_1^{\log(r_3/r_2)} \ M_3^{\log(r_2/r_1)}.$$

This inequality has numerous applications, and in particular plays an important role in modern operator theory.

The year 1896 saw the appearence of Hadamard's sensational paper with the solution of the long-standing problem of the distribution of prime numbers [I.45].[2] These numbers are placed chaotically among the integers but already Legendre in 1798, analysing tables of primes, conjectured that their density is approximately $\mathrm{const}/\log x$. Also from looking at tables Gauss predicted the more accurate approximation

$$\pi(x) \approx \int_2^x \frac{dt}{\log t}, \tag{2.3}$$

---

[2]For the history of this problem see, for instance, the book [III.118] and the very readable articles [III.151], [III.359].

where $\pi(x)$ is the number of the primes not exceeding $x$. In 1850 Chebyshev rigorously proved the inequality

$$0.9219\frac{x}{\log x} < \pi(x) < 1.10555\frac{x}{\log x} \tag{2.4}$$

and also checked that the constant in Legendre's law can not differ from 1.

The next contribution was made in the eight-page article by Riemann in 1859. He tried to show that $\pi(x)$ is the sum of a series with $\int_2^x \frac{dt}{\log t}$ as a principal term. Although Riemann's arguments were incomplete, his methods, based upon the study of the function $\zeta(z)$ as a function of a complex variable, proved to be very important and strongly influenced all the subsequent events. It is in this paper that Riemann stated the above mentioned hypothesis on zeros of the function $\zeta(z)$.

Finally, more than thirty years after Riemann's paper, the mysterious Gauss' law (2.3) was justified simultaneously, independently and with different proofs by Hadamard and de la Vallée-Poussin. They showed that the zeta function has no zeros on the line $\mathrm{Re}\, z = 1$, which is, of course, much weaker than Riemann's hypothesis, but proved to be sufficient to obtain the prime number theorem. "It is this result, in what one can call, as

*Charles de la Vallée-Poussin*

*(1866-1961)*

Vessiot does, transcendental arithmetic, which made Hadamard's name well known to mathematicians of all countries, when he was only twenty-seven years old," wrote Fréchet [II.13, p. 4083].

In 1896 Hadamard received the *Prix Bordin*[3] for his work *Sur certaines propriétes des trajectoires en dynamique* [I.48]. He was interested in the topic from the first year of his teaching of mechanics in Bordeaux. The subject proposed by the *Académie des Sciences* was *Perfectionner en un*

---

[3]Bordin was a notary who, in his will, left a sum of money to the *Institut de France* so that the interest would be used for prizes. According to the conditions in the will, each of the five constituent Academies had to announce the theme of the competition every year, and the theme should serve the general good and the progress of science and art.

*point important la théorie des lignes géodesiques.* By a geodesic one means a curve on the surface, which minimizes arc length between any two of its sufficiently close points. The geodesics in the plane are straight lines, and on the sphere they are great circles.

*Jacques Hadamard in 1890s*

The subject of geodesics had by that time a rich history going back to Gauss and even earlier. As an example, we mention Gauss' theorem on the orthogonality of geodesics emanating from a point to the geodesic circle centered at this point (1828). Several chapters of Darboux's *Leçons sur la théorie générale des surfaces* contain a lot of information about geodesics.

What was the reason for the *Académie* to choose that particular topic? The answer is that the facts about geodesics obtained previously concerned only their properties "in the small", whereas their global structure remained *terra incognita.* Already Jacobi's explicit formulae for geodesics on an ellipsoid (1838) had shown that their behaviour is far from being simple.

Have "the dynamic trajectories" in the title of Hadamard's paper anything to do with the geodesics the *Académie des Sciences* had proposed to study? The answer is yes, and the explanation lies in the deep analogy between geodesics and trajectories of point masses on the surface:[4] in fact, both are governed by similar differential equations. Therefore, by studying the global behaviour of solutions of a certain class of non-linear differential equations, Hadamard could obtain results of interest both for mechanics and geometry.

His work was influenced by Poincaré's papers *Sur les courbes définies par les équations différentielles* (1885), where Poincaré created the qualitative theory of ordinary differential equations studying properties of solutions without finding them explicitly. By using geometrical methods he described all possible forms of trajectories in the case of the first order system with two unknown functions.

Because Poincaré's approach was not applicable to the second order equations Hadamard dealt with, he invented a new one based on the study of extremal values of an auxiliary function. By choosing this function conveniently one could come to conclusions about solutions of dynamic systems. As an example, the following theorem was proved in Hadamard's memoir: Two arbitrary closed geodesics on a closed surface of positive curvature (i.e., the boundary of a convex body) have common points.

However, it was the method itself and not the results (not very numerous and by no means final) that deserved the appreciation of the Committee. Poincaré, the *rapporteur*, wrote:

> The Committee has judged that the author has shown great ingenuity of mind, having put forward plenty of new ideas which, as their every appearance would suggest, will one day be prolific; only lack of time has inhibited him from taking more advantage of it. The small amount of precise results which are formulated in this paper is enough to leave no doubt in this respect [III.334, p. 1111].

In two years Hadamard confirmed these expectations in his memoir [I.62] devoted to geodesics on surfaces of negative curvature (such as the hyperboloid of one sheet, for instance). Here his method gave a complete description of all types of geodesics on such surfaces.

---

[4]For the history of the geometrization of mechanics see Lützen's paper [III.252].

*A.M. Liapunov (1857-1918)*

In order to solve this classical problem of differential geometry dealing with smooth surfaces and curves Hadamard had to use concepts of Cantor's set theory which was considered at that time largely as a tool for only treating pathological functions.

It was through his *Prix Bordin* memoir that Hadamard came into contact with Liapunov, well known for his works in the theory of stability of mechanical systems, and his studies of equilibrium flow of a liquid. Liapunov also made important contributions to potential theory, which was a topic under intensive development at that time. Hadamard wrote to Liapunov when he learned that one of his theorems, fortunately not a major one, was proved in Liapunov 's thesis *A general problem of the stability of motion* written in Russian.[5]

---

[5]Here are the translations of two letters of Hadamard to Liapunov [IV.37] kept at the Saint Petersburg section of the Archives of the Russian Academy of Sciences.

Bordeaux, December 29, 1896

Monsieur, I dealt with the instability of equilibrium position of a free material point in my memoir, which was awarded a prize by the *Académie des Sciences* last year, and then learned in Paris about the existence of your important treatise (published in 1892), in which you considered the same problem and, undoubtedly, obtained many other results on trajectories in dynamics. May I ask you to give me the exact title of your work so that we can order it in our department.

With many thanks in advance, J. Hadamard.

*Université de Paris*, August 9, 1907

Monsieur and Highly Esteemed Colleague!

I would be very glad to see your remarkable paper of 1892 on the stability of motion translated into French. I hope to have it translated. But first of all I would like to know if you agree to give your consent to this. Will you be so kind as to give me an answer on this matter. I would be most grateful and assure you of my deepest feelings of respect and devotion.

You will also give me much pleasure by indicating other memoirs of yours in East-European languages, which have not been translated, and whose translation you would consider desirable. Jacques Hadamard.

Hadamard spent the summer of 1897 in Cenon, near Bordeaux, with his wife, and he did not go to the First International Congress of Mathematicians in Zurich. An explanation can be found in his letter to Hurwitz, who was a member of the Organizing Committee:

*Adolf Hurwitz (1859-1919)*

Dear Sir,

I am very grateful for the honour you show me in asking me to speak at the Congress and I would like you to know that I would not decline unless I found it absolutely impossible to give my all.

However, I am unable to promise you anything because I am expecting the birth of a child at the beginning of August, and my decision will necessarily be influenced by my wife's state of health.

Once more, please accept, together with my assurance of the efforts I shall make to join you, my deepest regards.

J. Hadamard

I take the opportunity to tell you of the interest I take in your research, announced in an article by Mr. Franel, on the extension to more general series of the fundamental relation which holds for $\zeta(s)$ and Dirichlet series, and of my wish to see them published soon [IV.49].

*Louise with Pierre and Étienne*

A few days after July 26, when his second son Étienne was born, Hadamard sent to Hurwitz the text of his short talk *Sur certaines applications possibles de la théorie des ensembles*, where he noted the utility of the study of sets of functions with properties different from those of numbers or *n*-dimensional vectors. Hadamard's talk was read by Picard in the analysis and function theory section on August 10.[6]

---

[6]Note Bourbaki's statement: "The official recognition of set theory occurred at the First International Congress of Mathematicians (Zurich 1897) at which Hadamard and Hurwitz gave many examples of its applications in analysis" [III.58, p. 44].

## 2.4 Influence on Borel

Besides his already mentioned paper on geodesics on surfaces of negative curvature, in 1898 Hadamard published a note on integral invariants in optics and a number of other papers. Among them was the article *Théorèmes sur les séries entières*, where a theorem on the multiplication of singularities of analytic functions was proved. We recall that Hadamard had become interested in these singularities in his student years, and here he turned to them again.

Perhaps his thought was provoked by the following simple observation. Together with two geometric series

$$\sum_{n\geq 0} \alpha^{-n} z^n \quad \text{and} \quad \sum_{n\geq 0} \beta^{-n} z^n$$

consider the third one obtained by multiplication of their coefficients

$$\sum_{n\geq 0} (\alpha\beta)^{-n} z^n.$$

By summing up each of these geometric progressions one readily finds that the product of the poles $\alpha$ and $\beta$ of the first and the second series is equal to the pole $\alpha\beta$ of the third one. It may appear a pure coincidence, valid for this particular example, that multiplication of coefficients leads to the multiplication of poles. However, Hadamard was able to discover a general law here by showing in essence that the singularities of the series

$$\sum_{n\geq 0} a_n b_n z^n \tag{2.5}$$

are contained in the set $\{\alpha\beta\}$, where $\alpha, \beta$ are singularities of

$$\sum_{n\geq 0} a_n z^n \quad \text{and} \quad \sum_{n\geq 0} b_n z^n.$$

This result inspired a flow of subsequent studies of many well-known mathematicians. An immediate response to it was given by Émile Borel, who, in the same year 1898, showed how the character of singularities of the product series (2.5) depends on the singularities $\{\alpha\}$ and $\{\beta\}$.

In fact, Borel was the first and the most brilliant successor of Hadamard's in complex function theory, and their names often stand side by side in monographs and textbooks on the subject. They met when Borel, five years younger, was a student of the *École Normale* and Hadamard taught at the *Lycée Buffon*. Both returned to Paris in 1897: Hadamard from Bordeaux and Borel from Lille, where he had been *maître de conférences* at the University since 1893.

Borel had published his first two notes in 1890, when only a *conscrit*. He recalled many years later: "At the time when I had finished my studies at the *École Normale Supérieure de Paris* in 1892, the theory of analytic functions was one of the scientific domains where all of the important discoveries had been made during the preceding decades, but where much, certainly, still remained undone. I was attracted by this field, and especially by the study of the influence of singular points on the properties of functions" [III.52, p. 2096]. In 1894 Borel defended his thesis *Sur quelques points de la théorie des fonctions*. Although less than fifty pages, the thesis attracted general attention by its depth and originality, many of Borel's subsequent discoveries in function theory and measure theory being anticipated in it.

*Émile Borel (1871-1956)*

Perhaps better than anybody else, Borel knew Hadamard's early work and made spectacular use of it on several occasions. For example, already in his first paper on meromorphic functions, the note *Sur une application d'un théorème de M. Hadamard* (1894), he used a property of determinants from Hadamard's thesis to prove that non-rational meromorphic functions are not representable by power series with integer coefficients – a quite unexpected phenomenon at first sight.

Moreover, by using his own methods, Borel could push forward some of Hadamard's results on singularities and entire functions. An opportunity presented itself in connection with a theorem in Hadamard's thesis where, among other questions, he treated the following: when does the circle of convergence consist of singular points? In other words, when does the domain of analyticity defined by a power series coincide with its disc of convergence? Before Hadamard, several examples of power series with this property were constructed (by Weierstrass, Darboux, J. Tannery, Fredholm and others), and all of them were gap series, i.e., series with infinitely many large gaps between non-zero coefficients. In his thesis Hadamard proved the first general theorem explaining this property. He showed that no power series $\sum_{k\geq1} a_k z^{n_k}$ with $n_{k+1}/n_k \geq \text{const} > 1$ admits analytic continuation across its circle of convergence. However, the theorem was not exhaustive: it covered all known examples but a single one, due to Fredholm (1890). In 1894, Hadamard said to Borel that his result could possibly be improved to include Fredholm's function, although his method seemed insufficient for that. Soon Borel succeeded with a quite different method in the spirit of his theory of divergent series. A bit later Fabry obtained an even better condition by skillfully modifying Hadamard's approach.

Borel improved Poincaré's and Hadamard's theorems on the relation between coefficients, the distribution of zeros, and the growth of an entire function by introducing a notion of order for such a function. This enabled him to prove an assertion about the Weierstrass formula (1.3), which is the converse, in some sense, to that obtained by Hadamard, and nowadays the Hadamard-Borel factorization formula is a standard term in complex function theory.

Another topic of common interest for Hadamard and Borel was the classical theorem of Picard (1879) which states that a nonconstant entire function omits at most one value (for instance, $e^z$ takes any value but zero). In his proof Picard used elliptic modular functions introduced by Hermite. For twelve years attempts to prove Picard's theorem without modular functions (a mysterious trick drawn from a different branch of mathematics) were unsuccessful. This was a major theme in complex function theory, and Hadamard was the first to advance in this direction in his *Grand Prix* memoir of 1893. He gave an "elementary" proof of the Picard theorem, valid for a special class of entire functions. In 1896 Borel published the first proof for the general case which became a great event at the time. It is worth mentioning that Borel's argument relied upon Hadamard's "real part theorem" of 1892 (an inequality between the maximum modulus and the maximum real part of an analytic function).

New aspects of the Picard theorem were revealed later by Hadamard's younger friend Paul Montel, who graduated from the *École Normale* in 1897 and taught at different *lycées* until 1911. In 1907 Montel defended his thesis *Sur les suites infinies de fonctions*, where he introduced the so-called normal families of analytic functions, closely related to the principle of compactness studied by Ascoli and Arzelà. The notion of normal families made a strong impact on various branches of complex function theory. By showing that the family of analytic functions, which omit two fixed values, is normal, Montel arrived at a new proof of Picard's theorem.

*Paul Montel (1876-1975) was professor at the Faculté des Sciences in Paris in 1911-1946. He was elected to the Académie des Sciences in 1937, became its president in 1958. Besides complex function theory he worked in geometry, mechanics, and the history of mathematics.*

In the 1900s Hadamard and Borel ceased active work in complex function theory. Hadamard gradually switched his main interests to differential equations, whereas Borel, who had already obtained classical results in the theory of divergent series as well as in the theory of measure and integral, made a change of course in the direction of probability theory, statistics and the theory of games.

## 2.5 The Dreyfus affair

In the middle 1890s France was divided into two camps, Dreyfusards and anti-Dreyfusards, over the Dreyfus affair. Ministers, political parties, the

National Assembly, army officers, well-known scientists and writers, the press, and financial circles were, over a period of several years, engaged in a battle with both moral and political overtones. An enormous amount was written about this case, from leaflets to the seven-volume book *The history of the trial of Dreyfus* by Reinach. Roger Martin du Gard wrote about it in his novel *Jacques Baroit*, and plays and films have been based on it.

The events began in September 1894, when the French intelligence service, through its agents at the German Embassy in Paris, came across an unsigned note to the German military attaché, promising some secret military documents. Suspicion fell on Alfred Dreyfus, an officer at the *Deuxième Bureau* in the French General Headquarters, because his handwriting was similar to that in the note.

Dreyfus firmly denied the accusation, but he was arrested immediately and brought to trial. Despite the dubious nature of the allegations (only two of five experts attributed the handwriting in the note to him) he was unanimously found guilty, sentenced to penal servitude for life and sent to Devil's Island off the coast of Guiana.

*Alfred Dreyfus (1859-1935)*

Dreyfus was a Jew, and this was decisive for the army leadership, which was chauvinistic and anti-semitic.  All attempts to set him free were in vain, despite desperate efforts by his family and a small group of friends who believed in his innocence.  After the trial it became known that the judges had illegally used information from a secret dossier of which the defence was unaware.  Such details, however, aroused doubts about Dreyfus' culpability only in a narrow circle of people, and the general public was convinced of his guilt by the unanimous verdict of the court-martial.  The journalist Léon Blum, later to be the Prime Minister in the Popular Front Government, wrote: "It had been some three years since Captain Dreyfus had been arrested, condemned, degraded, deported.  The drama had moved opinion for several weeks, but very quickly it had been forgotten, absorbed, dismissed.  Nobody had any longer thought of Dreyfus in the meanwhile, and to reconstitute the events which his name evoked, an effort of memory, already quite difficult, was needed" [III.43, p. 17].

Then in 1896, Colonel Georges Picquart from General Headquarters found a document containing strong evidence that Dreyfus was innocent and that another officer, a Major Esterhazy, was the guilty person.  He reported this to his superior officers, but they were unwilling to reopen the matter.  For his persistence, Picquart was transferred to a remote part of Tunisia.  In November 1897, Mathieu Dreyfus, a brother of the accused captain, learned by chance that the original note was very similar to Esterhazy's handwriting, and he wrote to the Minister of War about it.  On January 13th, 1898, two days after Esterhazy was court-martialled and immediately acquitted, Émile Zola published, in the newspaper *L'Aurore*, his famous open letter *J'accuse* to the President of the French Republic, naming generals and other officers and accusing them of falsifying documents and twisting facts in order to have Dreyfus found guilty.  Zola's article had the desired effect: proceedings were instituted against him for insulting the French army, and in this way the Dreyfus affair was again the centre of public attention.

In August 1898 it was discovered that an important document used as evidence against Dreyfus had been falsified.  An intelligence officer, Major Henry, confessed that it was he who had forged it, and then committed suicide after being arrested.  Esterhazy fled to England in September 1898 where he confessed his espionage.

After a long inquiry, the court of appeal ordered a second court-martial for Dreyfus, which took place in Rennes in 1899.  On September 9, Dreyfus was again found guilty by five votes to two, but this time with "extenuating circumstances."  He was promptly pardoned by the president, Emile Loubet.  The fight to clear his name continued, more fabricated evidence was revealed, and finally in 1906 the verdicts of the two previous courts-martial were overturned.

*The reading of the verdict in front of the council of war in Rennes, 1899*

At the beginning of the affair, the Hadamards were in Bordeaux, and they were shocked at Dreyfus' arrest. Their emotional involvement was especially deep as Jacques was a distant cousin of Dreyfus' wife Lucie, née Hadamard. She was the daughter of Amédée Hadamard's cousin David, a rich diamond merchant. However, the anti-semitic nature of the affair seemed to come as a surprise for Hadamard. In 1956 he recounted:

> On hearing the news of the arrest, my colleague at the Bordeaux University, where I was teaching at that time, the physicist Duhem, an ardent antisemite with whom, paradoxically, I had excellent relations, said to me: 'Why does one dare to maintain that Jews do not understand one another, when one sees this sea of anger?' I answered: 'What sea of anger?' [I.399, p. 80]

As with many others, Hadamard had no reason to suspect that a grave miscarriage of justice had been perpetrated: "The 'Intellectuals', and notably the University academics, were very slow in moving. At Bordeaux, where I was teaching, many did not see the flaw in the judgement" [I.399, p. 87].

Only after leaving Bordeaux for Paris in Autumn 1897 did Hadamard hear arguments against Dreyfus' conviction. Jacqueline Hadamard wrote:

> One day, my father was invited to a reception given by the
> *Administrateur* of the *Collège de France*. He said that he would
> not go. My mother protested 'You should not be ashamed of
> showing yourself', so my father went, alone. Then my mother
> began getting anxious: he hadn't come home. Then finally he ar-
> rived back and explained to her that he had met eminent people
> who were convinced that Dreyfus was innocent! [IV.1, p. I(10)]

Once he had been persuaded by Mathieu Dreyfus of his brother's innocence, Hadamard became an active Dreyfusard, and his views on the matter
quickly became well known in mathematical circles. On a visit to Hermite,
Hadamard was met by the words "Hadamard, you are a traitor." Hadamard
turned pale, but Hermite continued: "You have left analysis for geometry."
Hermite may have had in mind the then recent *Prix Bordin* memoir by Hadamard on dynamical trajectories [I.48]. "A typical bad joke of the time
when treachery and treason were key words of social life in France," comments J.-P. Kahane [II.29, p. 25]. With this incident in mind, Hadamard
remarked [I.372, p. 109]: "He had a kind of positive hatred for geometry and
once curiously reproached me with having made a geometrical memoir".

At first Painlevé also believed that Dreyfus was guilty but he later came
to doubt the verdict of the court-martial and became a Dreyfusard. An
important role in his conversion was a conversation he had with Hadamard
in the spring of 1897, which he described in a letter he wrote to Mittag-
Leffler on April 26, 1899:

> I will tell the facts in a few words: Jacques Hadamard, a
> cousin by marriage of Dreyfus, was led to speak with me of the
> trial of 94 (several months before the Esterhazy affair). For al-
> most an hour, he tried to convince me of Dreyfus' innocence, and
> at the end, faced with my disbelief, he did his best to make me
> understand the intrinsic value of his arguments and his complete
> lack of passion and sentimentality, telling me that Dreyfus was a
> stranger to him, that he had seen him once in his life and whose
> face he disliked, but that he based his belief in his innocence on
> facts.
>
> This conversation, about which I spoke at the beginning of
> the Esterhazy affair, was reported to General Gonse, but in the
> following way: 'Hadamard had told me that the Dreyfus family
> now had proof of Dreyfus' treason'.
>
> At General Gonse's request, I went to see him, I explained to
> him for more than half an hour that Hadamard had not ceased
> ardently to defend Dreyfus' innocence. He replied that in this

case, this conversation told them nothing and was of no importance to them. Then, behind my back, without warning me, he wrote a note saying that after a conversation I had with J. Hadamard, a cousin of Dreyfus, a conversation he had learned of from my lips, some members of the Dreyfus family believed in Dreyfus' guilt [III.301, p. 812].

Asked the question: "Did you not find yourself personally involved in the Affair, being related to the prisoner?" in an interview he gave in 1956, Hadamard replied:

> In fact, during the period which followed Zola being sentenced and the judicial manoeuvres against Picquart, I knew only what everybody knew. But it was at that time that rumours about me were circulated: they had me saying that the Dreyfus family had serious doubts about his innocence: a complete and crude lie, as I told you. These lies were put into the mouth of my comrade and friend Paul Painlevé, who was told about them by d'Ocagne, our colleague at the *École Polytechnique*. As for me, I experienced them at the time of the first inquiry ordered by the Court of Criminal Appeal; it delegated a judge as a rogatory commission, to question all three of us. I must say, by the way, that this questioning left me with a disturbing and deplorable impression of the methods of our criminal procedure: when the examining judge takes a deposition, he does not have anything written under the dictation of the witness, it is he himself who then speaks and dictates the text of the deposition to the clerk of the court...
>
> I was at that time *maître de conférences* at the *Sorbonne*, I had weighed up carefully the words I would use, and I had the greatest difficulties in defending my deposition, to have written what I had said (and then not quite exactly) — and not something else. As far as most witnesses are concerned, I would say that the system used gives the judge the possibility of making them say what he would like [I.399].

At the inquiry Hadamard recollected his conversation with Painlevé and reaffirmed his conviction that Dreyfus had been condemned without sufficient proof and by a violation of the law.

*Hadamard's articles about the Dreyfus affair*

The struggle for Dreyfus' exoneration was also a fight for the honour and dignity of the individual in the face of chauvinism. This commitment to protect the individual led the senator Jacques Trarieux to found the *Ligue des Droits de l'Homme* during the trial of Zola. Trarieux became the first president of the *Ligue*, with Émile Duclaux, director of the *Institut Pasteur* and the chemist Édouard Grimaux as vice-presidents. By April 1, 1898, the *Ligue* had 269 members, and its inaugural meeting took place on June 4. Hadamard became active in the work of the *Ligue* from its inception. Many years later, he would still passionately recall the Dreyfus affair. At the ceremony on the occasion of the sixtieth anniversary of the founding of the *Ligue*, held in 1958 at the *Sorbonne*, the ninety-year old Hadamard said: "It is due to this nobility of values that the *Ligue* continues to exist, and, even today, to proclaim that there is a Dreyfus affair each time the law is violated, each time that injustice and arbitrariness succeed." [I.402, p. 4]. Elected as a member of the central committee of the *Ligue* in 1909, he remained in this post until the last years of his life, to be succeeded by his daughter Jacqueline. The Hadamards' third son, Mathieu-Georges, born on February 27, 1899, was named in honour of Mathieu Dreyfus and Georges Picquart.

Laurent Schwartz wrote: "It is the Dreyfus affair which was in this sense (defence of justice) the great affair of his life. From the moment when he understood the enormity of the injustice perpetrated against a man in the name of reason of state, and the consequences which anti-semitism could have, he devoted himself passionately to the review of the trial. This affair marked his life" [II.58, p. 320]. From then Hadamard's life took on a political involvement, once, in 1924, even reaching to the League of Nations.

## 2.6  In Paris again

On October 29, 1897, Hadamard obtained a position as *maître de conférences* at the *Faculté des Sciences de l'Université de Paris*, and in November he also became the *suppléant* to Maurice Lévy in *Mécanique Analytique et Mécanique Céleste* at the *Collège de France* [III.78]. "Until the 20's there was no pension and people carried on until their death. Only when they thought themselves a little enfeebled did they name a *suppléant*. This *suppléant* received half their salary; he was always very happy to accept the position of *suppléant*: moreover, he hoped to succeed the incumbent. People accepted this half salary to get to the *Collège de France*, since before the war it was infinitely difficult to be appointed in Paris, even more difficult at the *Collège de France* than at the *Sorbonne*. It was absolutely obligatory that everybody went first to the provinces. To be appointed was very difficult." [III.265, p. 26] Hadamard had left a full professorship in Bordeaux for an apparently humbler position: such was the difference in the prestige of appointments in the provinces and in Paris.

The teaching duties of professors at the *Collège de France* are small, but the content of their courses is required to be new. During his seventieth birthday celebrations, Hadamard recalled the unique atmosphere in this institution: "To follow one's imagination freely is part of the tradition of the *Collège de France*" [II.27, p. 55]. Not only the content, but also the titles of courses are determined by the lecturers. It requires merely the formal approval of the title at the general meeting of the professors, where only a few understand the subject in question.[7] Lebesgue described the procedure as follows:

> The Administrator reads the titles: *Le Cakuntalà et la Mrc-chakatikâ; Différenciation du thymocyte, apparition de granulo-cytes et de plasmocytes, Les interprétations de K'ong Ngankouo,*

---

[7]There are fifty-three professors at the *Collège de France* today, of which four are in mathematics.

*Li-king ou Yili* and *Tch'ouen ts'ieou fan lou*. This is the terrible exercise imposed by our *docet omnia*! It seems that the reader becomes confused by the syllables; one even thought sometimes that he did not exactly know the meaning of all the words. The reading ended, and we voted in secret scrutiny for these titles we did not understand.

Well, *Messieurs*, this colourful control is effective, efficient. Everyone of us knows that he will be understood by two or three colleagues and that these, being his close friends, are too jealous of the dignity of the *Collège* to tolerate even the slightest deviation without immediately alerting all the professors who would gather against the delinquent; and the rule is respected. What rule? To do one's best to prepare, to facilitate tomorrow's discovery; for that, to teach the new science, in formation; also the science forgotten or little-known; above all, not to repeat oneself, to go forward. [II.27, p. 10].

*The Collège de France. A postcard from around 1920*

From the time when Hadamard became the *suppléant* to Maurice Lévy at the *Collège de France* he mostly turned his research interests to the domain of mathematical physics. His lectures reflected his recent mathematical results and provoked new questions which became subjects of his new studies. In his lectures on kinematics given during 1898-1899 Hadamard developed a theory of discontinuities in the rectilinear motion of gases, continuing the earlier investigations of Riemann, Christoffel, and Hugoniot. In the next academic year he spoke about the extension of this theory to the three-dimensional wave motion in gases and elastic bodies. In particular, he gave a classification of compatibility conditions for two motions of the medium, before and after the wave front, separating kinematical and dynamical phenomena. The full value of his results has been revealed only comparatively recently, especially with the development of gas dynamics and dynamic elasticity theory.

The year 1898, the first volume of *Leçons de Géometrie élémentaire – Géométrie plane* was published. The second volume devoted to three dimensional geometry appeared in 1901. There were subsequently numerous editions with various additions. The books were written on the initiative of Gaston Darboux, who edited the collection *Cours complet pour la classe de Mathématiques*. Courses on arithmetic by Jules Tannery and on elementary algebra by Carlo Bourlet had been published previously in the same series. In contrast to today's prevalent practice (*nomina sunt odiosa*), the names of authors were printed in large letters whereas the editor's name was in small letters, in proportion to the work done.

Many mathematicians were given a love for geometry through Hadamard's treatise. Gaston Julia recalls:

> An adolescent, born shortly before this century, one day receives the *Leçons de Géométrie élémentaire* as a mathematics prize. In the vacations which follow, he devours the book, he works furiously at solving the innumerable problems given. The book becomes an everyday companion. This book, and the teacher of elementary mathematics who gave it to him, orient his vocation, he will become a mathematician, for the love of the richness of imagination which the book reveals to him, for the taste of work well done [II.56, p. 730].

The impact of Hadamard's *Leçons de Géometrie élémentaire* on mathematical education in France at the beginning of this century is described by A. Weil:

> At that time, the textbooks used in secondary education in France were very good ones, products of the "new programs" of 1905. We tend to forget that the reforms of that period were no less profound, and far more fruitful, than the gospel (supposedly inspired by Bourbaki) preached by the reformers of our day. It all began with Hadamard's *Elementary Geometry* and J. Tannery's *Arithmetic*, but these remarkable works, theoretically intended for use in the "elementary mathematics" (known as

"math élém") course during the final year of secondary school, were suitable only for the teachers and best students: this is especially true of Hadamard's [III.418, p. 22].

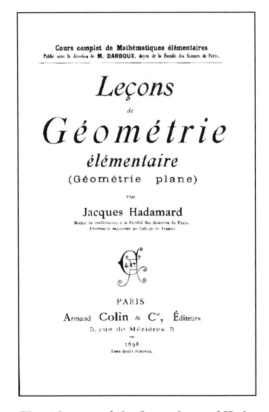

*The title page of the first volume of Hada-mard's Leçons de Géometrie élémentaire*

Hadamard's research was highly regarded, and in 1898 a commission consisting of Hermite, Bertrand, Poincaré and Sarrau awarded him the *Prix Poncelet*[8] for his work over the preceding ten years. He was undoubtedly held in esteem by his older colleagues. Poincaré, in his work *Les méthodes nouvelles de la mécanique céleste*, refers to Hadamard's theorems on analytic functions and geodesics and Hadamard's results became standard in university courses, while Picard and Goursat included them in their treatises on analysis.

---

[8]An annual prize awarded by the *Académie des Sciences* for works in pure and applied mathematics, established in 1867 in the testament of J. Poncelet, creator of projective geometry.

In 1899 Hadamard was awarded an honorary doctorate by Göttingen University. That was the university of Gauss, Dirichlet and Riemann. In 1886, then thirty-seven, Klein became a professor in Göttingen. By that time he had already done his principal work, securing his place at the mathematical Olympus. Enormously productive in his youth, in search of inherent links between different domains of mathematics and natural sciences, he obtained first class results in geometry, group theory, algebraic equations, elliptic and automorphic functions. In 1895 he was joined in Göttingen by Hilbert, who was already famous for solving fundamental problems in the theory of invariants and worked at that time on general laws of the theory of algebraic numbers. Headed by these two outstanding personalities Göttingen became the leading scientific center of Germany and the rival of mathematical Paris.

*Felix Klein (1849-1925)*

When addressing Poincaré in Göttingen in April 1909, Hilbert included Hadamard's name in a short list of the best French and German mathematicians:

You know, highly honoured colleague, as do we all, how steady and close the mathematical interests of France and Germany have been and continue to be. Even when we recall only quickly the developments of the recent past, and out of the rich and many-voiced concert of mathematical science we take hold of the two fundamental tones of number theory and function theory, then we think perhaps of Jacobi, who had in Hermite the outstanding heir to his arithmetical ideas. And Hermite, who unfolded the flag of arithmetic in France, had our Minkowski, who brought it back to Germany again. Or if we only think of the names Cauchy, Riemann, Weierstrass, Poincaré, Klein, and Hadamard, these names build a chain whose links join one another in succession. The mathematical threads tying France and Germany are, like no two other nations, diverse and strong, so that from a mathematical perspective we may view Germany and France as a single land [III.347, p. 75-76].

*David Hilbert (1864-1945)*

The Hadamards were personally acquainted with Hilbert and his wife Köthe, as is witnessed by the following undated letter from Hadamard to Hilbert[9]:

Saint Quay, Côtes du Nord

Dear Sir,

I was most touched, together with Madame Hadamard, by your most friendly letter, and I went with great interest through your talk, of which I only knew the abstract. I hope very much, as you tell me, to talk again about these and many other questions with you and to listen to the very original views you have given on the questions which we are all occupied with. We will be very happy, my wife and I, to continue together with Madame Hilbert and you, the friendship which we were only able to touch upon during your visit to Paris.

Soon, when I have done some of the quite considerable work I still have to finish, I also expect to send you an article for the *Mathematische Annalen*, happy that you have so amicably offered me hospitality there.[10]

Please give all my regards, and remember Madame Hadamard, to Madame Hilbert.

Yours sincerely,

J. Hadamard [IV.50]

## 2.7 The *savant Cosinus*

Hadamard impressed everybody with his prodigious memory. It seemed that he never forgot anything he had been taught at school and at the *rue d'Ulm*. "Even at an advanced age, he was able, in conversation, to cite passages from Greek and Latin." [II.8, p. 34]. Languages were never a problem for him: he spoke German and English from childhood. Jacqueline Hadamard refers to his linguistic ability: "In 1919, he went to give a series of lectures in Madrid. During the Easter vacations, we joined him. He told us proudly that he had succeeded in giving his last lecture in Spanish!" [IV.1, p. III(2)]. Indeed, Hadamard was a veritable fountain of knowledge in geography, history, philosophy, psychology, and, especially, botany, in which

---

[9]This letter is kept at the library of Göttingen University together with two more letters from Hadamard to Hilbert.

[10]None of Hadamard's papers was published in *Mathematische Annalen*.

he was almost a professional. Surprised by Hadamard's memory, Fréchet wrote: "...quite recently a well-known chemist remarked to me how he had been deeply impressed by Hadamard's knowledge of chemistry. His publisher told me that, while preparing his book on analysis, Hadamard returned to the bookshop where he had been correcting the proofs, saying 'I have just realized, in the street, that I forgot to correct an error in the third line on page 169' " [II.13, p. 4082].

However, what made him conform to the traditional image of a scientist (apart from his immense knowledge) was his incredible absent-mindedness in everyday life. "He thinks of things other than cars, when he crosses the street and he has never been able to make a knot in his tie," wrote J.-P. Kahane [II.29].

At the turn of the century, a series of comics was very popular with the French intelligentsia, the hero being a quick-witted and extremely absent-minded scientist, *Cosinus*. In outward appearance, *Cosinus* did not resemble Hadamard, but, as Jacqueline Hadamard observed: "...this creation by Christophe is, for people of our generation and class, inseparable from the image of my father, who perhaps gave the idea for these famous comics. In fact, Mr. Colomb (the real name of Christophe [11]) was a friend of the family, and was our teacher in natural sciences" [IV.1, p. 1].[12] Georges Colomb (1856-1945), Deputy Director of the botanical laboratory at the *Sorbonne*, was the author of numerous works popularising science, and he is considered as the inventor of comic strips, one of which is *L'idée fixe du savant Cosinus*, first published during the years 1893-1899 [III.81].

*Cosinus* is the sobriquet of Christophe's hero, which he earned after becoming *un monsieur excessivement savant*, his real name being Brioché, and with the Christian names *à la fois harmonieux, poétiques et distingués de Pancrace, Eusèbe, Zéphyrin*. His specialities were mathematics and mechanics, and, like Hadamard, he was passionately in love with botany and travelling. However, unlike Hadamard, all attempts by the savant *Cosinus* to leave Paris proved unsuccessful, due to his terrible absent-mindedness. While his luggage crossed borders, *Cosinus* and his dog *Sphéroïde* got into the most unpleasant predicaments in Paris. Here and on pages 110 and 166 we reproduce some of these cartoons from Christophe's book.

---

[11]Notice the joke: Christophe Colomb is the French for Christopher Columbus.

[12]We happened to hear a different opinion that it was Poincaré who had inspired Christophe. The *savant Cosinus*, like Poincaré, graduated from the *École Polytechnique*, and Poincaré's absent-mindedness was the talk of the town.

Of a pugnacious temper, Zéphyrin never missed the opportunity to apply to the nose or the eyes of his closest friends, convergent series of punches. There was, moreover, reciprocity. This is what he wittily called "the multiplication of blows". (There is a possible connection with Hadamard's theorem on the multiplication of singularities.)

Unfortunately, when he is preoccupied, Cosinus is generally absent-minded. This explains the strange way he is dressed when he is on the way to the prefecture and, at the same time, the great commotion which he causes among the people on his way

As well as the story of the sister forgotten on the glacier, and the engagement ring forgotten in his pocket, there are other anecdotes in the same vein. To quote Jacqueline Hadamard:

> The Cosinus side of my father always made us laugh – and not just us, but all his friends who knew of the great gap between him and daily life. Painlevé once met him on the boulevard Saint Michel and said to him: 'I see that Madame Hadamard has already gone on holiday'. 'But how can you see that?', asked my father. 'Your tie is behind your right ear'.
>
> In fact, I believe that once he married, he never dressed by himself! We always wondered how he managed when he went away to give lectures or to a conference. He gave a special demonstration of this on the day of the formal inauguration of

the *Bureau International du Travail*, which was held at the *Hôtel du Louvre*.

When he was climbing the steps, a young man, quite intimidated, stopped him: 'I'm sorry, Monsieur, you have no tie'. Taken aback, he asked the young man: 'What does one do in such a case?' The young man suggested that he go to the big *Magasins du Louvre*, a minute away, and buy one. 'What a marvellous idea, I would never have thought of that.' (As my mother bought everything for him, he had never been into a shop other than a bookshop.) 'But it seems to be rather big, how would I find a tie there?' The young man (decidedly a fountain of knowledge) explained to him that he only had to ask for the tie department. 'Sensational!'. He went after having thanked this precious young man, and found the tie department. The salesman asked: 'What sort of tie? For every day? For dressing? A bow tie?' 'I don't know. As you can see, it appears that I do not have what is necessary to go to an official event.' The salesman chose an appropriate tie: 'There you are, Monsieur'. But my father hadn't finished: 'You couldn't put it on for me? You see, I wouldn't know...' [IV.1, p. III(27-28)].

Anecdotes about Hadamard's eccentricities are still being recounted many years after his death, mostly without any foundation in facts. Recently we happened to disabuse a well-known French mathematician who told us the following 'true' story: "Hadamard was a widower three times. When his third wife died, he came to the funeral with an umbrella although the sky was cloudless. Asked 'Why?', he answered: 'You know, it rained at the funerals of my previous wives.' And, in fact, it soon started to rain." When the story-teller learned that Hadamard had been married only once, he asked us not to mention his name.[13]

---

[13]A variant of this story, with Boussinesq as the main character, can be found in [III.358, p. 156].

# Chapter 3

# Mature Years

## 3.1  New theme

The thirty-four year old Hadamard met the beginning of the twentieth century full of energy and plans. One could only admire the depth and diversity of the papers he had published. The forthcoming years would bring him growing personal recognition and new mathematical breakthroughs.

About 1900 Hadamard began active work on a new theme which had interested him for a long time, the theory of partial differential equations. The impetus was given by his recent studies in gas dynamics.

Unlike ordinary differential equations, partial differential equations contain derivatives of an unknown function of several variables. They describe various processes of mathematical physics including electromagnetism, elasticity, heat conduction, hydrodynamics, gravitation, and a lot more. Usually the same equation models different physical phenomena. For example, harmonic functions, i.e., solutions of the Laplace equation

$$\frac{\partial^2 u}{\partial x^2} + \frac{\partial^2 u}{\partial y^2} + \frac{\partial^2 u}{\partial z^2} = 0$$

can be interpreted as a stationary distribution of temperature, a potential of the electric field, a hydrodynamic potential etc. The so-called wave equation

$$\frac{\partial^2 u}{\partial t^2} = \omega^2 \Big( \frac{\partial^2 u}{\partial x^2} + \frac{\partial^2 u}{\partial y^2} + \frac{\partial^2 u}{\partial z^2} \Big), \tag{3.1}$$

where $\omega$ is a real constant and $t$ is the time, appears in acoustics, optics, and hydrodynamics. Transient diffusion processes are governed by the heat equation

$$\frac{\partial u}{\partial t} = \omega^2 \Big( \frac{\partial^2 u}{\partial x^2} + \frac{\partial^2 u}{\partial y^2} + \frac{\partial^2 u}{\partial z^2} \Big).$$

Attempts to solve these three equations had been undertaken in the
second half of the eighteenth and the beginning of the nineteenth centuries,
and during the nineteenth century other partial differential equations of
mathematical physics were discovered and investigated.[1] By the end of the
century the theory of partial differential equations was a vast collection of
facts and ingenious methods without, however, much order and system.

The difficulty of treating such equations stems from the fact that their
general solution depends on arbitrary functions and not on arbitrary constants, as in the case of ordinary differential equations. In order to find a
unique solution one adds some conditions at the boundary of the domain
where the equation is considered. For example, for the Laplace equation the
boundary values of the solution can be prescribed (this is called the Dirich-
let problem). Together with the wave equation the initial data $u|_{t=0}$ and
$\frac{\partial u}{\partial t}\big|_{t=0}$ form the so-called Cauchy problem.

The choice of boundary value and initial conditions is usually dictated
by the physical problem and it was observed that in such cases the problem
often has a unique solution. For example, the Dirichlet problem for the
Laplace equation and the Cauchy problem for the wave equation are uniquely
solvable. On the contrary, as Hadamard noticed in 1900 [I.88], the Cauchy
problem for the Laplace equation is generally unsolvable, and, from the
very beginning, he sought for the explanation of this difference between the
equations. In the *Princeton University Bulletin* of 1902, he published the
article *Sur les problèmes aux dérivées partielles et leur signification physique*,
where he introduced a concept of the well-posedness of a boundary value
problem, which had great consequences for the future development of the
theory of partial differential equations and functional analysis and is known
as the notion of the well-posedness of a problem in the sense of Hadamard.

The above three classical partial differential equations are of second order, that is they contain derivatives of order two and none of higher order.
More general second order equations with variable coefficients appear in
modelling of physical processes in inhomogeneous media. In a posthumous
paper of 1889 du Bois-Reymond introduced the classification into elliptic,
hyperbolic, and parabolic equations, generalizing the Laplace equation, the
wave equation, and the heat equation respectively. By the end of the nineteenth century little was known about properties of second order partial
differential equations with variable coefficients, especially when the number
of the space variables is greater than or equal to two. Perhaps the first attempt to advance in this direction was undertaken by Picard, who, starting
in 1890, extended some properties of the Laplace equation to general elliptic
equations.

---

[1]See Demidov [III.98], Lützen [III.251].

About 1900 Hadamard became attracted by general second order hyperbolic equations with variable coefficients which describe wave propagation in inhomogeneous media, and he was to return to this topic repeatedly. His main concern was with the solution to the Cauchy problem and Huygens' principle, by which Hadamard understood the following property of wave phenomena: once a wave has passed through a given point, this point remains at rest.

In order to solve the Cauchy problem, Hadamard gave his now famous construction of the fundamental solution. By fundamental solutions one means those with singularities of a special form. They permit one to solve the Cauchy problem with arbitrary initial data by simply taking integrals. For hyperbolic equations the singularities of fundamental solutions lead to divergent integrals of the form

$$\int_\Omega \frac{f(x)}{g(x)^\alpha} dx, \quad \alpha > 1,$$

where the integration is extended over a domain $\Omega$ with a part of its boundary situated on the smooth surface $g(x) = 0$. Hadamard overcame this difficulty by introducing the concept of the finite part of a divergent integral (1902), the idea closely connected to distribution theory which was to appear many years later.

From his construction of the fundamental solution for the Cauchy problem Hadamard deduced, for example, that if Huygens' principle holds, then the dimension of the space is odd. Since this is only a necessary condition, he formulated the problem of describing all the equations satisfying Huygens' principle. This problem proved to be very difficult and it has not been solved up to now although considerable progress has been made.

Partial differential equations dominated Hadamard's work during half a century. Paul Lévy wrote:

> One of Hadamard's admirers once told me that it was in the theory of partial differential equations that he gave the most striking proof of his genius. He devoted numerous articles and two large books to this subject, one book on the propagation of waves and the equations of hydrodynamics (1903), the other on the Cauchy problem. On the subject of these works, I will first of all note that he was not interested in the problems one poses because they seem easy to solve: I do not believe that he ever studied cases where elementary integration is possible. On the contrary, he devoted himself to problems which are posed

naturally, and especially those from physics: for these, even a little progress seemed important to him [II.36, p. 7].

## 3.2   A chronicle of the years 1900-1914

From 1900, International Congresses of Mathematicians have been held every four years, with the exception of two world wars. As Tricomi ironically remarked at Hadamard's centenary, they "were a matter of quite important meetings because they had not yet degenerated into chaotic (and essentially useless) gatherings of the masses which they have become today" [II.5, p. 24]. The Second International Congress of Mathematicians, held in Paris in August 1900, is known for Hilbert's formulation of his famous twenty-three problems. Starting with this Congress Hadamard was present at all of them, except for those in Toronto in 1924 and in Oslo in 1936. His contribution to the Paris Congress was the short communication [I.88], where he introduced the so-called mixed problem for the wave equation with Dirichlet data on the boundary of the domain at any time, and the Cauchy data at the initial time.

*Georges Humbert (1859-1921)*

Also in 1900, on December 3, Hadamard was on the list of candidates to the *Académie des Sciences* presented by the geometry section, but it was Painlevé who was elected this time. Then quite soon after that, due to Hermite's death in January of 1901, he was on the list again, but was ranked only third by the geometry section, after Georges Humbert and Édouard Goursat, and equal with Émile Borel.

Simultaneously with the second volume of his *Leçons de Géométrie élémentaire* Hadamard's little book *La série de Taylor et son prolongement analytique* appeared in 1901, and it quickly became a manual for all interested in the area.

In October 1901 Hadamard made his first trip overseas to represent the *Faculté des Sciences* of the *Sorbonne* at the 200th anniversary of Yale University as well as to have the title of *Doctor honoris causa* conferred on him. In 1902 he was again in the United States, to give a series of lectures on the theory of elasticity, methods of mathematical physics, and geometrical optics, at Princeton University.

The year 1903 saw the appearance of his book *Leçons sur la propagation des ondes et les équations de l'hydrodynamique* [I.109] based on lectures Hadamard had given at the *Collège de France* in 1898-1900.

In the same year Hadamard was made an honorary member of the Kharkov Mathematical Society. In a letter dated November 25, 1903 he wrote to Steklov:[2] "I am extremely honoured to be elected a member of the mathematical centre, to which we are indebted for such remarkable discoveries during the last few years" [IV.38].

The title of Hadamard's report to the Third International Congress of Mathematicians in Heidelberg (1904) was *Sur les solutions fondamentales des équations linéaires aux derivées partielles* [I.117] and a detailed exposition was given in his memoirs in the *Annales Scientifiques de l'École Normale Supérieure* [I.114], [I.119].

In 1906 he was elected President of the French Mathematical Society, and the *Académie des Sciences* awarded him the *Prix Petit d'Ormoy*.

The next year Hadamard submitted his large memoir [I.145] for the *Prix Vaillant* announced by the *Académie des Sciences*. Here he considered the mechanical problem of the equilibrium of a thin plate clamped along its edge, which is modelled by a fourth order partial differential equation supplemented by two boundary conditions. He concentrated mostly on the study of Green's function $G(x, y)$ which is interpreted as the deflection of the plate at the point $x$ under the unit normal load applied at the point $y$. This function enables one to express the solution of the boundary value problem in the form of an integral. Hadamard obtained many interesting results in the memoir, and his main achievement was a formula for the variation of $G(x, y)$ under an infinitesimal variation of the domain. By the decision of jury he was awarded three-quarters of the prize, and the rest was given in equal parts to Boggio, Korn, and Lauricella.

---

[2]V.A. Steklov (1864-1926) worked mainly on problems of mathematical physics, especially in potential theory. When constructing solutions of boundary value problems in the form of expansions in eigenfunctions, he introduced the important notion of a closed orthogonal system, i.e., a system whose span is dense in the space $L^2$. He was the first to propose a certain procedure of averaging of functions. After 1917, Steklov was the vice-president of the Academy of Sciences (1919-1926), and in 1921 he founded and headed the Institute of Physics and Mathematics, from which the Mathematical Institute of the Academy of Sciences of the U.S.S.R. was separated in 1934. This Mathematical Institute now bears his name.

In 1908 Hadamard obtained the *Prix Estrade Delcros*, awarded by the *Académie des Sciences* for his work of the preceding years. The same year in April, he participated with two communications *Sur certaines particularités du calcul des variations* and *Sur certains cas intéressants du problème biharmonique* at the Fourth International Congress of Mathematicians in Rome. Together with Langevin, Prandtl, Steklov and others he served as a member of the commission on the unification of vector notation which was organized by the Congress.

Shortly after retirement of Maurice Lévy in 1909 Hadamard left the *Sorbonne* and succeeded him at the *Collège de France* as *Professeur de mécanique.* At the *Collège de France* Hadamard continued his course on entire functions, the

*Jacques Hadamard in the 1900s*

Riemann zeta function and its applications to number theory, which he had started at the *Sorbonne* in 1907. His lectures at the *Collège de France* on Calculus of Variations became the basis of the treatise [I.159] published in 1910. It filled a gap in the French mathematical literature of that time.

In October 1911 Hadamard delivered four lectures at Columbia University, New York, in which he touched upon various problems in the theory of differential equations, in particular, the role of topology in the study of ordinary differential equations and smooth mappings [I.194].

On January 5, 1912, he succeeded Camille Jordan as *Professeur d'analyse* at the *École Polytechnique*. The most eminent mathematicians in France – Poincaré, Picard, Goursat and others, unanimously acclaiming Hadamard's scientific and pedagogical achievements, put his name forward for Jordan's position. Poincaré's report said that "Hadamard is recommended as a great scientist who has done first-rate work in numerous branches of mathematics, in particular in the theory of elastic waves" [II.27, p. 19].

On July 17, 1912 the French and world scientific communities suffered a great loss: Henri Poincaré died suddenly, aged only fifty-eight. He left hundreds of articles devoted to asymptotic and qualitative theories of ordinary differential equations, function theory, infinite determinants, group theory, non-euclidean geometry, hydrodynamics, topology, number theory, mathematical physics, celestial mechanics, probability theory, foundations of mathematics, and philosophy of science.

*Henri Poincaré*

In some of his works Hadamard was directly influenced by Poincaré's studies: for instance, this was the case with the memoirs on entire functions, dynamic trajectories, and with excurses in topology and probability. The many statements made by Hadamard about Poincaré are full of marvel and praise for his talent. He very quickly wrote articles in expressing these feelings: towards the end of 1912 he published two surveys of Poincaré's work in the journals *Revue de métaphysique et de morale* and *Revue du mois* and he prepared a long paper *L'œuvre mathématique de H. Poincaré* for *Acta Mathematica*.

Paul Lévy wrote about these articles: "... he had undoubtedly written them within the space of two or three months. I do not need to remind you what an extraordinary genius Poincaré had been. If we only mention mathematics, he renewed all parts of it. One had to be Hadamard to dare undertake the exposition of all of his immense work which dealt with so many different areas, and to finish it in one summer. If I add that one of these expositions is forty-one pages long and the other eighty-five pages long, you will realize that these were very deep studies" [II.5, p. 15].

HENRI POINCARE

L'ŒUVRE SCIENTIFIQUE
L'ŒUVRE PHILOSOPHIQUE

PAR

**Vito VOLTERRA**
Professeur à l'Université de Rome, correspondant de l'Institut.

**Jacques HADAMARD**
Membre de l'Institut,
Professeur au Collège de France et à l'École Polytechnique.

**Paul LANGEVIN**
Professeur au Collège de France.

**Pierre BOUTROUX**
Professeur à l'Université de Poitiers.

LIBRAIRIE FÉLIX ALCAN
1914

*In 1914 the book Henri Poincaré, L'œuvre scientifique, L'œuvre philosophique appeared, consisting of articles by Volterra, Hadamard, Langevin, and Boutroux*

The following story of Lebesgue is indicative of Hadamard's deep respect for Poincaré's prodigious ability: "After lectures he had delivered, some scientists gave a banquet in honour of Hadamard. In his speech, it occurred to somebody to compare him to Poincaré. Hadamard rose, indignant, and shouted: 'No! There are some stupid things one doesn't do!' " [II.27]. According to Montel, the scene took place in 1930 when a delegation of French mathematicians visited the U.S.S.R. Montel said: "When we left the Academy of Sciences in Kiev, where the President, in his wellcoming speech, compared him [Hadamard] with Henri Poincaré, he was sincerely angry, since he modestly placed himself below the great genius" [II.27, p. 12].

Jacqueline Hadamard wrote:

> For my father, Poincaré was always the supreme genius, the person he called *mon Maître* and whom he admired not only for his mathematical work, but also for the beautiful language he used in order to write it.

In fact, my father, who had great difficulty in translating his thoughts into words, was one of those people who could not suffer mediocrity in this and the greater part of his working time was taken up by this hard labour. He had infinite patience for it [IV.1, p. III(2)].

On the day after his forty-seventh birthday in 1912, Hadamard was elected to the *Académie des Sciences*, succeeding Poincaré in the geometry section. In his report to the Secret Committee of December 2, 1912, Appell placed Hadamard "to the first rank of geometers of the epoch" and said:

Mr. Hadamard's work is varied, powerful and rich: it encompasses almost all the areas of mathematics: it deals with the theory of functions and the theory of numbers, the integration of the differential equations of mechanics and mathematical physics, the study of the newest problems of functional calculus.

In all these areas, Mr. Hadamard has obtained general results of great importance, using powerful methods which are his own and which have opened the way to much research in France and in other countries [IV.22].

*Athena, the emblem of the Académie des Sciences*

This time he was first on the list, followed by Borel and Goursat, second, with Guichard and Lebesgue placed third. Hadamard won the first round by a vote of thirty-six out of fifty-seven.

A few weeks after Poincaré's death the Fifth International Congress of Mathematicians started in Cambridge (England). "Even the English skies seemed to mourn the loss of the great mathematician, for it rained almost steadily throughout the Congress" [III.6, p. 14]. Hadamard was vice-president of the Congress and gave a talk entitled *Sur la série de Stirling*. By invitation of Mittag-Leffler it was decided to have the next meeting in Stockholm in 1916.

We have already mentioned Hadamard's treatise on elementary geometry. It was an indication of his constant interest in the problem of improvement in the teaching of mathematics. In 1914 he was an active participant, and later the leader, of the French group of the International Committee on Mathematical Education, which was established in 1908 at the International Congress of Mathematicians in Rome. Representatives of eighteen countries took part in the work of this committee, its first chairman being Felix Klein.

Hadamard was one of the organizers of the International Conference on Mathematical Education, which took place from the 1st to the 4th of April, 1914, in Paris, and gave a talk there [I.188]. In the discussions he warned against formalism in education, and proposed that one should appeal to common sense in introducing new material. He talked about the teaching of mechanics in secondary schools and its connections with mathematics on the one hand and with physics on the other. He asked the general question: "What can be done to give teachers a sufficient sense of applied mathematics?" He recalled his experience as an examiner, and on questions of methodology (how one should teach Taylor's series, and so on). He also discussed the relationship between intuition and rigorous proof (the dictum "logic merely sanctions the conquests of intuition" is due to him).

Immediately after this conference there was another one, on mathematical philosophy, from the 6th to the 8th of April in Paris, and here Hadamard gave three talks on different topics: the intrinsic properties of space, the calculus of functionals – analysis and synthesis, and on mathematical principles and mathematical reasoning.

The following undated Hadamard's letter to Hilbert seems to be written not long before, in connection with the first talk:

> Dear Colleague,
>
> I hope that your tiredness of which, as Mr. Xavier Léon tells me, you still complain, is well on the way to improving. It would be quite bad of me to importune you before you have recovered from it. However, I beleive that I can ask you a question whose answer, it seems to me, will not require any trouble of you.
>
> I intend to mention the question of Klein-Clifford space at the Easter Conference. I take the liberty of sending you a provisional version of the note I am going to read at the Conference, although, as you will see, this note contains absolutely nothing original and is limited to the classic statement of the problem.
>
> If I am not mistaken, you have not touched this question in your fundamental research on the principles of geometry. What I would like to know is precisely if that is really the case or if, on the contrary, the passage which you devoted to it eluded me.
>
> I again hope that I am not being importune in asking you for this information, and in thanking you in advance, I assure you of my cordial sentiments.

J. Hadamard [IV.50]

Could the tiredness Hadamard mentioned at the beginning of this letter be the onset of Hilbert's pernicious anaemia, which would have killed him had not a remarkable new treatment not come along in 1926?[3]

## 3.3 Hadamard and Volterra

In Hadamard's *Prix Vaillant* memoir, as in some of his other works, one can see conceptual connections with the calculus of functions of curves and surfaces, developed by Vito Volterra. Hadamard and Volterra had much in common. Both came from Jewish families with a rich genealogy. They showed a gift for mathematics at an early age: Volterra read Bertrand's *Traité d'arithmétique* and Legendre's *Eléments de géométrie* at the age of eleven, and when he was thirteen, impressed by Jules Verne's *De la terre à la lune*, he tried to find the equation of the trajectory followed by the heroes of the novel. Furthermore, both were *normaliens*: Volterra was a graduate of the *Scuola Normale Superiore di Pisa*, the Italian equivalent of the *École Normale Supérieure*.

They were both active in public life, with similar political views, and shared an interest in mathematical problems arising in the natural sciences. Volterra, like Hadamard, enriched mathematics by solving important problems while still young. He was the originator of new directions in research in analysis, for which he also found various applications. In 1899 he was elected to the *Accademia dei Lincei* in Rome. The next year he married Virginia Almagià, with whom he was to live until his death, and moved from Turin to Rome, where he had been invited to take the chair at the university which had fallen vacant on the death of Beltrami.

Here is the first letter from Hadamard to Volterra, kept in the collection of the library at the *Accademia dei Lincei*. It is undated, but was most probably written at the end of 1900, soon after the Second International Congress of Mathematicians in Paris.

25 rue Humboldt

Dear Monsieur,
  I write to you in Turin, since I do not know whether you are already installed in Rome, and I hope that my letter will reach you in any case.

---

[3]See sec. 2.3 in the book by C. Reid [III.340].

Its aim is to ask you to please send to me, or, better still, to Mr. Dupouy, the secretary of the Congress, 162 boulevard Péreire, in Paris, the text of the report you made on equations with real characteristics, which we would like to have for the proceedings of the Congress.

I am happy to take this opportunity to remember myself to you and I ask you to accept, with all my regards for Madame Volterra, my best wishes.

J. Hadamard [IV.44]

*Vito Volterra (1860-1940)*

Volterra was one of four plenary speakers at the Paris Congress, the others were Poincaré, Mittag-Leffler and the historian Moritz Cantor. Unlike the International Congresses of today, these lectures were intelligible for all participants, with topics taken from the history or philosophy of mathematics. Volterra's talk, entitled *Betti, Brioschi, Casorati – Trois analystes italiens et trois manières d'envisager les questions d'analyse* [III.408], commemorated three colleagues who died during the previous ten years. He also presented a shorter report to the Congress: *Sur les équations aux dérivées partielles* [III.409], mentioned in the above letter from Hadamard.

Whenever their work had common ground, Hadamard extended Volterra's ideas. That was the case with his work on the equations of mathematical physics, particularly in the theory of waves, and with Volterra's theory of functions of curves, which were called *functionals* by Hadamard. He was the first to describe, in 1903, linear functionals on a function space. Considering the space of continuous functions on the interval $[a, b]$, Hadamard showed that every functional is the limit of a sequence of integrals:

$$U[f] = \lim_{n \to \infty} \int_a^b h_n(t) f(t) dt.$$

This theorem heralded the beginning of a large domain of functional analysis. The non-uniqueness of the sequence $\{h_n\}$, which is a disadvantage of Hadamard's representation, was avoided in 1911 by F. Riesz, who found a canonical representation of $U[f]$ as a Stieltjes integral – "In mathematics simple ideas usually come last", as Hadamard once observed.

Two years before Volterra gave a lecture course on functions of curves at the *Sorbonne*, in 1910 Hadamard had presented the elements of Volterra's theory in his book *Leçons sur le calcul des variations*, including in it some of his own results. This fundamental work had its origins in the lectures he gave on the subject at the *Collège de France*. In 1912 Hadamard published his article *Le calcul fonctionnel* [I.174], where he proposed to make a general study of functions whose arguments have an arbitrary character. Fréchet later wrote that it was this "prophetic article" by Hadamard which inspired him to create the theory of abstract spaces [II.13, p. 4084].

Other important contributors to the development of the ideas of functional analysis, begun by Volterra and Hadamard, were Hadamard's pupils René Gâteaux and Paul Lévy.

## 3.4   Home life

By 1900, the Hadamards had moved into a house at *rue Humboldt 25*, between the *boulevard Arago* and the *boulevard Saint-Jacques*. The chemist Victor Auger lived in the house next to them. Later, one of Louise's sisters bought a house in the vicinity, and some time afterwards the philosopher Gaston Milhaud became a neighbour. Jacqueline Hadamard wrote: "All these houses, well hidden behind a block of apartments, were (and remain) an ocean of peace with a farm at the end of the road, with pigs, turkeys, hens and ducks, and even some cows. Unfortunately, this farm on the *rue du Faubourg-Saint-Jacques* has been replaced by a large apartment block and a big garage".

The Hadamards already had five children, the two daughters being born in the new house: Cécile in 1901 and Jacqueline in 1902. At that time, some of the university professors wanted to educate their children at home. The Perrins, Langevins, Curies, and Hadamards decided to do this. Hadamard wrote: "In two years, by teaching my own children, I elaborated a series of ideas on the teaching of elementary geometry: a question '*à l'ordre du jour*' at least in France, but my ideas are pretty divergent from those which are presently in favour" (from a letter to E.O. Lovett, July 12, 1912 [IV.32]).

During daytime, it was quiet at home only when Fräulein took the youngest to the *Jardin du Luxembourg*. It was an ideal time to work and receive visitors – Hadamard did this at home since there were no offices at the *Collège de France*. Hadamard's way of working is described by Jacqueline Hadamard:

> He practically never wrote a word. He always told me that he thought without words and that for him the great difficulty was to translate his thoughts into words. He only scribbled down equations, not at a table, but at a high wooden plinth of the kind that were normally used, at that time, to put a bust on (at my grandmother's, it was a bust of Beethoven, of course). In the hall, he could write down his mathematical formulas, while walking up and down. So for many years, I used to hear my mother taking down dictated sentences of the kind: 'We integrate – poum –, we see that the equation – poum, poum, poum – equals zero, takes the form – poum, poum, poum, poum'. The number of "poums" indicated the length of the space to be left for the formulas.
>
> My mother, of course, was unable to write these, since her education had never gone further than simple arithmetic. In this, she surpassed my father, who had never been able to count beyond four; after that came $n$.
>
> He couldn't count and neither could he write, and it was my

mother who took care of this. For instance, he once received such a charming letter from his friend Duhem (who had remained in Bordeaux) that my mother told him: 'This time, you have to reply'. My father agreed. One hour later, my mother came to look: 'How far have you got?' – 'Not far'. In fact, he had written: 'My dear Duhem...' and that was all! [IV.1, p. I(13)]

He himself recounted about his working habits in the book [I.372, p. 34]: "Helmholtz's and Poincaré's statements seem to suggest that they often used to sit at a work table. I never do so, except when I am obliged to effectuate written calculations (for which I have a certain reluctance). Except in the night when I cannot sleep, I never find anything otherwise than by pacing up and down the room. I feel exactly like the character of Émile Augier who said: 'Legs are the wheels of thought'."

Interestingly, Hadamard invented his own system of stenography to write non-mathematical texts which Louise mastered perfectly. He dictated, she took down shorthand and then, without his participation, prepared the final text.

Louise's kindness, her attractive appearance and musicality were far from being an exhaustive list of her qualities. She was "a generous and intelligent woman who supported and helped him throughout his career", wrote Fréchet [II.13, p.4081]. Their marriage, which lasted for 68 years, was a model of harmony between two distinguished and gifted people.

*Tante Lou*, as she was called by her family and friends, protected Hadamard from the little things in life, easily adjusting herself to the peculiarities of life together with a great mathematician. Jacqueline Hadamard wrote:

> What a wonderful life we had under the extraordinary charm of my mother, who was able to instill her brood with total respect for my father's work. When I was two years old, I was snubbed (it seems) by my older brothers and sister for a whole day because I had touched Papa's papers!
>
> It must be said that it was difficult to avoid these papers, because, on top of everything else, my father was afraid of someone putting them in order. Otherwise, he claimed, he wouldn't be able to find anything again. So the papers were (everywhere) in great heaps.
>
> My father had a room and a large desk; or, rather he had one at one time. But to my knowledge, he hadn't used it for a long time. In fact, I only remember the desk being covered with five layers of papers, separated from each other by layers

of newspapers which he put there when he changed the subject of study. There wasn't a sixth layer, because this would have been too high for it to be convenient to use. I am convinced that nobody (not even himself) knew what there was in the lower layers, which were uncovered only in 1935 when we moved.

This deluge flowed over into the living room, which created problems when friends or famous foreign scientists came, because we had to clear the chairs and sometimes even (in summer) the stove. During the times of great activity, it even tried to take over the dining room table: it was actually a temptation, this beautiful surface being clear for hours! It obviously presented difficulties because we had to clear it at mealtimes (oh, what a drama!). My sister even claims that once or twice the table cloth was put over the papers.

Naturally, this allergy to order didn't go without causing problems, because I always remember my father complaining about the whole of humanity, that they had moved his papers. Later, I did try to make things better by offering him a vast desk with eighteen drawers, but without success: there was only more room for him to spread his papers. But the strange thing is that in spite of that, he almost never lost a thing ... except once when some memoir was found at a costermongers who sold his vegetables in the papers in question.

Le chien Sphéroïde.

*Here you see ... the dog Sphéroïde, called so because he is vaguely of the Bulldog breed [Boule dogues, literally, Ball dogs] (from Christophe's L'idée fixe du savant Cosinus [III.81]).*

... Mathematics, naturally, dominated in the house, and even extended to the animal world: a stray dog found by us was christened 'X-prime'. Later, another dog was named 'Zêta de s' (a function which my father was studying very much at that time). The last one was called 'Logarithme népérien', which quickly became 'Log' [IV.1, p. I(14)-I(15)].

Hadamard loved music passionately. He resorted to the violin when he encountered difficulties in his mathematical work. Tricomi recalled: "In love with music and able to play the violin, he once confessed to me that he did not know what gave him more satisfaction – the violin or mathematics" [II.61, p. 378]. Louise was a good pianist. The children were brought up with the belief that one had to learn to play at least one instrument. Their eldest son Pierre could play the violin, Cécile's instrument was the viola, Jacqueline played the violin, while Étienne, the second son, was an extraordinarily gifted pianist.

The children spent some of their time with Jacques' mother, who, having attained a certain prosperity, had bought a large country house in *La Queue en Yvelines*, fifty kilometres from Paris. There, they would play in the old garden and tour the area on bicycle.

*En famille, 1913*

The whole family went on summer vacation, which lasted two and a half months, mostly to Brittany but sometimes to the *Massif Central*. Jacqueline Hadamard recalls:

> It was in the Corrèze that one of my father's colleagues, Mr. Périer, a professor at the *Sorbonne*, initiated my father (and us as well) to the passion for mushrooms. They were so little sought after at the time, that at the end of summer we revolted against

the abuse of ceps!  But the initiation into the diagnosis of the
species of a mushroom did not go without playing a trick or two.

So it was that one day my father, having
overlooked the remark "slightly purgative"
about a certain goatsbeard, put some in the
omelette, which meant that the whole family benefited all night from this "slightly"!
What would it have been like if it had been
marked "strongly"!

In Brittany we never stayed in a hotel and my mother rented
a house.  We took the train and my father had the annoying habit
of, once we were seated in our compartment, alighting from the
train to go and find out something or other.  Once the train
departed without him when he had the family's tickets and the
household purse!  My mother was not calm until he had returned
to sit down [IV.1, p. I(18)-I(19)].

Hadamard's house was always open to guests.  Among his close friends
were many outstanding people such as Volterra, Einstein, Painlevé, Montel,
Lebesgue, Langevin, Perrin, Bédier.  Jacqueline Hadamard recounted:

I remember his [Einstein's] first visit when I was little, his
clear expression, almost that of a child.  Of all my distant recollections, the most vivid is that of his great laugh which could
be heard in the whole house.  When he came he did something
I have never seen anyone else do:  he would take the footstool,
put it on his knees and lean on it so that he could talk about
mathematics in comfort.

Another visitor we liked a lot was Rivet [a famous anthropol-
ogist and ethnologist] who told us about his travels.  I remember
the horrific account he gave of a trip to South America where he
had collected a sample of poisonous spiders (which killed more
people in Brazil than snakes did, he said).  So he returned to
Rio de Janeiro with a large box containing about a hundred of
these spiders and registered in a hotel.  His assistant picked up
the box and dropped it; it broke, and there were all these beasts,
each of them a carrier of death, escaping in a hotel full of people!
The story he gave about the *corrida* to recapture them gave us
shivers down our spines.  In fact, they found them very quickly,
except for two!  He spent two days in anguish before succeeding
in catching these last two, waiting to hear any minute of the
death of people, whose death he would have felt responsible for.

Happily, everything finished well and he could finally sleep with
a clear conscience.

Painlevé, a mathematician who later made a political career
often came to us. I remember in particular his theory on the
perversity of inanimate objects which play the most rotten tricks.
He said to my father: "You haven't noticed that collar buttons
hide by jumping from one drawer to another? I've seen them do
it."

We also often saw the Milhauds, who lived in a house near
ours. Gaston Milhaud, a philosopher, was an extremely kind
man and his wife had kept the freshness of a child to the age
of sixty plus. She demonstrated this later by starting a very
successful career as a painter, having passed sixty. The eldest of
their sons was the playmate of my brothers and I had the joy
(being then seven years old) of giving the feeding bottle to the
youngest.

Of course, we saw the Augers a lot. Victor Auger, Professor
of Chemistry at the *Sorbonne*, was a born philosopher. He was
content when something happened and happy even when nothing
occurred. But I was very impressed by one of his reactions: he
did not allow one to be served a cracked plate and when he saw
one on the table, he would break it in two, which as a result made
his wife angry. Their children, about the same age as us, were
our playmates, and the first to finish their homework called the
others. We used to go to the [*Jardin du*] *Luxembourg* together
when our parents thought we had made enough noise or damage.

But our favourite visitors, for us children, were Lebesgue and
Montel, two mathematicians, very different but full of life both
of them [IV.1, p. III(31)].

Here is how Steklov describes his impressions of Hadamard in a letter to
Liapunov of September 17, 1903:

Hadamard somehow managed to find me himself. Once he
failed to catch me at home, the next day he appeared at half
past eight, when we had barely woken up. He came to Paris
for a couple of days (he examined for the *baccalauréat*) and on
the very day of his departure for the provinces, where he spends
the vacations, he dropped by before the exam. He spent half an
hour with me and said more than one can during a whole day. A
typical Parisian, incredibly lively and quick, he behaves himself
as if we have been acquainted for a long time and only happened
not to have seen each other for a short time. In general, it is

a surprising ability of French people to behave, during the first encounter, in such a way that one forgets that one is seeing the man for the first time, and after two or three meetings one finds oneself an *ami* [IV.39].

It was an exceptionally happy time for Hadamard: new theorems, lectures, a loving wife and children, friends, and a comfortable house filled with the sound of music.

## 3.5   The First World War

At the age of seventy Hadamard said that he was blessed with the beauty of life from 1892 to 1916 "... after which no joy was really pure" [II.27, p. 5]. The first date was the year he got married and the second was during the First World War, which resulted in personal tragedy for him.

On July 27, 1914, Austro-Hungary declared war on Serbia. Within one week the tragic events had involved Germany, Russia, France, Great Britain and Belgium in a conflict with which no previous war can be compared in scale or consequences (the number of people affected, the length of the fighting fronts, and the technology used). It was a war for the re-division of an already divided Europe, and its impact was felt far outside its borders. On the side of the Entente were Great Britain, France and Russia, to be joined later by Italy and, in 1917, the U.S.A. By the end of the "war to end all wars", thirty-four countries had entered the Entente. They were opposed by Germany, Austro-Hungary, Turkey and Bulgaria.

Hadamard's sons studied at the *Lycée Louis-le-Grand*. The older, Pierre, was accepted at the *École Polytechnique* in July 1914. "Towards his brothers and sisters he felt himself to be the 'big brother', with a secret awareness of his role as the eldest and of the moral responsibility which this imposed on him," wrote Hadamard [IV.2]. In August Pierre was called up. Jacqueline Hadamard wrote:

> In 1914, we were on vacation at Tréboul, near Douarnenez, with our friends, the Augers, when the catastrophe was triggered off. Returning from a walk, we learned of the mobilisation and that my parents and my elder brother Pierre (who had been accepted at the *École Polytechnique*) had gone to Douarnenez to take the train to Paris. We ran after them and succeeded in catching up with them before they got onto the train, and were quite bewildered, like everybody else.

*Louise Hadamard as a nurse, 1915*

Our parents came back after a few days and announced that my brother had joined up as a sub-lieutenant in the artillery. He soon went to the front. We returned to Paris after the victory of the Marne, won to a large extent thanks to the G.7 taxis which transported the troops.

But all the North was occupied and the refugees rushed towards Paris. It was necessary to organize their reception, to take care of the single children. A friend of my mother, Madame Grumbach (an amazing woman), organized a charity and my mother helped her [IV.1, p. II(1)].

The following letter was written by Hadamard to Volterra on January 15, 1916:

25 rue Humboldt

Dear Friend,

Your very cordial note gave me real pleasure, as you may imagine. It is quite an agreeable sensation, and very necessary at this moment, to feel, even at a distance, the precious friendships which have been valued at all times, but which are more so, and which one knows to be keener and closer in the middle of the present crisis.

It is quite natural that our destinies resemble each other at this moment. I am, like you, kept fairly busy, and by questions close to those to which you have devoted yourself. And as a consequence also I teach rare courses for very rare students at the *Collège de France* – I don't mention, of course, the *École Polytechnique*, whose students are all fighting, and which has been converted into a hospital.

My wife sends her best regards to Madame Volterra and says that she must be happy to have such young children at this present time. Up to now, fortunately, our eldest son has come through the war without damage, even though he has taken part in some of the fiercest actions.

I thank you so much for your good wishes, and send you all of ours together with my respects to Madame Volterra, my best regards.

J. Hadamard. [IV.44]

Hadamard was engaged in theoretical work of military significance, in the *Direction des Inventions* (Department of Technical Inventions) organized by Painlevé and Borel within the *Ministère de l'Instruction Publique, des Beaux-Arts et des Inventions intéressant la Défense Nationale*, where Painlevé was Minister in 1915-1916.[4] In Hadamard's dossier at the *Archives Nationales* one can find the following document confirming his participation in military research:

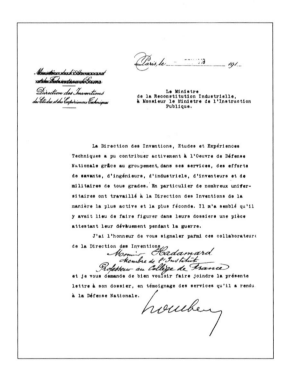

From the Minister of Industrial Reconstruction
to the Minister of Education

The Department of Technical Inventions, Studies and Experiments was able to contribute actively to the task of national

---

[4]This ministry was created specially for Painlevé; he became Minister of War in 1917.

defence, thanks to the coordination in its services of the efforts of scientists, engineers, industrialists, inventors and soldiers of all ranks. In particular, many university academics worked in the Department of Inventions in a most active and fruitful way. It seemed to me that there were good reasons for a document to be put in their dossiers which would testify to their devotion during the war.

Among these colleagues of the Department of Inventions, I have the honour of calling to your attention

*Monsieur* Hadamard
*Membre de l'Institut*
*Professeur au Collège de France*

and I request you to add the present letter to his dossier, as a witness to the services he has rendered to national defence. [IV.26]

The Department of Inventions also used the services of Volterra, particularly on the problem of location of sound. Volterra, a senator, did a great deal for Italy's entry into the war on the side of the Entente powers. When this occurred, in 1915, he enlisted as a volunteer in the army, being assigned to the air force (then a part of the army) as an officer of the engineer corps. He worked on developing artillery for aircraft, took part in dangerous military operations, and was sent on military missions to England and France, visiting the Franco-German front with Painlevé and Borel.

In the beginning of 1916, Hadamard wrote to Volterra:

My Dear Friend,

Painlevé has just informed me of your desire to have a French professor in Rome, and adds that you had me in mind. I do not need to tell you how much I appreciate your appeal. You know, also, the great pleasure with which I would have accepted in ordinary times, and that it has always been my most ardent wish to return to your country as often as possible.

You will understand the objections, and you will have, I believe, forseen them. It is not only that, as everybody, I am unable to think of anything other than that which concerns the war – and it is true that your idea stems from this, so that this would not be an objection – but I have the joy and pride of taking part in projects directly useful to defence. I am attached to the new Department of Inventions, at the head of which, as you know, are Painlevé and Borel.

I cannot tell you that I am considering, without ulterior motive, the idea of interrupting the work I am pursuing. Would not such a radical interruption (which would be incurred by the distances and necessity of adjusting my mind – so far away, for the moment, from pure science – to teaching such as I would like to do at the University of Rome) cause me to lose contact with the problems I have begun to deal with, at the same time as to remain ignorant of those which will continue to present themselves?

Furthermore, I understand, as you do, all the importance which an affirmation of solidarity in the domain of science would have for our two countries. I would very much like to be able to reconcile this duty with the ones I have just told you about.

The main question, for me, in these conditions, is that of the length of my stay in Rome. I am obliged to insist on shortening this duration, as much as I would have wished to extend it at another time, and as much as that will be reasonably possible.

So I begin by asking you what sort of teaching you have in mind and what is the minimum time you thought of allotting to it?

I add, with the same point of view, that I would be happy to be able defer my leave of absence as far as possible, so that I may continue my collaboration until such a time as this work will have begun to find an application.

I fervently hope that we shall come to an agreement which satisfies you and, once again with our warmest and most heartfelt thanks, I assure you of my sincere friendship.

J. Hadamard. [IV.44]

Hadamard arrived in Rome on May 2, in the morning. From the fourth to the seventeenth of May, he gave lectures at the university, on the Cauchy problem for hyperbolic equations. He again met Volterra, who had managed to come to Rome from the army for short time. Soon after Hadamard's departure from Rome, Virginia Volterra received a telegram from France, in which it was announced that Pierre Hadamard had been seriously wounded. She contacted the mathematician Fano in Turin, with the idea of informing Hadamard through him. Fano sent her the following telegram:

TELEPHONED GENOA YESTERDAY EVENING
WHEN LEARNED HADAMARD STILL THERE NOW
INFORMED LEAVING TODAY. FANO [IV.44]

## 3.6   Deaths of Pierre and Étienne

In the spring of 1916 Pierre Hadamard was given the command of an artillery battery near Verdun, which was stubbornly defended by the French. The fortress at Verdun played a central role in the plans of the German high command on the western front, at the beginning of 1916: its capture would open up the road to Paris. The battle which began here in February 1916 was so awful in their slaughter, that even now, long after the more horrifying Second World War, Verdun is a gruesome symbol of mass human tragedy.

On his return home from Italy, Hadamard learned the terrible truth:

> *Sous-lieutenant* HADAMARD, Pierre, of the 103rd Battery of 58 of the 49th Artillery Regiment.
>
> A young officer-bombardier of great courage and with a very deep feeling of duty, sent on May 18, 1916, to a German trench, a few hours after its capture, to observe the shooting of its mortars. He left it, having accomplished his task; returned to take part in a counter-attack. Fell gloriously, fighting with the infantry which he encouraged by his example.

General Commanding the 2nd Army

Signed: Nivelle. [IV.1]

On May 29 Hadamard wrote to Virginia Volterra:

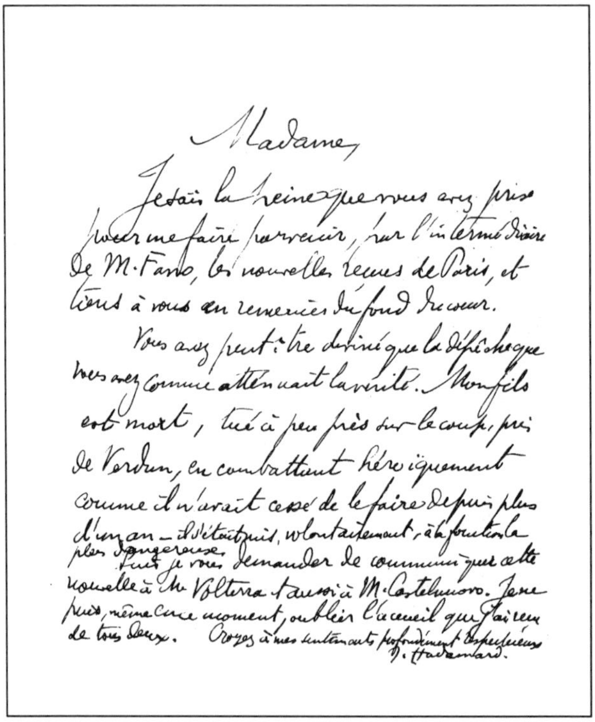

*Facsimile of Hadamard's letter to Virginia Volterra*

Madame,

I know the trouble you took to get the news from Paris to me, through Mr. Fano, and I thank you for this from the bottom of my heart.

You have perhaps guessed that the dispatch you read attenuated the truth. My son is dead, killed almost instantly, near Verdun, fighting heroically, which he had not ceased doing for more than a year – he had voluntarily taken on the most dangerous duty. May I ask you to convey this news to Mr. Volterra and also to Mr. Castelnuovo. I cannot, even at this moment, forget the reception I had from both of them. Please accept my deepest regards.

J. Hadamard. [IV.44]

Soon he got a telegram:

> Hadamard
> 25 rue Humboldt
> Paris
>
> WITH DEEP CONDOLENCES SEND EXPRESSIONS OUR
> ADMIRATION YOUR SON FALLEN HEROICALLY FOR
> THE CAUSE OF JUSTICE
>
> VITO VIRGINIA VOLTERRA[5] [IV.44]

When the war began, the Hadamards' second son Étienne, who showed remarkable ability in mathematics, was only seventeen and studied at the *Lycée Louis-le-Grand.* In June 1916, after receiving the news that he had passed the entrance examinations to the *École Centrale,* Étienne joined the army as a volunteer, and within a month was sent to Verdun. He was assigned to a battery four kilometers from where his brother Pierre had been killed two and a half months previously. The day after his arrival, he was badly wounded in his first battle, with shell splinters in both legs. There was a lack of transport, so that he was operated on two days after being wounded. His parents and the youngest son Mathieu managed to arrive in time to find him conscious. Étienne lived for two hours longer.

Paul Lévy recalls Hadamard's sad words: "...what I did in mathematics is nothing compared with what he [Étienne] could do if he were alive today." [II.37, p. 3]. A similar statement by Hadamard is recalled by J. Nicoletis, a military engineer who attended Hadamard's lectures as a young man: "As a mathematician he himself 'didn't exist' compared with this boy who had been accepted into the *École Centrale*". During the war, Nicoletis served in the same regiment as Pierre, who saved his life. "He pulled me, half-dead, into the trenches at Artois", wrote Nicoletis [II.49, p. 10]. The Hadamards became attached to Nicoletis, his presence reminding them of their departed sons. Jacqueline Hadamard recalls: "The shock is terrible for my parents. But they were able to endure it by trying not to let it weigh too heavily on us, the young ones (which I realized only much later). When my father died in 1963, I found that he always carried with him the letter that my brother Pierre's Commander had sent him, describing how he had died doing more

---

[5]The following touching story told by Hans Lewy proves to be erroneous: "Jacques Hadamard was giving some lectures in Rome, and then Hadamard left. And afterwards came a telegram for Hadamard which Volterra opened. And it said, "Your son has been killed in the war". That was the first World War. Hadamard had just left, so Volterra looked at the train schedule, found a faster train and caught him at the border so that he could rather tell him personally about his son" [III.7, p. 186-187].

than his duty and the friendship everybody felt for him." [IV.1, p. II(2)-II(3)].

Eight months after his visit to Italy Hadamard was made an Officer of the Order of the Crown of Italy and on April 4, 1918, the *Accademia dei Lincei* elected him a foreign member.

In 1918, the Germans employed their Big Bertha gun to shell Paris, but without much success, as Jacqueline Hadamard recounts:

> It wasn't very effective, and, except for one shell which hit *Saint Gervais*, during mass (I think), there weren't any victims.
>
> But when the alert was sounded, our parents made us go down to the shelter. My sister was studying for her *baccalauréat* and my father found that these hours spent in the cellar was a chance to make her work. He didn't realize at all that she was tired and thought that she didn't make enough effort. The neighbours, also in the cellar, were taken aback by his severity and, from time to time, intervened.
>
> Our parents, on finding out that the trajectory of the Big Bertha shells was on a north-south axis passing over us, wanted us to go to the countryside. We protested and refused categorically. So the family, as well as the cousins living near us, went to some other cousins who had a large apartment in the 16th *arrondissement*, where we settled in after a fashion. I remember I slept on the billiard table [IV.1, p. II(35)].

Towards the end of the war, Hadamards' youngest son Mathieu, who had passed the entrance examinations to the *École Polytechnique* in 1918, was also mobilized. He was sent to the quieter Italian front, from which he returned alive.

The war ended on November 11, 1918, and on June 19, 1919, the Versailles peace treaty was signed. Jacqueline Hadamard wrote: "...the end of the war came with the armistice. There could not be great jubilation for the great number of families (as ours) in deep mourning. But it was at least the end of anguish for so many. It was the moment of detente for those of us left of our generation" [IV.1, p. II(37)].

The war became catastrophy for French mathematics, whose death toll was terrible, because unlike the Germans and the British, the French did not protect young mathematicians but sent them to the Front, and as a result appalling numbers were killed, thus severely weakening French mathematics in the post-war years. From 240 pupils of the *École Normale* who were called up on August 2, 1914, only twenty-three returned intact and 120 were killed [III.195, p. 174].

*The names of Hadamard's sons on the memorial plaque at the Lycée Louis-le-Grand*

Among Hadamard's colleagues who also suffered losses were the Picards who had lost a daughter and two sons and the Borels whose adopted son had been killed.

Jacqueline kept a copybook, written in 1923 in Hadamard's hand, containing detailed biographies of Pierre and Étienne and how they died. A faded wild flower from the herbarium of one of the dead boys is sewn into a page of this copybook, which begins as follows:

*The beginning of Hadamard's copybook with reminiscences about Pierre and Étienne*

To our grandchildren

For you, dear children, whom we may perhaps never see, we write these words from the depth of our despair. They will tell you about those you should have known and who should have been able to witness your first steps: your uncles; but neither you nor we will be able to imagine them in that respect.

Look at their portraits: they are schoolboys like you, when you see this for the first time. They were not even schoolboys who were more serious than others. At home, their upbringing had been done in an atmosphere of freedom (we did not know any other), and we sometimes, jokingly, accused them of taking their principles from Kipling's *Stalky and Co.*, since this book was their constant delight up to the day of their departure.

It was an accusation as a joke: as schoolboys they were as attentive to their homework (from the very first months after entering the *lycée* they were among the best) as they were adored by their schoolfriends and their teachers, many of whom would later speak about them to their younger colleagues.

This work, from which they did not flinch, was done, as with everything else, with joy: a joy which never ceased to illuminate their childhood and youth, and which they relished so much more when together. There was no finer union than theirs, and we always remember the musical improvisations '*à deux*', when, seated at the same piano, they needed only to agree on the genre of the 'piece', the time and pitch, and they were away, anticipating each other all the time, keeping in harmony in all the improvisations dictated by their fantasy.

It is those schoolboys who have sacrificed the beautiful lives they should have lived for their country. And now they are only something in our memory and in that of some friends, ours and theirs, who will, I sincerely hope, be able to tell you about them; and a little in the fine, but so vague, memory which France offers its legions of dead.

This recollection will not be enough for you. You will not let them die for good as long as you are alive. You owe it to them, to know them, to know what they were, all their promise, which, of their own accord, they left, to save us [IV.2].

## 3.7 Robert Debré about Hadamard

For a Frenchman, there is no need to present Robert Debré, a doctor, after whom a Paris hospital and many pediatric departments all over France are named. His son, Michel Debré was a long-time associate of De Gaulle, becoming Prime Minister in the first government of the Fifth Republic, and later Minister of Finance, Foreign Secretary and Minister of National Defence.

Robert Debré was Louise Hadamard's nephew and close friend of her family, being the *garçon d'honneur* at the Hadamards' wedding. The following lines are his unpublished reminiscences about Hadamard, written at the request of his cousin Jacqueline Hadamard:

In answer to the question you ask me, what springs immediately to mind is the really unforgettable memory of the atmosphere of your parents when they were young. I have great experience (in my very long life) of the intimate life of many couples, because I have often been called out because of the anxiety, the anguish for small sick children, and I was aware of the atmosphere in quite a number of homes.

*Robert Debré (1882-1978)*

I have met a good many happy couples. But really the couple
Jacques and Louise Hadamard were something exceptional and
have left me with a unique image, superior to all others. Because
they knew how to appreciate every small particle of happiness:
just a simple little bouquet which he saw her arranging, and he
admired every flower, even each leaf. A simple joke about one of
their friends would give them joy. It was the ability to enjoy each
morsel, an ability which rested in the end on complete happiness.

And that was combined with a gentility which she radiated
completely. It wasn't a blind gentility: they saw clearly their
faults and even the faults of those around them. But this was in
a thriving atmosphere which made them a model of life.

My uncle Jacques accepted his work with a smile and accomplished the tasks in an impeccable way. But besides that there
blossomed a delight in everything: reading, music, art, nature,
friendship and family life. And the astonishing thing was that
nothing of it was naive, but it was accompanied with perfect
judgement.

Both of them had an extraordinary taste for memories, and
each opportunity to recall a past episode made them equally
happy.

At the same time, a remarkable admiration for the intelligence, bordering on the genius. A critical intelligence, but penetrating, and in a number of fields, including history and contemporary events (one must remember that at that time the evolution of thought was not guided by any preconditions).

Jacques Hadamard had an untiring instinct for justice, applied to everyone in society; but the social field was still far away for him and I think that he could not endure it. He had a deep feeling for liberty, a complete respect for the human conscience – something rarer than one may believe.

All that manifested many characteristics from childhood and adolescence, persisting into old age. Consequently a certain enthusiasm: the generosity of the adolescent. That contrasts happily with the attitude of the most *blasés*. But at the same time he had a very dignified insouciance towards practical problems. It wasn't a form of contempt because he did not have contempt for anything.

In all this stage of their life, what stands out are the intellectual qualities: the erudition, the judgement, the good sense in educational methods, with a great seriousness in contemplation. He had an early interest in the education and forming of man.

In summary, the recollection I have of that period is the irradiation of a glowing intellectual power.

A second stage was marked in his life by the Dreyfus affair, which triggered off in him a judgement about people based solely on the criterion of knowing whether they were pro- or anti-Dreyfusards. He was literally obsessed by it, it filled his entire life, on the look-out for the least word, the least episode, the least newspaper article. Around them, people such as Seignobos, Séailles, [6] the *Sorbonne* group of Dreyfusards. But at home more than anywhere else, a loss of proportion which lasted about five years, after which they returned to normal.

Then came the moment, with the start of the 1914-1918 war, when their hopes of peace and internationalism took a bruising. The ideal collapsed. However, at this time, the French were very divided. As all of us, the Hadamards did not have a specific opinion.

At the beginning of the war I didn't see them, of course. When I saw them again, I had a shock: they were not at all the

---

[6]Ch. Seignobos (1854-1942), historian, the author of *Histoire de la civilisation* and several books on the history of France. G. Séailles (1852 - 1922), philosopher who worked on the theory and history of art as well as on problems of moral philosophy. Both became members of the Committee of the *Ligue des Droits de l'Homme* from its foundation.

same. They were searching for a philosophy, a discipline, while keeping (him) a scientific serenity. Of course, he had kept all the clarity of his intelligence. But the joy of life had gone for ever.

In ending, I would say that his influence on me was, and is, fundamental [IV.1, p. I(20)-I(21)].

# Chapter 4

# After the Great War

## 4.1 The twenties

On July 24, 1919, Hadamard wrote to Mittag-Leffler: "For my part, I can barely get back into scientific life, and life as a whole". However, he soon resumed his work with the same intensity, publishing a series of papers each year. "To bear such misfortunes without being irredeemably overwhelmed, mathematics was a great help to the grief-stricken father by forcing him to cut himself off from reality", wrote Denjoy [II.8, p. 34].

In 1920, Hadamard succeeded Appell to the chair of mathematical analysis at the *École Centrale*, although he still kept his positions at the *École Polytechnique* and the *Collège de France*. His pedagogical activities reached their widest scope: while he gave a comparatively elementary course at the *École Centrale*, his lectures at the *École Polytechnique* were considerably more difficult. According to Paul Lévy "...the harder parts were picked out and explained in a way which aroused the admiration of the audience" [II.37, p. 4].

An interesting detail concerning Hadamard's teaching at the *École Polytechnique*, though related to the pre-war years, is mentioned in the survey article [III.24, p. 184]: "Hadamard had elevated demands regarding the contents of his courses, introducing the system of notes at the end of the page on the sheets distributed among students and multiplying the references to recent mathematical publications and articles. The consequences are implacable for the students: 1000 pages of lithographic courses for two divisions 1912-1913 and 1913-1914; this is a record of the time".

Mahieux, *le Général commandant l'École Polytechnique* from 1965 to 1968 and former student of Hadamard, confessing that he "...never left the last dozen of the first hundred of his year", said at the celebration of Hadamard's centenary:

> The *taupin* who saw Jacques Hadamard enter the lecture theatre, found a teacher who was active, alive, whose reasoning combined exactness and dynamism. Thus the lecture became a struggle and an adventure. Without rigour suffering, the importance of intuition was restored to us, and the better students were delighted. For the others, the intellectual life was less comfortable, but so exciting... And then, above all, we knew quite well that with such a guide we never risked going under [II.5, p. 8].

Mandelbrojt recalled at the same jubilee:

> For several years, Hadamard also gave lectures at the *Collège de France*: lectures which were long, hard, infinitely interesting. He never tried to hide the difficulties, on the contrary he brought them out. The audience thought together with him; these lectures provoked creativity. The day after a lecture by Hadamard was rich, full and all day long one thought about the ideas.
>
> It was in these lectures that I learnt the secrets of the function $\zeta(s)$ of Riemann, it was there that I understood the significance of analytic continuation, of quasi-analyticity, of Dirichlet series, of the role of functional calculus in the calculus of variations [II.5, p. 25-27].

In 1920 Hadamard used invitations to give lectures at the Rice Institute (Houston, Texas) and at Yale University. The first invitation was extended by Edgar Lovett whom he knew from the 1900s. Lovett, a mathematician and astronomer, was the first president of the Rice Institute founded in 1908. Twenty eight years later he wrote: "We recall with gratitude the counsel and encouragement which Professor Hadamard accorded the plans of the Rice Institute before pen had been put to paper or spade in the soil. With characteristic unselfishness, he solicited and secured the cooperation of his colleagues in our undertaking" [II.27, p. 68]. At the Rice Institute Hadamard spoke about the early scientific work of Poincaré [I.220].[1]

---

[1] He also lectured on this topic during his courses at the *Collège de France* in 1922.

HE KNOWS MATHEMATICS

This is Professor Jacques Hadamard, of the Department of Mathematics of the French Institute and College of France, and the French Polytechnic School. He comes to this country to lecture on higher mathematics at Yale University and the Rice Institute of Texas. He is considered one of the foremost mathematical geniuses and expounders of the present day

Hadamard's portrait and the newsclipping from the New York City *Town & Country*, April 1, 1920 [IV.33]

When the Rice Institute opened, it comprised only a few buildings, the most impressive of which is now called Lovett Hall

Rice's student newspaper *The Thresher* reported on Hadamard's visit in the following article of March 19, 1920:

### NOTED FRENCHMAN IS
### GUEST OF INSTITUTE

### Dr. Hadamard, Famous Mathematician,
### Gives a Series of Lectures

Doctor Jacques Hadamard, one of the foremost mathematicians in France, has been a guest of the Rice Institute for the past few days. This is the third time Doctor Hadamard has been in America, but on his former trips he did not come as far South as Houston. He has, nevertheless, been interested in Rice for some years, as he was one of the men whom Doctor Lovett consulted before the building of the institution started.

*Edgar Lovett (1871-1957)*

Doctor Hadamard delivered three lectures on mathematics and also a public lecture, the subject of which was "France's

Effort in the War". The chief topic considered was the use of mathematics in the anti-aircraft warfare. "When one shoots at a flying bird with a small gun and allows for the speed of the bird in shooting ahead of it, one illustrates a fundamental principle in anti-aircraft artillery", said Doctor Hadamard. The greatest difficulty is in getting the instruments used as "eyes" at a great enough distance apart and at the same time having reports come from them instantaneously. This was finally accomplished, Doctor Hadamard said, by having a row of men extending from one "eye" to the other and each passing on the readings taken at the eyes. Another topic discussed was the ever changing conditions of warfare. There were many machines used one month which would be out of date and ready for the scrap pile the next.

The lecture was delivered in English without the use of technical expressions, which, together with Doctor Hadamard's wit, made the talk very enjoyable to those present there. [IV.34]

*Old print of Yale College and State House, New Haven, Connecticut*

His Yale lectures were supported by the Silliman foundation. This is the name of a family which left a legacy to the University in order "to establish an annual course of lectures designed to illustrate the presence and providence, the wisdom and goodness of God, as manifested in the natural and moral world... It was the belief of the testator that any orderly presentation of the facts of nature or history contributed to the end of this foundation more effectively than any attempt to emphasize the elements of doctrine or of creed; and he therefore provided that lectures on dogmatic or polemic theology should be excluded from the scope of this foundation, and that the subjects should be selected rather from the domains of natural sciences and history, giving special prominence to astronomy, chemistry, geology and anatomy" [I.223, p. vi].

The fifteenth in the series of Silliman lectures was given by Hadamard and devoted to his work on second order hyperbolic equations. It summarized his many years of research. Hadamard has previously lectured on this topic at Columbia University (1911), and at the Universities of Rome (1916) and Zurich (1917).

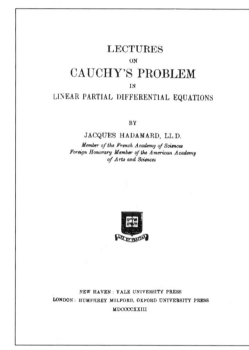

Poincaré's work on partial differential equations was the topic of Hadamard's lectures given in Barcelona in 1921 [I.225]. In 1922 Hadamard's probably best known book *Lectures on Cauchy's problem in linear partial differential equations*, based upon his Yale Lectures, was published in America.

Hyperbolic equations were also the subject of his two talks at the International Congress of Mathematicians in Strasbourg in September 1920: *Sur la solution élémentaire des équations aux dérivées partielles et sur les propriétés des géodésiques* and *Sur le problème mixte pour les équations linéaires aux dérivées partielles*. The Congress had a distinct political flavour. Even the choice of the place was provocative because Strasbourg belonged to the territory (Alsace-Lorraine) lost by France in 1872 and returned after the First World War. Mathematicians from twenty-seven countries participated in the Congress, though those from Germany and its allies did not, since they were not invited. As a result there were many debates whether this meeting could be considered as a regular International Congress of Mathematicians. In particular, Mittag-Leffler insisted that the choice of Strasbourg contradicted to the decision of the previous Congress to have the next one in Stockholm. Hardy, Littlewood, Mittag-Leffler, the president of the American Mathematical Association, D.E. Smith, and others were in opposition to the Congress and did not come. The Congress was presided over by Émile Picard, and Camille Jordan was its Honorary President (see [III.6]).

We do not know if Hadamard had his doubts before the meeting but as the twenty-six year old Wiener, who came to Strasbourg, recalled:

> Professor Jacques Hadamard of Paris played a great part at the congress. He was then only in his middle fifties, but his reputation had been well established before the end of the nineteenth century, and to us fledglings he was a great historical landmark. Small, bearded, very Jewish-looking in the *fin* French way,[2] he occupied a unique position in the affections of his younger colleagues [III.422, p. 67].

When, in 1924, Hadamard took part in the conference marking the seven-hundredth anniversary of the University of Naples, the young Tricomi gave a talk on his recent studies of partial differential equations of mixed type, being elliptic in one part of the domain and hyperbolic in the remainder. The equations are indispensable in transonic aerodynamics, as was discovered much later, but at that time their importance was not obvious. Hadamard immediately grasped the significance of Tricomi's new theory, even upbraiding one of the participants for raising quibbling objections during the discussion which followed. As Tricomi wrote in [III.399, p.14], the talk had infinite value for him since it gained him Hadamard's esteem.

Being invited again to the Rice Institute by Lovett in 1925, Hadamard spoke about Poincaré's later work [I.311]. Lovett wrote : "Twice in our history he delivered under our auspices courses of lectures which he allowed us to publish and distribute, under our imprint, to men of science in this country and abroad. And in the course of this fleeting quarter of a century he has received in the hospitality alike of his home and of his seminar in the halls of the *Collège de France* students and instructors of this institution" [II.27, p. 68].

Hadamard participated in the affairs of the rapidly growing Hebrew University of Jerusalem inaugurated on April 1, 1925. The University was administrated by an international Board of Governors headed by Chaim Weizmann. The Board elected the president of the University, discussed the budget, authorised the establishment of chairs and faculties, appointed professors and lecturers. Since self-governement of the University during its early years

---

[2]It is interesting that in the Russian translation of *I am a mathematician*, Moscow, Nauka, 1964, Wiener's words "very Jewish-looking" were omitted.

was impossible, a Provisional Academic Council was created consisting of its professorial staff and eminent foreign scientists.

Hadamard became a member of both the Board and the Council in September 1925, together with Albert Einstein, the philosopher Martin Buber, the psychiatrist Sigmund Freud, the mathematician Edmund Landau, and others. They met once a year in different European cities [IV.36]. At the second conference of the Academic Council, held in Zurich in August 1929, Hadamard supported the inclusion of applied mathematics in the curriculum. He said that "it was most important to fight against the tendency of young students to erect mental walls in their minds separating the various disciplines. An exclusive study of theoretical mathematics had often been described as dangerous and he thought it was essential that the teaching of mathematics should be supplemented by that of applied mathematics" [IV.35].

*Einstein played an important role in the Board of Governors and in the Academic Council of the Hebrew University of Jerusalem. On this photograph of 1923 he is planting a tree on Mount Scopus, where the first buildings of the University were erected.*

In 1928 Hadamard spent a week in Czechoslovakia. The following account of his visit appeared in *Časopis pro pěstování matematiky a fysiky:*

On May 22, an outstanding French mathematician, Jacques Hadamard, professor at *Collège de France* in Paris and a member of the French Academy of Sciences, arrived with his wife in Prague. He was invited by the Faculty of Natural Sciences of the Charles University to give three lectures that he delivered on May 23 in the morning and in the afternoon and on May 24 in the morning.

As the topic for his lectures he chose Huygens' principle. First, he formulated Huygens' principle in three different theorems, after which he gave an explanation of these theorems and talked about their mathematical significance. A large part of the lectures consisted of the results of his own scientific work. He had been led to the study of Huygens' principle by his work on partial differential equations. His exposition fully revealed the great French mathematicial tradition, it was filled with a great richness of mathematical ideas and it was remarkable in its spirit and clarity. In his lectures, Mr. Hadamard never omitted emphasizing those points in which the material discussed had not yet been satisfactorily solved, and which can therefore become starting points of new research. With his perceptive intellect he was able to reveal a great number of such points.

In honour of Mr. Hadamard and his wife, the Faculty of Natural Sciences organized a dinner in the Municipal Halls, which was attended by the chairman and several other members of the Union [of Czechoslovak Mathematicians and Physicists]. Mr. Hadamard left for Brno on May 24 to deliver a lecture on the same subject the following day at the Masaryk University. On May 29, he took part in the final meeting of the VIth Congress of Czechoslovak Natural Scientists, Physicians and Engineers in Panteon. After an excursion to the Tatra mountains, he returned to Paris [III.429].

The first and the second volumes of Hadamard's *Cours d'Analyse de l'École Polytechnique* were published in 1927 and 1930. In writing this course, he did not intend to create a new treatise similar to those written by his teachers Jordan, Picard and Goursat, where many of the chapters had the character of monographs. Hadamard's course had a pedagogical purpose and was intended for physicists, astronomers, and engineers. The text contained elements of the theory of functions of a complex variable, mathematical physics, the calculus of variations, and probability theory.

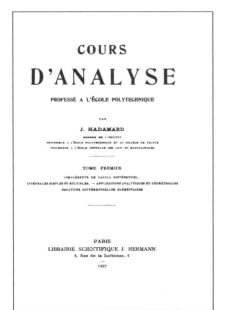

*The title pages of the two volumes of Hadamard's Cours d'Analyse*

In 1926 the second edition of *La série de Taylor et son prolongement analytique* [I.253] appeared, which was prepared by Hadamard in co-authorship with the twenty-seven-year-old Szolem Mandelbrojt. The book reflected recent progress in the domain. At the same time, Hadamard devoted several papers to quite another area, the theory of Markov chains in probability theory. The topic of his report to the International Congress of Mathematicians in Bologna in 1928, was *Sur le battage des cartes et ses relations avec la mécanique statistique.* In Paul Lévy's opinion "this work is above all important because of the impulse it gave to the theory of Markov processes, which today is one of the principal parts of the calculus of probabilities" [II.37, p. 17].

Hadamard travelled a lot. Besides his already mentioned trips to the U.S.A. in 1920 and 1925, to Spain in 1921, to Czechoslovakia and Italy in 1928, and to Switzerland in 1929, he visited Brazil in 1924, Argentina in 1930, and Egypt in

*Hadamard in Egypt, 1933*

1933.

He headed the French delegation to the 1932 International Congress of Mathematicians in Zurich, where he gave a talk entitled *Sur les équations aux dérivées partielles d'ordre supérieur*. In June 1935, Hadamard was a chairman at the cycle of international lectures on partial differential equations, organized by the University of Geneva, and gave a talk on well- and ill-posed boundary value problems. He published articles abroad: 1923 – in Barcelona; 1924 – in Naples and Rio de Janeiro; 1925 – in Madrid; 1926 – in Kazan; 1929 – in Prague; 1930 – in Kharkov; 1933 – in Tokyo.

*Hadamard during the Congress in Zürich, 1932 (to the right is Louis Kollros)*

During the 1920s and 1930s Hadamard continued to work on problems of education, both mathematical and general. He wrote many papers on teaching methods as well as articles and letters on programmes for mathematical education and improvements in the presentation of various topics. To help those who taught analysis, he addressed the articles *La notion de différentielle dans l'enseignement* [I.224], *Sur la théorie des séries entières* [I.251] and others, and he gave an exposition of some questions in algebra which are difficult to present: *La non-résolubilité de l'équation du cinquième degré* [I.323], *La théorie des équations du premier degré* [I.331], as well as in geometry *Sur un théorème de géométrie élémentaire* [I.216], *Les développables circonscrites à la sphère* [I.329]).

In [I.224] Hadamard endorsed Poincaré's opinion [III.319] that one should speak about derivatives in the course of analysis and avoid complicated explanations connected with the symbol $d$. His arguments were later used by some marxist historians of mathematics to support Karl Marx' philosophical speculations on derivatives and differentials [III.269]. There even exists a concept of the differential as an operator symbol in the sense of Marx-Hadamard [III.209], [III.150]. We refer to the last sentence of the article [III.150]: "The definition of the differential in Hadamard's course[3] demonstrates that mathematicians also begin to have the same understanding of the general nature of the differential calculus to which the dialectic in the hands of a philosopher-materialist led, half a century ago".

$$dy(x) = y'(x)\,dx$$

Hadamard published a polemical paper [I.252] in 1926 expressing his views on the new programme in mathematics. He was also interested in questions concerning the humanities, an interest which is demonstrated in his long article *À propos de l'enseignement secondaire* [I.210], dealing with the interconnections between the teaching of languages, literature, and history. His thoughts on the development of culture in the process of secondary education are presented in the paper *L'enseignement secondaire et l'esprit scientifique* [I.212].

In 1932 Hadamard was elected Chairman of the International Committee on Mathematical Education. For many years, he wrote for the journal *L'Enseignement Scientifique*. On his seventieth birthday, this journal published a detailed survey of his work on the problems of education [II.9].

## 4.2  *Jeux d'esprit*

Hadamard's scientific curiosity was insatiable, being just as undiminished as during his youth, as Jacqueline Hadamard recalls:

> The sessions at the *Académie des Sciences* were an extraordinary source of education for him. Our friend Magat[4] told me that he was present at one session. What struck him was that

---

[3]In Chapter 1 of *Cours d'analyse* [I.258].
[4]Michel Magat (1908-1978) was a well-known physical chemist.

most of the members sat quietly in their seats. But two of them, Jean Perrin and my father would flit about incessantly from one row to another. As for Jean Perrin, I can't say why, but as for my father, I know very well that for him it was a matter of asking one specialist colleague at what stage genetics was, another what was the name of the eagle he had seen during the vacations, and a third if such and such a chemical theory had made any progress, and so on.

*Jean Perrin (1870-1942)*

Coming out of the meeting where, like the others, he hadn't heard what was being said from the rostrum, he thought about the answers he had obtained, bought a cornet of fried potatoes and ate them while walking, enchanted by "his club". Unfortunately, having looked at his watch, he saw that he had a quarter of an hour before going home. So he thought he would have the time to go to the museum or to a library... and he forgot the time, he who was, despite this, astonishingly punctual when it was a matter of going to one of his classes. But otherwise he lost track of time. My mother, seeing that his return home time had passed an hour or two ago, would fret with anxiety. I had all the trouble in the world to prevent her from telephoning the police or the hospital.

And then he would come home, quite relaxed, completely surprised that it was that time already: "I only went to see the mushroom exhibition!" [IV.1, p. III(27)]

Although Hadamard defined philosophers as those who look for a black cat in a dark room, where there is none, and who find it there [III.358, p. 45], he became a member of the French Philosophical Society (*Société française de Philosophie*) from its inception, on February 1, 1901. He was a frequent participant at its meetings which took place several times a year. He also became involved in the work of the Society on the Philosophical Dictionary which started in 1902 and lasted as much as twenty years.

The meetings of the Philosophical Society began with a main lecture and this was followed by a discussion, the accounts of the sessions being published in the *Bulletin de la Société*. The audience, of course, varied according to the topic of the meeting. Over the years one could meet with philosophers such as Émile Boutroux, Henri Bergson, Xavier Léon, Émile Meyerson, the sociologist Émile Durkheim, the psychologist Henri Delacroix, the historian and philosopher Léon Brunschvicg, the poet and philosopher Paul Valéry, the mathematicians Jules Tannery, Émile Borel, Paul Painlevé, the physicists Paul Langevin, Jean Perrin, and many others. The topics of the pre-war meetings in which Hadamard took part in discussions were: *On the objective value of physical laws, The place and the character of philosophy in secondary education, Matter and motion, The significance of pragmatism*, and even *The learned horses of Elberfeld*.

One may wonder what horses had in common with philosophy. During the early years of this century there were many speculations about intelligent animals, especially about horse-computers, this latter due to the disciples of the trainer Krall from Elberfeld, a suburb of the city Wuppertal (Germany). Claparède, a well-known Swiss psychologist who had observed Krall's experiments was invited by the Philosophical Society

---

**Séance du 13 Mars 1913** [1]

**LES CHEVAUX SAVANTS D'ELBERFELD**

1. — M. Claparède a donné dans les *Archives de Psychologie* (vol. XII, p. 274), le récit de quatre séances qui lui ont été offertes par M. Krall, le 30 et le 31 août 1912.
Voici quelques *faits* extraits de ce récit :
1° M. Krall écrit à la planche :
$$\sqrt{36} \times \sqrt{49} =$$
*Muhamed*, l'étalon qui est son meilleur élève répond, d'abord *52*, puis, sur cette remarque que c'est faux, *42* (juste). — M. Krall inscrit alors le signe + au-dessous du signe × dans l'opération ci-dessus, et prie le cheval d'additionner les deux réponses. Celui-ci répond aussitôt *13*, qui est juste.
2° M. Claparède propose l'opération suivante : $\sqrt{61466}$. Réponse en quelques secondes : *28* (juste).
3° *Zarif*, l'émule de *Muhamed*, a donné une épreuve notée par M. Claparède comme excellente : M. Krall ayant placé devant lui divers cartons portant des chiffres de couleur différente (7, 1, 6, 5, 4, 3), lui demande d'*additionner les chiffres bleus* (soit 6 et 4). *Zarif* a donné immédiatement la réponse.
4° Comment t'appelles-tu? — *Garif* (confusion du *g*, donné par le nombre 33 avec le *z* correspondant à 53). — Quelle autre lettre aurais-tu pu mettre à la place du *f* (*v*, juste, car en allemand le *v* se prononce *f*). — Qui est *Zarif*? — Réponse : iig (c'est-à-dire, ich,

1. Présents à cette séance : MM. Abauzit, Beaulavon, Dʳ Beredska, Bouglé, Couturat, Cresson, Dagnan-Bouveret, Delacroix, Delbos, Drouin, Dumas, Hadamard, E. Halévy, H. Lachelier, Lalande, X. Léon, Menegaux, Meyerson, Milhaud, Parodi, Quinton, Rey, Roustan, Vendryes.

BULLETIN SOC. FRANÇ. DE PHILOSOPHIE. T. XIII, 1913.    9

to give an account at the meeting held on March 13, 1913.

*Karl Krall (1863-1929), a jeweller and animal trainer fanatically devoted to the idea of intelligent animals. Much interesting material about Krall and his experiments with intelligent horses and dogs is collected in Stadtarchiv Wuppertal. One can find there, in particular, his book Denkende Tiere. Der Kluge Hans und meine Pferde Muhamed und Zarif (1912, Verlag von F. Engelmann, Leipzig)*

Claparède said:

> What impressed me above all were the replies given by the horse when everybody, Krall included, were outside the room, and the replies were monitored only through a glass hole made in the large wooden door separating the shed from the courtyard. In these conditions, Zarif, and especially Muhamed, gave some astounding replies, such as: $\sqrt[4]{614656} = 28$. (The number was chosen by me from a list of powers; I did not know the answer) [III.234].

The participants of the discussion expressed quite varied opinions about this. Hadamard's attitude was one of caution: he only asked Claparède to give "some details of the experiments in which he had taken part, on the presence or absence of the trainer when the horses gave good and bad answers, on the way in which the questioners had proceded."

Claparède concluded the discussion with the following words, close to the heart of a computer scientist:

I had already, in my *Archives* article, called attention to this
fact that dexterity in calculating is not a proof of intelligence,
and I cited the example of the calculating prodigies. I am happy
to see that Misters Couturat and Lalande share this opinion. In
the December 1912 issue of the *Rivista di Psicologia*, Dr. Ferrari
asked if the horses did not have recourse to a number system
simpler than the decimal system, for example binary counting,
in use in China. But Krall's teaching was based on the decimal system, it would be quite impossible for the horses to have
substituted another system themselves! [III.234]

Hadamard was quite a competent participant of the sessions devoted to
the philosophical problems of physics, then in a state of rapid development.
(He kept abreast of new physical ideas at Langevin's "Tuesdays" at the
*Collège de France*, and in conversations with Langevin and Perrin.) When
Einstein's theory of relativity was the subject of the meeting held on April
6, 1922, Hadamard held a discourse on the relation of this theory to reality.
Einstein's reaction was:

*Paul Langevin (1872-1946)*

> I have only one word to say about Mr. Hadamard's remarks. Mr. Hadamard has said that a theory should first be logical, then agree with experimental facts. I do not believe that this is sufficient and, in any case, it is not evident *a priori*. To say that a theory is logical means it consists of symbols which are connected with each other by means of certain rules, and to say that the theory conforms to experience means that one has rules of correspondence between these symbols and the facts. Relativity emerged from experimental necessities; this theory is logical in the sense that one can give it a deductive form, but one must also know unambiguous rules which make its parts correspond to reality. There are therefore three postulates and not two, as Mr. Hadamard thought [III.227, p. 97].

Hadamard commented: "In other words, what one examines is a logical theory and the totality of rules which make it correspond to reality."

At another session of the Society, which took place on November 12, 1929, and was devoted to *Determinism and causality in modern physics*, Louis de Broglie spoke on the fundamental ideas of quantum mechanics, in the presence of Einstein, Langevin, Perrin, Borel, and Hadamard. In the subsequent discussion, Hadamard recalled his conversations with Durkheim in Bordeaux:

> Physics, we have just had the confirmation of this, finds itself at this moment more and more led to admit a certain indeterminism, one can almost say a certain free arbiter, in the interior domain of the atom. Now this fact, the fact that such a free atomic arbiter would be, in virtue of the law of large numbers, perfectly compatible with natural determinism which experience shows us, was remarked upon a long time ago by the metaphysicists. I should know, because it is from conversations I had with Durkheim between 1893 and 1896, so, it must be admitted, many years before dawning on the physicists, that this idea was communicated to me [III.102, p. 151].

On January 28, 1928, the subject of the meeting of the Philosophical Society was *Artistic creation*, and the speaker was Paul Valéry. Hadamard was in the audience and took part in the discussion afterwards [III.220, p. 18-20]. He had long been interested in the psychology of invention, stimulated by his own work and the memory of his mental processes, failures, and flashes of inspiration. He followed contemporary psychological work on the mechanism of creativity, and was impressed by Poincaré's lecture on mathematical invention, given in 1908 to the *Société de Psychologie de Paris*, the first study of this subject.

Eight months before his talk at the Philosophical Society, Valéry, who was often inspired by physical and mathematical associations, wrote to Hadamard:

> Monsieur and Illustrious Colleague,
>
> If I were not overwhelmed with work and under the pressure of an imminent departure to England, I would have come today myself to thank you for your friendly and obliging reply. At the first occasion of free time, I shall go and ask for the *Nouvelles Annales* in a library and I shall try to assimilate the principles of an analysis which I know to be so precious and so rich in applications in all contexts. I do not, in particular, see anything that might aid us more powerfully to draw up a map of our mind and of its internal relations whenever a question requires this.

*Paul Valéry (1871-1945)*

You see all the fancifulness of my idea immediately, and how it relates bizarrely or illegitimately to mathematics!...It is just that I frequently invoke your subtle gods when I struggle in a disorder of impressions and relations which defy vocabulary and syntax – and the possible reader. In certain researches (which I shall call "philosophical" to spare our time) I have wished a thousand times that a mathematical demon – not the one of Maxwell, but the one of Riemann – appear and aid me in constructing a scheme of my simultaneous ideas. Because of the simplicity of the initial conditions and the clear and perfect separation of operations, the geometer has the fortune of realizing, what the literary artist perceives and misses at each instant, because of the complexity of the objects, the impure and statistical character of common language and of its forms... I'll stop... I was going to write things that scandalise logic, and perhaps common sense. Analogy gives the most dangerous temptations. For you, happy geometers, it describes itself or manifests exactly through the identity of the forms or their transformations. But for us, it only happens in the domain (which is not at all analytical) of dreams.

Please accept, my dear and illustrious colleague, my renewed thanks and my devout respects.

Paul Valéry.

Also, please excuse the inversion of this page – which should be transformed into a surface with only one side! [III.404, p. 171]

*Möbius' band made of Valéry's letter, as he had wished*

Hadamard seriously addressed the problem of mathematical creativity when preparing his talk for one of the meetings of the *Centre International de Synthèse* in Paris. This was a society founded in 1924 by the historian and philosopher Henri Berr. The "weeks of synthesis" at the Centre were devoted to discussions of different aspects of human thought and activity. The topic of the ninth "Week", held between May 18 and May 22, 1937, was *Invention*, at which Louis de Broglie and Edmond Bauer spoke about thought processes in experimental sciences. Paul Valéry's talk dealt with creation in poetry, and Hadamard analysed mathematical invention. With his lecture, he laid the foundations for his future study of this topic.

## 4.3   Correspondence with Mittag-Leffler

*Gösta Mittag-Leffler (1846-1927)*

Gösta Mittag-Leffler graduated from Uppsala University in 1872 and the next year he went to Paris to continue his mathematical studies. Hermite met him warmly but surprised him by saying: "You have made a mistake, sir, you should follow Weierstrass' course at Berlin. He is the master of us all" [III.281]. Following this advice, Mittag-Leffler attended Weierstrass' lectures in Berlin during the academic year 1874-1875. In 1876 he proved his famous theorem on the decomposition of meromorphic functions, analogous to the decomposition of rational functions into partial fractions. By the end of the century he had won an international reputation for his work in the theory of analytic functions and differential equations. In a series of papers on meromorphic functions published after 1900 he studied the analytic continuation of a power series outside its disc of convergence. After spending the years 1877-1881 in Helsingfors as professor at the University, he moved to Stockholm where he took the newly founded chair of mathematics at the recently organized *Högskola*. A prominent teacher, he worked there until his retirement in 1911.

In 1882 Mittag-Leffler founded the journal *Acta Mathematica*. The proposal to start a Scandinavian mathematical periodical was suggested to him by Sophus Lie. Mittag-Leffler took up the idea enthusiastically, and transformed it into that of a major international journal. A high-ranking mathematician with broad connections in the mathematical world, a pupil of Hermite and Weierstrass, ambitious and charming, he appeared to be a right person to realize the project. The story of the foundation of *Acta Mathematica* is described in the interesting paper by Y. Domar [III.108]. We mention that Hermite proved to be very helpful. He wrote to Mittag-Leffler about mathematical life in France and, in particular, about achievements of his talented young pupils Appell, Picard, and Poincaré, whom Mittag-Leffler immediately attracted to the journal. Especially important was Poincaré's promise of his long mathematical papers for the first and subsequent issues of *Acta Mathematica*.

To obtain a positive attitude from Weierstrass and Kronecker toward the journal was a more delicate task, partly because the new journal would be a rival to *Journal für die reine und angewandte Mathematik*, which they were in charge of. In order to overcome this difficulty, Mittag-Leffler obtained and cleverly used the support of the Swedish King Oscar II, so that the German mathematicians, who were previously awarded with a Swedish royal order, responded benevolently.

Now, when the idea of the journal was approved both by French and German mathematical communities, Mittag-Leffler was able to launch a really international publication. In fact, during all his life he, a representative of the neutral Sweden, played a mediator's role in the European mathematical world split first by the Franco-Prussian and then by the First World War. Mittag-Leffler's success with *Acta Mathematica* "came like a bolt of lightning from the blue, and the way he achieved it reflected not only his imagination and daring but also his unusual talent for scientific diplomacy" as D.E. Rowe put it [III.348, p. 599].

During many years Mittag-Leffler's permanent concern was to attract rising mathematical stars to his journal and, of course, Hadamard appeared in the list in due course. The beginning of Hadamard's correspondence with Mittag-Leffler dates from 1893, just before Hadamard's move to Bordeaux. There are thirty-five letters from Hadamard and nineteen from Mittag-Leffler, kept at the Mittag-Leffler Institute in Djursholm.

The first letters concern Mittag-Leffler's proposal to publish something from Hadamard in *Acta Mathematica*. One reads that Hadamard first refused the offer, and then hesitated as to whether he ought to refuse a second time, because he had only a paper which did not seem deserving publication in such an authoritative journal. Finally, he promised to submit this article

but took his time in sending it. "You are doubtless somewhat surprised at
still not having my manuscript," he wrote later in 1894, "This is because,
when writing, I added some developments which make the paper somewhat
longer than I had thought at first. However, I think that I shall be able to
send you my work soon" [IV.46]. The paper on series with positive terms
[I.19] appeared in *Acta Mathematica* the same year.

Mittag-Leffler and his wife Signe visited the Hadamards in August 1900,
during the Second International Congress of Mathematicians in Paris. From
then on, Mittag-Leffler began his letters to Hadamard with *Mon cher ami*
instead of *Cher Monsieur* as before. (Hadamard, twenty years the younger,
carried on with *Cher Monsieur* in his letters to Mittag-Leffler.)

*Mittag-Leffler in his house in Djursholm, a suburb of Stockholm*

Many of the subsequent letters from Mittag-Leffler also deal with Hadamard's contributions to *Acta Mathematica*. The following is the beginning of
the letter, dated March 24, 1907, in which Hadamard's large memoir on the

Cauchy problem [I.148] and the famous library of the future Mittag-Leffler Institute are mentioned:

> My Dear Friend,
> I am sorry about the delay in publishing your memoir and I can most definitely state that such a delay will not occur again. The explanation is that I am in the process of rebuilding my house completely so as to get as much space as possible for my library. During the rebuilding, everything has been left in a frightful disorder and it is only after having received your letter that I found your memoir, which was immediately sent to the printers with instructions to typeset it without delay and to stop work on articles already begun. Please forgive me and be assured that this will not happen a second time. There is always room for you and you will be printed without delay and before all others [IV.45].

In the following letter from 1913, Mittag-Leffler writes about Hadamard's article for *Acta Mathematica* giving a survey of Poincaré's work:

> Tällberg, August 8, 1913
>
> My Dear Friend,
> Your kind letter came to me here in the country, where I have a small property.[5] I include a photograph of my house. I think that the most practical thing will be if you finish your article as soon as possible. It is of some importance for me to have it soon. Otherwise everything will become complicated with its publication. That doesn't prevent you from being able to make some changes afterwards, when you are back in Paris.
>
> It is true, it is a difficulty which Poincaré himself made the basis of the principal aspects of his work. But two different people neither see nor judge the same subject in the same manner, even when it concerns mathematics. You have doubtless your own view about Poincaré's work and there is no living geometer whose opinion would be as interesting as yours.

---

[5]Carleman wrote: "Apart from his mansion in Djursholm, Mittag-Leffler built a magnificent summer residence Tällgården in Dalecarlia, situated in great natural beauty at the southwest slope of Tällberget, with a free view over the lake Siljan and its immense framing of mountains dressed with forests. Here he spent the summers every year and as a rule also the Christmas and New Year celebrations during the two last decades of his life [IV.47].

I am myself also very tired, and I absolutely must rest for a time. Madame Mittag-Leffler joins me in asking you to remember us to Madame Hadamard, your kind and untiring secretary. [IV.45]


It seems that by autumn 1913, Mittag-Leffler received two chapters of Hadamard's manuscript devoted to function theory and to ordinary ifferential equations. Only the third chapter was missing, which was devoted to Poincaré's work on partial differential equations, and Hadamard did it very quickly:

Djursholm, October 11, 1913

My Dear Friend,

The third part of your study on Poincaré's work is now in my hands. I have not yet had the time to read what you have written, only a chance to glance through some parts. I already have the impression that you have – as I expected – erected an imperishable monument to your great predecessor. I thank you profoundly and for my part I shall do everything I can to give you prominence.

Please give my respectful regards to Madame Hadamard and thank her kindly from me for everything she has done to facilitate your work.

When I have read your article entirely, I shall immediately send it to the printers. That will be in two or three days.

Please be assured, my dear friend, of my great appreciation of your mathematical genius and of my deep gratitude. [IV.45]


During the hard years of the First World War Mittag-Leffler did his best to keep mathematicians from belligerent countries together by publishing their papers in *Acta Mathematica*. G.W. Dauben reproduces the following abstract from Mittag-Leffler's letter to Fejér dated November 29, 1917:

As you have no doubt noticed, I am always able to publish articles from various countries side-by-side. Mathematics is the science which is the least national, and I hope that mathematicians will be able one day to take over the direction when the time again comes to reestablish international scientific contacts. [III.93]

In 1916, the year of his seventieth birthday, Mittag-Leffler and his wife transferred the ownership of their house in Djursholm to the Royal Swedish Academy of Sciences. It became the building of the future Mathematical Institute,[6] which was opened formally in 1919. Mittag-Leffler served as its director until his death in 1927 and was succeded by Carleman.

The correspondence of Mittag-Leffler and Hadamard, interrupted by the war, resumed in 1919:

Tällberg, April 8, 1919

My Dear Friend,

I have just received the *Comptes Rendus* of the 17th of March, where I find your extraordinary note *Remarques sur l'intégrale résiduelle*. Would you not like to write a lengthier article on this subject for my *Acta Mathematica*?

My volume dedicated to the memory of Poincaré, in which your article is among the longer ones, was printed a long time ago. However, I didn't want to publish it before peace was concluded and a little tranquility reigned over the world. My hope is to be able to come to Paris to present it to the *Académie des Sciences* myself.

I hope that you, as well as your family, did not suffer too much from this terrible war which France endured with so much heroism and glory.

Remember us both, Madame Mittag-Leffler and myself, to Madame Hadamard, my dear friend.

Hoping to be able to shake your hand once again in a not too distant future.

Your very devoted,
Mittag-Leffler. [IV.45]

After learning of the terrible losses Hadamard had suffered, Mittag-Leffler answered with the following letter. (It is also interesting because

---

[6] At present the Mittag-Leffler Institute carries out annual research programs in mathematics by inviting about twenty active specialists in a certain field. The rich library of Mittag-Leffler is systematically completed and his archive contains a priceless material for historians of mathematics. The Institute publishes two journals: *Acta Mathematica* and *Arkiv för matematik*.

of his mention of young Carleman.)

Tällberg, September 9, 1919

My Dear Friend,

I thank you kindly for your letter of July 25. It caused me great pain to learn that you also had to suffer, as so many of my other friends, the hardships which this tragic war has caused all of humanity.

For me personally, these four years have been quite a hard time, because of the state of my health. But I do not want to speak of that now. However, I permit myself to send you two small communications which, I hope, will interest you. They were given to me by Mr. Carleman, the first to be made *boursier* at my Institute. I find that the choice of first *boursier*[7] I made was not bad and I expect much from Mr. Carleman in the future. I would like, moreover, to send him to you in Paris, as soon as the difficulties which still remain after the war have disappeared, and when such a journey will be profitable for him. You will then see if you share my opinion.

The Poincaré volume of *Acta Mathematica* is already printed. Only an introduction is lacking, put aside by the war. As soon as the journey will not be too difficult, I hope to be able to come to Paris myself and to present it to the *Académie*.

Meanwhile, please remember us, Madame Mittag-Leffler and myself, to Madame Hadamard, and be assured of my old and faithful friendship and my admiration for the discoveries with which you have already enriched our science, which, I dare hope, will be followed in a not too distant future by many others.
[IV.45]

It was not until 1921 that the volume of *Acta Mathematica* dedicated to Poincaré appeared, with, amongst other contributions, a long survey of Poincaré's mathematical work, written by Hadamard before the war. "The present volume had just about been printed five years ago, but because of the weight of the misfortunes which afflicted the different peoples in the world, we felt unable to publish it until now," Mittag-Leffler wrote in the preface.

The last two letters from Hadamard were written in 1924. In the first of them, Hadamard discussed Mittag-Leffler's proposal to publish a survey of Hadamard's own work.

---

[7]scholarship holder

Paris, February 9, 1924

Dear Monsieur,

I send you the corrections to my 1908 memoir, as you requested. It could be useful for readers if you indicate that this theory is taken up again in my Yale work.

I now come to your proposal concerning an analysis of my works. We talked about this with Mr. Lebesgue, to whom you naturally made a similar proposal. We both think that it should not be only French mathematicians who ought to participate in your project, and so we would be grateful if you would inform us of the foreign mathematicians who have joined in.

Independently of this question of principle, there is in fact a difficulty which I should not hide from you. I am not, I hope, too old, of which you express a slight fear; unfortunately, I have been completely snowed under with tasks of varying order for some years now, which comes to the same thing. This is to tell you that the analysis in question will not be possible unless

1. You are not in a hurry.

2. I may take as a starting point my old *Notice* for the *Académie des Sciences*, and even then making only very few changes (a little different stress in some passages, made necessary by the new circumstances in which I shall be). It is impossible for me to consider the idea of taking on all this writing from the beginning.

In wishing to give you satisfaction, I hope that these conditions, which I am obliged to require of you, are not such that they will stop you.

Yours sincerely,

J. Hadamard. [IV.46]

Mittag-Leffler evidently accepted Hadamard's conditions, but almost a year passed by without the survey having been completed, and Hadamard wrote on December 21, 1924, explaining why:

I have not forgotten your request, a great honour for me, to speak about my work to your readers – which, it is true, is not without some difficulty, as it is always embarrassing to talk about oneself. Returning just now from a long journey in Brazil that compelled me to put aside many things which I find again on my return, I am obliged to ask you once again for a little patience. However, I can assure you that I have not forgotten my promise and I shall begin writing as soon as it is possible. [IV.46]

We do not know whether Hadamard ever began working on the survey, but it has never been published.

Mittag-Leffler died on July 12, 1927. Among the numerous newspaper articles written by Hadamard there is one entitled *French and Swedish research in close cooperation*, which appeared in Swedish translation in the October 27, 1939, edition of the daily *Svenska Dagbladet*, together with articles by Volterra and two astronomers, W.S. Adams and E. Strömgren, on the occasion of the bi-centenary of the Royal Swedish Academy of Sciences. Hadamard's contribution begins with an appreciation of the work of his deceased Swedish friend:

> I find it a pleasant duty to remember everything that the Swedish mathematical school means to mathematical science in general, for my country's science and for my own scientific life.
>
> When I think of the problems which have been the subject of my own studies, one great name above all comes to mind: Mittag-Leffler. While I was still at the school-desk, he formulated a classic theorem, a beautiful generalization of a result of Weierstrass. But from 1900, he came with completely different investigations, discoveries of a much more personal nature, more unexpected and profound than the theorem just mentioned, bringing a completely new development of the theories to which we devoted our main articles [III.99].

## 4.4  Hadamard and André Bloch

Hadamard's name appears in some stories about André Bloch (1893-1948), a brilliant mathematician with a tragic life. Some recent sources of information about Bloch are two papers in *The Mathematical Intelligencer*: one by D.M. Campbell [III.65] which contains reminiscences of other mathematicians, and the second paper by H. Cartan and J. Ferrand [III.71] giving a documented account of Bloch's life.

In 1917, returning to Paris from the front with a contusion, Bloch killed his brother, his uncle and his aunt in order to perform "his eugenic duty: he had to eliminate a branch of his family that he considered defective" [III.71]. He was placed in the Saint-Maurice psychiatric hospital, where he spent thirty-one years until his death. During all this time, Bloch worked on mathematics. As well as some results on the theory of holomorphic and meromorphic functions, his papers deal with number theory, geometry, algebraic equations, kinematics, and the teaching of mathematics. He was awarded the *Prix Becquerel* by the *Académie des Sciences* just before his death.

Cartan and Ferrand write:

> André Bloch corresponded with several mathematicians, among whom some were unaware of his condition, for he simply gave his address as 57 Grande rue, Saint-Maurice, without specifying that it was a psychiatric hospital. In order not to give himself away, he avoided meetings on the pretext of ill health. We can appreciate the nature of his confinement from a letter he wrote to P. Montel on 16 January 1940, in which Bloch offered to take the necessary steps to obtain permission to see Montel: 'I was allowed out one afternoon fifteen years ago' [III.71].

Hadamard was among those to whom Bloch wrote and sent his papers. Some of Bloch's articles in the *Comptes Rendus de l'Académie des Sciences* of 1924-1925 were communicated by Hadamard and he included Bloch's paratactic circles into the *Leçons de Géométrie élémentaire*. Did he know about Bloch's situation? Did he meet Bloch?

Mordell, in his *Reminiscences of an Octogenarian Mathematician*, answers these questions affirmatively, referring to Hadamard himself:

> He said to me that as editor of a mathematical journal, he received rather good papers from someone unknown to him, so he invited him to dinner. His correspondent wrote that owing to circumstances beyond his control, he could not accept the invitation, but he invited Hadamard to visit him. Hadamard did so and found to his great surprise that his author was confined to a criminal lunatic asylum. Apparently he was quite sane except for the murder of his aunts. His name was A. Bloch, and he was a very good mathematician [III.284, p. 953].

## 4.5 Meetings with Einstein

In March and April of 1922 Einstein gave his first lectures in Paris. He was invited to the *Collège de France* on the initiative of Langevin, who had to overcome the opposition of people who still held anti-German feelings.

These were the years of Einstein's biggest triumph. In 1917, the well-known English physicist and astronomer, Eddington, proposed a method for the experimental verification of Einstein's prediction that light has a gravitational mass. Two years later, two expeditions organized by Eddington proved the deviation of solar rays near the sun, after making photographs of the sky at the moment of complete solar eclipse. Eddington's report and the reaction of scientists produced a world-wide sensation. It became fashionable to talk of the "heaviness of light" and of the "curvature of space". Here is how Infeld explains the reasons of Einstein unprecedented prominence:

*Einstein at the Collège de France, 1922 (Hadamard's bald head can be seen in the front row)*

It was after the end of the World War I. People had had enough hatred, killing and international intrigues. Trenches, bombs, murders had left a bitter taste. Books dealing with the war were neither read nor bought. Everybody expected an era of peace and wanted to forget about the war. Thus this phenomenon was able to impress the human imagination. Eyes were raised from the grave covered earth to the starry sky. An abstract thought carrying man far from the sadness of everyday life to the mystery of the Sun eclipse and the power of human intellect. Romantic scenery, a short darkness, the picture of curving light rays; everything so different from the overwhelming reality. And one more reason, probably the most important: the new phenomenon has been foreseen by a German scientist and verified by English scientists. Physicists and astronomers, belonging not so long ago to two opposite sides, are now working together! Is it not the beginning of a new era, an era of peace? This human longing for peace was, I believe, the main reason of the increasing popularity of Einstein [III.191].

*Élie Joseph Cartan (1869-1951) graduated from the École Normale Supérieure in 1893, was a professor of the Sorbonne from 1912, member of the Académie des Sciences from 1931. The main directions of his work are the geometry of Riemannian spaces, group theory, theory of invariants, differential geometry, and relativity theory. In 1922 he introduced the concept of the space with absolute parallelism (a space without curvature). Einstein used the term "Fernparallelismus" which is equivalent to the commonly used "distant parallelism".*

Einstein gave four lectures in the largest lecture theatre at the *Collège de France*, each being followed by a discussion on special and general relativity. The first lecture, on March 31, was particularly overcrowded. André Weil wrote: "Admission was by ticket only: I received mine thanks to Hadamard, I suppose. Everyone who moved in scientific or philosophical circles or in high society came, in such droves that the *Garde Républicaine* had to be called in to control the crowds" [III.418, p. 30]. On April 6th there was a meeting at the *Sorbonne* of the French Philosophical Society, already mentioned in Section 4.2, where Langevin spoke about philosophical aspects of Relativity Theory and Hadamard participated in the discussion together with Einstein, É. Cartan, Painlevé, P. Lévy, Perrin, Becquerel, Brunschvicg, Bergson, Le Roy, and Meyerson. Discussions also took place away from the lecture theatres: "I even remember trying, at Mr. Hadamard's home, to give you the simplest example of a Riemannian space with *Fernparallelismus...*" É. Cartan wrote in a letter to Einstein [III.70, p. 5].

Hadamard's attitude to Einstein's ideas was somewhat special: for him there was no question of accepting or rejecting them, as his own work in mechanics, mathematical physics, and geometry enabled him to understand them. He wrote:

> At least, I was from that time on already prepared to accept the relativistic point of view; in any case, I was disposed before-hand to respond rather badly to those – I swear to you that I heard it once more recently from recognized men of science – who not only assert their disbelief in relativity, which would be their right, but who also call it senseless, 'contrary to common sense', and who cannot understand how so weird an invention could have been accepted by anyone [I.239, p. 67].

In the preface to the book by Juvet on tensor calculus, published in 1922, Hadamard expressed his great expectations of the influence of general relativity on the future development of differential geometry:

> There was thus, in our opinion, a crisis in infinitesimal geometry, and it is this crisis that the advent of Relativity has just solved. It has from the beginning thrown an unexpected light on the previous geometrical work in showing the fundamental importance of this absolute differential calculus of Ricci and Levi-Civita, which remained almost without reaction from their appearance. Whatever opinion one holds about the well-foundedness of the new hypotheses, they give geometry a new role and open up a new period for it, not this short lived novelty which can too often only influence the mathematician left to his own devices, but this infinitely fecund novelty which springs from the nature of things [I.222].

Hadamard considered that one of his failures was not discovering the special theory of relativity. In 1924, he gave a talk at the International

Philosophical Congress entitled *Comment je n'ai pas découvert la relativité* [I.239], in which he tried to explain why the researcher often fails to notice facts which are at hand. He remarked that already at the beginning of his study of wave theory, he came upon transformations which left invariant the equation of the light cone, but he considered them to be physically meaningless (Hadamard had in mind the Lorentz transformations which were studied by Poincaré and played a decisive role in Einstein's special theory of relativity). He wrote: "I really believe that I owe a very personal *mea culpa*, and that I must have been extraordinarily stubborn to elude the consequences of research I had myself carried out in continuation of the classical results of Kirchhoff and Volterra" [I.239, p. 66].

After giving a thorough analysis of his reasoning, he judged himself harshly:

> To think that this Kirchhoff line did not have a physical significance – that was a leap too bold for me. Like all mathematicians, I admired the vast and ever-expanding body of work that was being produced by physicists; this admiration mingled with respect imposed by my own incompetence. I had not yet clearly understood that one should be able on occasion to fail to show respect to physics.
>
> And that is how I, a mathematician with an impoverished imagination, found myself totally unable to interpret concretely the conclusion that mathematical theory was irresistibly forcing upon me; I contented myself to bow respectfully before Kirchhoff's point of view. The moral of this story, then, is in one's own field the scholar should respect nothing – one should not heed the word of a Kirchhoff any more than Copernicus respected the work of Aristotle or Ptolemy; that is what we all do ever since Einstein. It is the story of the egg of Christopher Columbus; I wonder whether it is not true of a number of mathematical discoveries as well [I.239, p. 66-67].

Such statements with admissions of failure are quite rare in mathematical literature, both before and after Hadamard. It is customary to present only the author's successes. Twenty years later, Hadamard mentioned this and other opportunities he had missed, in his book on the psychology of mathematical thinking [I.372].

On April 16, 1922, a week after Einstein left Paris, Hadamard wrote to him: "The glory of Paris is definitely yours" [III.120, p. 117]. To this, Einstein replied: " I am very content with my stay in Paris, happy to have made the acquaintance of Parisian mathematicians and physicists, and hope that I have contributed to the restoration of friendly ties between French and German scientists" [III.120, p. 117].

During his 1930 speech in Buenos-Aires at the Academy of Science, Hadamard said: "I would like to recall Einstein as I remember him in our famous sessions at the *Collège de France*, developing, making precise, defending his brilliant ideas which are a renewal not only of modern physics, but also of all of our philosophy, and as I would also see him under my roof during informal meetings, with his powerful mind and his childish smile, with this generosity of spirit and ideas which have made him our guide, not only from the scientific point of view, but also from the moral point of view" [II.54, p. 79].

## 4.6   Home orchestra

Whenever he was in Paris, Einstein would visit the Hadamards, and as Paul Lévy recalls, the two men would speak more about music than about relativity [II.5, p. 23]. He also played the violin in an amateur orchestra organized by the Hadamards in their home.[8] Jacqueline Hadamard gives a description of this home orchestra:

How the idea to bring together an orchestra came to my parents, I do not know. But we began with some relatives and friends. It was not a small task. There was no problem with the scores: the *École Polytechnique* and *École Centrale* had orchestras and we borrowed the music; if necessary, we would have it sent from the *Maison Wolf* of Strasbourg.

As for the conductor, we fortunately knew some young composers who gladly accepted to practice what could be their future profession. The first was Maurice Franck; he no doubt did not imagine the difficulties posed by amateurs, who are naturally not so good players as professionals, but above all, less hard working and passive. Quite a variety of people came to our house and we even succeeded in finding wind instrument players. We were lucky to have Georges Duhamel[9] as our first flutist, very good,

---

[8]Fréchet mentions that Hadamard and Henri Villat, his colleague and also a member of the *Académie des Sciences*, participated in a quartet [II.13, p.4].

[9]G. Duhamel (1884-1966), a famous French novelist.

who later mentioned our orchestra in one of his books. Once, even, we were able to find trumpeters (for a work which required them), whom we shut up in a junk room to soften the force of their playing.

My mother stood in for the missing instruments on the piano, as, alas, rehearsal days were days of anguish for us: who was going to telephone and cancel at the last minute? Because you can't demand of amateurs what you demand of professionals. I remember one evening when we were only two violinists when there were five cellists, as many as at the Colonne[10] concert!

But the rehearsals struggled along all the same, the good players showing up the weak ones (of whom I was one). My great fear (and I was not alone) was the evenings when the conductor had the instruments practice one at a time: because when the turn came to the second violins, it had to be admitted that, apart from the andantes, I played only one note in four! On the other hand, I could be counted on when it involved keeping time, which was not the case for my father, too much the romantic, and who was unable to prevent himself from speeding up or slowing down, according to his mood.

So our orchestra must have been terrible to hear and I felt sorry for the conductor. Fortunately, there was no audience and, as we lived in a detached house, no immediate neighbours. In an apartment, the neighbours would certainly have complained! The important thing was we all took great pleasure in it and I don't think that one can really know a symphonic work without having massacred it in this way.

We had lucky days: that was when Einstein came to Paris, bringing his violin since he was delighted to participate and played very well. Everyone, I am sure, applied themselves ten times more on those days.

Because he was afflicted with the same "weakness" as my father, regarding tempo, our conductor decided that on those days, as an exception, of course, I was promoted to first violin, it being well understood that, sitting beside Einstein, and even if I played only one note in eight, my task was to keep him in time [IV.1, p. III(24)].

---

[10]Édouard Colonne (1838-1910) was a violinist and conductor, the founder in 1871 of the *Concert national*, which was later named after him.

## 4.7   How many snakes did Hadamard hunt?

Hadamard was an indefatigable traveller. Physical activity helped him keep up his prodigious scientific output. When he was over sixty, Hadamard climbed Mont Blanc, and gathered botanical specimens in the Mexican mountains, surprising his companions with his endurance and daring.

The French and Swiss Alps held no secrets from him, and he was not afraid of any climb. Montel, who once accompanied him to ascend the Rhône valley towards the Aletsch glacier, remembered that Hadamard refused to make the usual hourly pause, saying that he would not rest until they reached the destination [II.5, p. 21]. Tricomi, who was with them, wrote: "Everything was going fine, but in the evening our return lasted a little longer than intended, because Hadamard stopped every ten paces to turn round and admire one more time the peak of the Jungfrau, made golden by the last rays of the setting sun" [II.5, p. 24].

The scale of Hadamard's travel plans can be judged by the following extract from his letter to E. O. Lovett of January 31, 1911: "May I ask you to give me a help which, I hope, will not give you too much trouble? This would be to let me know if you hear of somebody intending to take a trip, on end 1911, in the tropical part of America: Mexico, West Indies, or perhaps Columbia ... I should like to join, if possible, some companions, as agreeable as possible, not too imprudent, and, above all, of course, coming there as I do to enjoy nature and not for snobism" [IV.32].

Nicoletis, who was Hadamard's companion in Brazil in 1924, recalled: "It was around the bay of Rio that he initiated me to the joys of mountains. Hadamard was an untiring mountaineer. He climbed, he swam, he walked and how many times have I not heard him insisting on the indispensable equilibrium one should maintain between body and mind" [II.49, p. 10].

Paul Montel and Jacqueline Hadamard give two accounts of Hadamard's trips to South America. Montel writes: "Invited to make a lecture tour in Argentina, he landed in Brazil in order to see orchids in full blossom. He went into the tropical forests and with a stone tried to kill a rattle snake, whose skin he had promised, and after refusing to go by sea, he rode on horseback to Buenos Aires, accompanied by his wife,

who was exhausted by this new mode of transport" [II.5, p. 21].

Jacqueline Hadamard gives the following version: "Going on a lecture tour to Rio de Janeiro, he had decided to go and see the source of the Iguaçú river where, it appears, there were some amazing ferns and where he could see some marvellous orchids and humming birds. For this, he did not go directly to Rio, but decided to cross the Parana forest on horseback. That was how he made my mother spend whole days on horseback, her first experience of this form of transport" [IV.1, p. III(3)].

Although the geographic details differ, the general atmosphere and other points in each story are the same. If the two stories describe the same trip, then what was the final destination, Rio de Janeiro or Buenos Aires? Jacqueline Hadamard gives one more story about a trip to South America: "Another time, he went to South America and then had to go to the

 United States afterwards. Of course he took an unorthodox route, lost his luggage on the way, forgot that he would go from the southern to the northern hemisphere, that is to say from summer to winter, and arrived in New York in the middle of a snowstorm in a white suit, but triumphantly carrying the tail of a rattle snake which he had killed by throwing stones at it" [IV.1, p. III(3)-III(4)].

This story raises a new question: is the snake the same one mentioned in Montel's story? These unresolved problems show clearly the difficulties met by biographers of great men.

## 4.8 Ferns and fungi

During his long life, Hadamard visited many countries. He gave lectures in the United States, the Soviet Union, Italy, Switzerland, Spain, Portugal, Germany, Great Britain, Belgium, Czechoslovakia, Rumania, Canada, Egypt, Israel, China, India, Brazil, and Argentina. These trips appealed to Hadamard in another, quite unusual way: he was, from his youth, passionately interested in botany. Paul Lévy recalls: "During the final years of the last century, Hadamard, my father's pupil [11] visited us frequently, and

---

[11]Lucien Lévy (1853-1912), the father of Paul, taught mathematics at the *Lycée Louis-le-Grand* during several years. He was Hadamard's teacher in the class of *rhétorique*. During twenty-three years he was an Examiner at the *École Polytechnique*. His geometrical memoir was awarded by the Royal Academy of Belgium. He also wrote several textbooks.

once, while we were sitting at the table, I saw that he pulled a magnifying glass out of this waistcoat pocket and explained: 'I always take it with me to examine unknown plants.' Then I thought: this man is a botanist. I later learned that he was a mathematician, and probably the greatest of his generation" [II.5, p. 9].

*Cosinus had hardly gone ten paces in the ruins when he discovered a veritable virgin forest and an abundant and variegated flora made him completely forget Sphéroïde.*  *But he comes back, at the sight of a new species. "I shall call it Briocheiaparisiensis Br.", he says.*

In his book on the psychology of invention, Hadamard remarked that "especially biology, as Hermite used to observe, may be a most useful study even for mathematicians, as hidden and eventually fruitful analogies may appear between processes in both kinds of studies." [I.372, p. 9]. He was mostly interested in ferns and fungi. According to Fréchet, even botanists consulted him about them [II.13, p.4082]. Roger Heim, a *Membre de l'Institut* and the *Directeur du Laboratoire de Cryptogamie au Muséum National d'Histoire Naturelle*, said in 1967:

> In 1927, when I was still a young man, I went with Professor Jacques Hadamard to the narrow valleys of the Gapençais mountains, to look for a rare fern which he did not have in his herbarium. He, who was one of the greatest mathematicians of our age, was carried away in a youthful passion by the study of Nature, in an area of botany on which he was a remarkable

expert. Mathematical analysis was the expression of his genius, but it was incomplete for him. He added to it the passionate love and brilliant knowledge of ferns, not those that grow in greenhouses, or in gardens, but where they live, in their environment, the steep rocks, the embankment of a path, the coping of a well, just as the farmer lives in his field, or the fisherman in his boat. He could have been an eminent pteridologist and done mathematics for enjoyment [III.173, p. 142-143].

His collection of ferns grew with every trip, and it was considered to be the third best in France, after the ones of *Muséum d'Histoire Naturelle* and of Prince Roland Bonaparte. However, Hadamard would not allow anyone else to gather specimens for him, as is evident from what Jacqueline Hadamard writes about her trip to Indochina:

Of course, I didn't forget my family during these two months of travelling, and especially my father's passion for ferns. I finally succeeded in detaching a very small foliole of a tree fern: it was hardly one and a half metres long. But to my great disappointment, my father disdainfully pushed aside this present which had caused me so much trouble. Any fern he had not picked himself was not worthy of going into his collection [IV.1, p. III(22)].

She then goes on to describe how Hadamard dealt with his specimens:

He had infinite patience for his beloved ferns. He had a negative talent for material things and would quite naturally try to open a suitcase by its hinges; but he would spend hours sticking each fern onto a large piece of cardboard, smoothing out each foliole. They were held on the cardboard by hairs stuck at the two extremities (hairs which he got from me because he had hardly any left), so that one could admire every detail of them. When he did this, the family was forbidden to open a window or a door for fear of draughts of air [IV.1, p. III(22)].

In the summer of 1936, after four months in China, Hadamard and his wife decided to return home. They took advantage of the opportunity to travel from East to West by the Transiberian Railway. Sixty years ago, the trip by railway from Vladivostok to Moscow took far longer than it does today, mainly because of long stops at intermediate stations and sidings. Hadamard used these stops to enlarge his botanical collections, and this made his wife very anxious in case he should be left behind. According to Montel, Hadamard "discovered in Russia a new species of fern to which his name was given: Hadamardus" [II.25]. Unfortunately, this name does not appear in any reference book on ferns which we have consulted.

## 4.9   Hadamard's puzzle

Hadamard was a humourous and witty man. "In honour of his friends, he would write charming poems which unfortunately disappeared during the last war" [II.5, p. 22]. "He loved humour and once read me a malicious little poem of his about a candidate to the Academy", Fréchet recalled [II.13, p. 4086].

George Pólya, in his book *Picture Album: Encounters of a Mathematician*, recounts: "At the Bologna Congress the meetings started in Bologna and ended in Florence. That's about a three-hour train ride and for this there was a special train. I recall we were in a compartment that was very noisy, and Hadamard was tired and wanted to have some peace. So he told the people in the compartment about a difficult problem, a puzzle. As soon as he told it, everyone started working on it, and it suddenly became quiet so Hadamard could sleep" [III.324].

Perhaps they worked on the following puzzle of Hadamard's (obtained from L. Schwartz). One is required to draw a picture where *le roi Pépin,*[12] *sans air, sans eau, sans lit, sans pain, ayant perdu le peu qui lui restait, gémit tout seul dans un coin.* The reader familiar with French will see that the solution is

This is a play on words and reads: King Pepin, without air, without water, without a bed, without bread, having lost the little that remained to him, moans all alone in a corner. It also means: the *roi Pépin*, without *r*, without *o*, without *the i*, without *pin* (pronounced *pain*), having lost the *pe* (prononced *peu*) which remained to him, *g* placed all alone in a corner.

---

[12]Pépin is the name of several kings in the 8th and 9th centuries, the most famous being *Pépin le Bref* (Pepin the Short) (715-768), the king of the Francs and the father of *Charlemagne* (742-814).

# Chapter 5

# *Le Maître*

## 5.1  *Le Séminaire Hadamard*

Hadamard's favourite creation was the seminar at the *Collège de France*, which he initiated in 1913. He began with his own surveys of Poincaré's works embracing various domains of mathematics. Then he introduced the practice of students' reports on mathematical memoirs.

*"Hadamard told us that the goal of his new seminar was de faire la gazette", remembered George Pólya, one of the first participants. (From the unpublished speech of G. Pólya at Hadamard's jubilee in 1936 [IV.22])*

These meetings were interrupted by the First World War, but they resumed in 1920, and were held regularly. The seminar very quickly gained a prestigious reputation and gradually its participants became active scientists rather than mere students.

The idea of a seminar was by that time not new: already in 1899, the young Hadamard took part with great enthusiasm in Marcel Brillouin's physics seminar at the *Collège de France*. There were some more or less specialised seminars abroad, but Hadamard's seminar became a unique phenomenon, not only for French mathematicians, but also for the world mathematical community. "Due to his erudition and to his ability to master any domain, Hadamard extended the seminar to include all areas of mathematics", wrote Fréchet in his obituary of Hadamard [II.13, p. 4085].

Hadamard's seminar continued for more than twenty years. Many French and foreign scientists gave talks there, among them such mathematicians as Émile Borel, Paul Montel, Henry Lebesgue, Paul Lévy, Maurice Fréchet, Jean Chazy, Henri Villat, Gaston Julia, Arnaud Denjoy, Georges Valiron, Élie Cartan, André Weil, Vito Volterra, Tullio Levi-Civita, Godfrey Harold Hardy, Edmund Landau, George Birkhoff, Sergeĭ Bernstein, Nikolaĭ Luzin, Mikhail Lavrentiev, Andreĭ Razmadze, George Pólya, Rolf Nevanlinna, Lars Ahlfors, Laurent Schwartz, as well as the physicists Max Born and Louis de Broglie.

André Weil describes the work of the seminar as follows:

> In Paris, while I was at the *École* and for a long time afterwards, there was only one seminar worthy of the name, and this was Hadamard's. At the beginning of the year, we met in the library of his home on the *rue Jean-Dolent*, where he handed out mathematical papers to be reported on. These papers were for the most part the offprints which he had received from all over the world - or at least those which appeared to him worthy of discussion. To these he added titles of various provenance, as well as titles proposed by others of us, for he was quite open to suggestions. Most of these works had been published in the last two or three years, but this condition was not hard and fast. As to the subjects covered, his aim was to provide as extensive a panorama of contemporary mathematics as possible. If it was not exhaustive, this was at least his goal. For every title he announced, he would ask for a volunteer, often explaining briefly why the paper excited his curiosity. Once the titles had been distributed, dates were set for reports to the group, and after some general chit-chat, we all left.

At that time the seminar took place once a week; later it met twice a week. Among its participants were highly accomplished mathematicians as well as beginners. Paul Lévy, who had been Hadamard's pupil, was among the faithful. Hadamard behaved as if the *exposés* were primarily for the purpose of informing him personally; it was to him that we addressed ourselves and especially for him that we spoke. He understood everything, as long as it was explained well; when the summary was not clear, he would request clarification, or else – not infrequently – supply it himself. He always reserved the option of adding his own remarks at the end, sometimes in a few words and at times in a more leisurely fashion. Never did he appear conscious of his own superiority: whoever was

*André Weil*

giving the *exposé* (it is on purpose that I do not use the word 'lecture', for in front of Hadamard it was impossible for it to come across as a lecture) was treated as an equal.

This was true even for me, callow student that I was when, shortly after entering the *École*, I was accepted as a participant – no mean mark of favor. I believed I already had some ideas on functions of several complex variables; I had made several observations (which I thought to be original, and which may perhaps have been so) on the domain of convergence of power series in several variables, generalizing Hadamard's classic theorem on series in one variable; but even more importantly, I had just discovered Hartogs' works in the library of the *École*. Although they had already been around for some time, they were little known in France, and had never been the subject of an exposé in Hadamard's seminar. I proposed this topic, and he received my suggestion with pleasure.

The *bibli*[1] and Hadamard's seminar, that year and the following ones, are what made a mathematician out of me [III.418,

---

[1]The library of the *École Normale*, the key of which Weil obtained while acting as assistant librarian [III.418, p. 35].

p. 39-40].

S. Mandelbrojt recollects how the topics of the talks were distributed:

At the beginning of the academic year, from October, Hadamard gathered people whom he thought able to speak at his seminar, Parisians and even those from the provinces, but not foreigners who could not come expressly. Madame Hadamard served delicacies. The room was full of memoirs and we discussed which ones we had to talk about and who would present them. We chose. 'And you, Mandelbrojt, do you want to do this?' Good, I chose myself. 'Chazy will do this, Valiron will do that'. As for the foreigners, we invited them. We asked Birkhoff, Pólya, Plancherel to come from far away in order to give a talk, most often on the subject they knew best and which they chose themselves [III.265, p. 17].

The main part of S. Mandelbrojt's speech at the centenary of Hadamard was dedicated to reminiscences about the seminar:

The sessions on Tuesdays and Fridays were certainly the most intensive that the collective mathematical thinking in France experienced between the two wars. Every mathematician, French or foreign, considered it an honour and an important proof of scientific esteem, to be invited by the illustrious Master to speak about his own research or simply to give an account and a comment on newly published research.

Vito Volterra came to the seminar to speak on nonlinear functionals, or on functional calculus in general. Levi-Civita talked about his research on Riemannian geometries, and absolute parallelism was often the subject of the seminar. Hadamard very much liked to listen to George Birkhoff discussing the ergodic problem. Pólya explained his beautiful results on analytic continuation and the distribution of the singularities of a Taylor series, a subject born with Hadamard's thesis. Landau spoke on number theory or on Dirichlet series. Serge Bernstein gave an exposition of weighted polynomial approximation and the theory of best approximation, which a considerable number of Russian mathematicians now deal with. It was at Hadamard's seminar that Rolf Nevanlinna made known his famous theory of entire functions, and Lars Ahlfors announced his distortion theorem.

J Hadamard

Everybody came to Paris for a few days to bring the fresh fruits of their latest researches to the Parisian mathematical public, and to other mathematicians passing through, who together made up the seminar audience, and above all to Hadamard.

I remember exciting sessions devoted to the theories of Brouwer, in which Lebesgue, Winter, Paul Lévy, and Wavre (from Geneva) took part. These theories puzzled us; imagine that the other young people (of that time!) were, like myself, frightened of going into it. They were very anxious to be able to continue their own research in less vertiginous areas.

Hadamard, who often saw the essence of the subject being considered better than the invited scientist who was, however, a specialist in that branch of mathematics, compared the results obtained with old results, sometimes seeing connections with a completely different mathematical area.

Thus we all worked in isolation, each of us thinking that one's own domain was the only interesting area or, at least, the most interesting.

Besides, every young mathematician had reasons to believe this: the inspiration which guided him, the difficulties one met, difficulties which nowadays one often dare not admit, afraid of seeing them treated by 'techniques', but they were difficulties one could overcome – the love and tenacity which one invested in one's research to reach the essence of the problem, to create one's poem – for how else can one call these pieces of pure imagination, isolated, but so rich?

General theories, those which gave the structures of mathematical phenomena, were few: at best they constituted an autonomous branch of mathematics, or were even a part of meta-mathematics which one did not dare touch.

It was here that Hadamard's broad, encyclopedic and, above all, profound mind intervened: he alone created the sort of synthesis which was lacking.

How many times did he state in the seminar a maxim which now appears to us evident, but which was less so at that time: 'generalize to simplify, or to understand better!'

He saw clearly and quickly that some theorem of the theory of analytic functions was nothing else than a version of a topological theorem, that another one could also be stated in terms of the theory of functionals, and could, under these conditions, give rise to a considerable number of versions in several branches of mathematics which were *a priori* unconnected.

Hadamard, without saying it explicitly, crystallized the structure of mathematical phenomena – the poems became a part of a grand epic. [II.5, p. 25-26]

The same tone of admiration is present in reminiscences of Paul Lévy:

During the seminar sessions I was sometimes unable to follow a difficult talk, and I was impressed by the fact that nothing appeared difficult for him. Always attentive, he often interrupted to clarify a point badly explained by the speaker, and if by chance something escaped him, he was never ashamed to say so, and to ask for further explanation [II.36, p. 4].

In his recollections S. Mandelbrojt remarks:

Nowadays one talks a lot about mathematics or other sciences done in a group. Bourbaki is done by about ten people. At that time, there was no collaboration and Hadamard's seminar was a kind of pre-Bourbaki; Hadamard was somehow the editor, if I may speak abstractly, the moral editor of Bourbaki. It was of capital importance, I believe, for French mathematics as well as for mathematics as a whole [III.265, p. 17].

In this chorus of praises, Dieudonné's words sound somewhat discordant:

Our only opening onto the outside world at this time was the seminar of Hadamard, a professor but not a very brilliant teacher, at the *Collège de France*. (He was a great enough scholar for me to be able to say this without harming his reputation). He had the idea (apparently taken from abroad, because this had never been done in France) of inaugurating a seminar of analysis of current mathematical work. At the beginning of the year he distributed to all those who wanted to speak on a subject, what he considered to be the most important memoirs of the past year, and they had to explain them at the blackboard. It was a novelty for the time, and to us an extremely precious one, because there we met mathematicians of many different origins. Also, it soon became a centre of attraction for foreigners; they came in droves. So it was for us young students a source of acquaintances and views that we did not find in the formal mathematics courses given at the University.

*Jean Dieudonné (1906-1992) in 1924*

This state of affairs lasted several years, until some of us – starting with A. Weil, then C. Chevalley, having been out of France meeting Italians, Germans, Poles, etc. – realized that if we continued in this direction, France was sure to arrive at a dead end. We would no doubt continue to be very brilliant in the theory of functions, but for the rest, French mathematicians would be forgotten. This would break a two-hundred-year-old tradition in France, because from Fermat to Poincaré, the greatest French mathematicians had always had the reputation of being universal mathematicians, as capable in arithmetic as in algebra, or in analysis, or in geometry. So we had this warning of the bubbling of ideas that was beginning to be seen outside, and several of us had the chance to go and see and learn at first hand the development that was going on outside our walls. After Hadamard retired in 1937, the seminar was carried on, in a slightly different form, by G. Julia. This consisted of studying in a more systematic manner the great new ideas which were coming in from all directions. This is when the idea of drawing up an overall work which, no longer in the shape of a seminar, but in book form, would encompass the principal ideas of modern mathematics. From this was born the Bourbaki treatise [III.104, p. 135-136].

Should one find Dieudonné's remark so odd? Perhaps not, because in spite of its diversity, Hadamard's seminar was mainly in analysis, understood in a very broad sense, whereas it worked in a period when topology, probability and abstract algebra were rapidly developing into major divisions of mathematics. Dieudonné was one of the founder members of the group Bourbaki which started off in 1934 just wanting to write a textbook in analysis.[2] Instead they created a many volumed treatise *Éléments de Mathématiques*, a survey of fundamental structures of mathematics written on the contemporary level of rigour and based on the most general principles. The above Dieudonné's remark which seems to be closer to the Bourbaki position of the 1940s and 1950s, strikes at the heart of what a universal mathematician is. Without trying to resolve this big issue we only quote A. Weil, another Bourbaki's collaborator:

> I had formed the ambition of becoming, like Hadamard, a 'universal' mathematician: the way I expressed it was that I wished to know more than non-specialists and less than specialists about every mathematical topic. Naturally, I did not achieve either goal [III.418].

We complete our selection of citations on the seminar with an enthusiastic account by S. Mandelbrojt and L. Schwartz:

---

[2]For the prehistory of the group see L. Beaulieu's article *A Parisian café and ten proto-Bourbaki meetings (1934-1935)* [III.21].

Those of us who have had the privilege of attending Hadamard's Seminar at the *Collège de France*, where he taught from 1909 to 1937,[3] would probably be unable to recall more inspiring hours of mathematical thought. Well-known mathematicians all over the world considered it as an honour, and sometimes as a redoubtable task, to be asked to state and to prove their recent results, or the results in their fields just discovered by others. But, without trying to diminish the contribution of the talent of the lecturers, we must say that the bulk of our feelings, of the richness of our inspiration, and our desire to continue to work, or at least to think, on the subject just treated in the Seminar, came from Hadamard's analysis of the lecture, from his critical views, from his interruptions, simple remarks and some prophecies on the future of the subject.

One of the characteristics of Hadamard's Seminar was its variety. It was not a Seminar on one branch of mathematics – it was one on Mathematics, pure and applied, on the philosophy of mathematics and numerical analysis as well. Often the lecture was an important mathematical event, where for the first time a very significant result was expounded. Sometimes new results were born at the Seminar, and published a few weeks later in the *C.R. Acad. Sci. Paris*.

Mathematical life in Paris in the twenties and early thirties was for the large part described by two words: *Séminaire Hadamard* [II.43, p. 117-118].

## 5.2   S. Mandelbrojt about Hadamard

The reminiscences of Szolem Mandelbrojt were published in 1985, as recorded by his nephew Benoit Mandelbrot, the creator of fractal geometry [III.265]. A brilliant story-teller, S. Mandelbrojt has much to say about Hadamard, who played an important role in his life and had a great influence on him as well as on his mathematics.

As a child growing up in Warsaw, Mandelbrojt developed an early passion for mathematics (he discovered the procedure for finding square roots when he was thirteen). From 1917 to 1919 he studied mathematics at the Warsaw Polytechnic and at the university, attending courses by Janiszewski,

---

[3]At the *Collège de France* Hadamard was *professeur suppléant* from 1897 to 1909 and *professeur* from 1909 to 1937.

Mazurkiewicz, Rajchman, and Sierpinski on group theory, general topology, trigonometric series and set theory. Mandelbrojt spent a cold and hungry year (1919-1920) in Kharkov, where his professor was Sergeĭ Bernstein.

*Szolem Mandelbrojt (1899-1983) worked in mathematical analysis, complex function theory, functional analysis, the theory of Dirichlet series, number theory, approximation theory. He entered the Académie des Sciences in 1972.*

Bernstein graduated from Paris when only nineteen, and within five years had solved Hilbert's nineteenth problem on the analyticity of solutions to regular variational problems and elliptic partial differential equations for the case of two independent variables. He included this in his thesis [III.32] presented to the jury, whose members were Picard, Poincaré, and Hadamard. Later, Bernstein's interests changed to probability and approximation theory, and in Kharkov he gave a course on analytic functions. After a couple of lectures, only one student remained, Szolem Mandelbrojt. He recalled: "He wasn't a brilliant teacher. Not at all! With each word, one saw that he suffered in order to bring out the right property,

*Sergeĭ Bernstein (1880-1968)*

to speak about the right function. Seeing him thinking, one felt his mind, a lively mind, vibrating, and I vibrated with him, he doing the talking and me thinking, he thinking and speaking. It was marvellous. I learned the theory of functions with Serge Bernstein" [III.265, p. 5-6].

However, Mandelbrojt's dream was to continue his mathematical education in France, and in 1920 he arrived in Paris, which he considered as a centre of world mathematical life. He was fortunate enough to win a scholarship, but it was so little that he ate mostly tomatoes which "were delicious in those days".

Mandelbrojt saw Hadamard for the first time in the autumn of 1921, at the inaugural lecture of Lebesgue at the *Collège de France*. He recounts:

> When he made his inaugural lecture, he came in through the Professors' door, accompanied by the big-wigs. There were ministers (at the time, an inaugural lecture at the *Collège de France* was a very important event) and there were professors of the *Collège* who were not mathematicians, and then there was a man, not very tall, with a goatee beard, a jolly little goatee beard, and I said to myself: "How dare he, that man there, come to the course by that door, as if he were important." It was only afterwards that someone said: "That's Hadamard." And I understood why he came into the room; however, I had never seen him before. After that, I followed his seminar, but at twenty-one, twenty-two, I didn't say a word to Hadamard unless he spoke to me. I must say that now one talks about mandarins. At that time I don't know if there were any mandarins, but there were people who believed that the professors should be mandarins. I wasn't one of them, perhaps because I came from Poland, perhaps because I came from far away, but the students had an enormous respect for the professors [III.265, p. 10-11].

Mandelbrojt would wake up at four in the morning and read Hadamard's thesis, his other works from 1896 and different papers on entire functions. He also studied Hadamard's little book on Taylor series, the works of Aristotle, and in the evenings he learned French, memorizing Rimbaud.

> And so that's how I began creating. I had always had results, to which, I must say, I did not attach any importance. In Hadamard's thesis, there are certain determinants involving the coefficients of a Taylor series. I said to myself: "Look, considering such determinants from one point of view gives something

interesting." A result of the kind one now calls lacunary. So I
noted it down. I said to myself: "Look, there's a theorem about
the multiplication of singularities". With that, I had other results, and I found theorems, other methods, I obtained further
results. But I considered them as exercises. I believed that to
do real work, you had to make a discovery like Newton, like Einstein, like the Lebesgue integral (because I knew that Lebesgue
had a doctorate). One had to do something infinitely important
[III.265, p. 11-12].

After preparing two articles for the
*Comptes Rendus*, he heard, to his surprise,
from Montel: "You know that you have
a thesis?", and so, in November 1923, he
became *Docteur ès Sciences* with *mention
très honorable*.

When Mandelbrojt showed him the
proofs of his thesis, Montel said: "You
know, you should see Hadamard", and
telephoned him. So Mandelbrojt went to
the Hadamards for the first time. Looking through the proofs, Hadamard was so
impressed that he immediately suggested
that Mandelbrojt write a joint book with
him on Taylor series. He meant the second edition of his own little book [I.80].
They began work in 1924, one part being
written by Hadamard, the other by Mandelbrojt, and the book appeared in 1926
[I.253]. It was reprinted in Mandelbrojt's
*Selecta* [III.264]. As one of the acknowl-

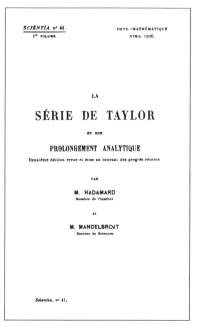

*The joint book of Hadamard and
Mandelbrojt*

edgements in his thesis, Mandelbrojt wrote: "I express my deep thanks to
Mr. Hadamard, who was the first to become interested in my researches
since my arrival in Paris" [III.265, p. 40].

During one of Volterra's visits to Paris, Hadamard presented Mandelbrojt to him, and soon afterwards Mandelbrojt made a trip to Rome, using
a Rockefeller grant. There he became acquainted with Lovett, who invited
him to visit the Rice Institute.

On his return to Paris in 1925, Mandelbrojt led the same life as before:

> I attended Hadamard's seminar before completing my thesis,
> only I didn't speak with Hadamard. But after 1923, he became

very friendly. On Thursdays, I would see Hadamard in his long dressing gown, talking quite ardently on several mathematical subjects, really very passionate. One day, I was a little ill, I lived in a hotel of the 20th category, on the fourth floor. Hadamard was no longer young, but knowing that I was ill he came to see me in my room. I needed a small operation to be done. He recommended me to one of his nephews called Schwartz [4] (the father of Laurent Schwartz. He was extremely kind to me. In his seminar, I learned masses of things. There was also a course given by Hadamard on Dirichlet series. Not as they are understood nowadays; what we now call Dirichlet series were called at that time general Dirichlet series. Most often it concerned functions resembling the Riemann zeta function. There again, it is very funny that I liked this course, which was not done brilliantly. I ended up by hating brilliant courses. Hadamard's courses, before coming from his mouth, had to pass through his body, through his heart. They had to pass through the whole of his brain. One reflected on, thought about and almost suffered the theorems [III.265, p. 16].

Mandelbrojt said that he changed the area of his studies several times, and the impetus was mostly given by Hadamard's seminar or by the works he read after the seminar meetings; often it was due to the way in which Hadamard viewed the results. "From this point of view", Mandelbrojt said, "I can say that I was Hadamard's pupil" [III.265, p. 18]. Did Hadamard consider any particular part of mathematics as the most beautiful, and the other parts uninteresting? "No. In his seminars we had hydrodynamics, mathematical logic, the theory of functions, topology, also things he didn't know at all, but wanted to learn. No, no, no, Hadamard considered his seminar as a place where you either learned things he liked, or things he didn't know. Moreover, he preferred things he didn't know, because he wanted to learn" [III.265, p. 18].

Did Hadamard's age become apparent with the years at the seminar?

No, no, certainly not. After all, you know that someone thirty years old considers a person of seventy as being old. And I, who was younger than thirty, didn't consider him as old. His remarks were always pertinent, extremely pertinent, very interesting. Oh, I remember many remarks Hadamard made which completely

---

[4] Anselme Schwartz was not Hadamard's nephew; he was a husband of a niece of Louise Hadamard.

undermined what the speaker had said, and which changed the things one had to publish. One speaker explained an article, Hadamard made some remarks, and two or three weeks later another speaker came to give a solution to the problem; or the same speaker came back to improve his theorem. It happened very frequently. It was really there (I don't mean anything pejorative with the word "cooking") that mathematics was "cooked", where mathematics was really done in France [III.265, p. 18-19].

Hadamard played a central role in Mandelbrojt's naturalization, as Mandelbrojt recalls:

*Stanislaw Zaremba (1863-1942)*

I want to tell you how I got myself naturalized in 1926. It was quite an amusing thing. Hadamard liked my work a lot. Quite simply (it doesn't matter if I brag) I was considered a very good mathematician. In 1923 when I did my thesis, Serge Bernstein was in Paris. Everybody knew that I knew him. Hadamard and Montel, being rather naive in these matters, told Bernstein: 'Now that Mandelbrojt has done such a beautiful thesis, he will get a professorship at the University of Warsaw'. I had never spoken of such a thing. Bernstein, who was Russian, not Polish, told them: 'Listen, there is a very deep-seated anti-semitism in Poland, and he will never become a professor, or at least, not quickly. You shouldn't count on it. There isn't one Jewish professor there.' After that, Zaremba, who was a good Polish mathematician, quite a lot more liberal than the others, and a Professor at Krakow, came to Paris. Hadamard said to him: 'Really, can't he get a professorship?' 'No! If you want, I shall make him assistant'. That was in 1924. Hadamard was really offended, Montel also. They asked Mme. Curie, because she was a Pole, to intervene. But she explained to them that there was no hope in Poland. So Hadamard told me to become naturalized...

So, a little later, when I was in Rome, I told Hadamard that I would be happy to become naturalized. Hadamard talked to someone at the Ministry of Justice. There was a special law which allowed naturalization of anyone who had rendered important services to the country. It was decided that I had rendered important services to the country. A propos this, I remember something very amusing. One had to go to the Council of State, otherwise you had to wait five or six or ten years. I went to the Council of State and a councillor of State called me. 'They tell me you are a very good mathematician. Are you like Pascal?' 'No, Monsieur, Pascal made a discovery when he was twelve years old, I wrote a thesis (a good one, it seems), but I was twenty-four'. 'Do you know Langeron?' At that time there was a prefect in Paris called Langeron. 'No, I don't know Langeron.' 'Every intellectual in France, every intellectual should know Langeron. He is one of the greatest physicists in the world'. 'You mean Monsieur Langevin'. 'Oh, yes, of course'.

I got engaged to Gladys in 1925, but I didn't want to become naturalized by marriage; once you're married it was easy, but I wanted to become naturalized before, because I could do it by the special law. I became naturalized on May 11th, 1926 and I got married on May 25th, 1926, as a French citizen. On the day of my marriage, I received a telegram to go to the Rice Institute. Lovett didn't know my address in Paris, but he knew Hadamard's address, because Hadamard had been there several times. Hadamard was a witness at my marriage and the same day, he brought me the telegram to go to Rice [III.265, p. 22-23].

Mandelbrojt's son, born in 1929, was named Jacques, after Hadamard.

*At S. Mandelbrojt's father's house in Warsaw, 1930. Sitting: S. Mandelbrojt, his brother (B. Mandelbrot's father), A. Denjoy, J. Hadamard, S. Mandelbrojt's father, P. Montel*

In connection with this photograph, we cite the following extract from the conversation Benoit Mandelbrot had with his uncle:

> *B.* I remember several of your visits to Warsaw. Most of all thanks to this family photograph, with Hadamard, Montel, Denjoy and you, and (in full view) what was said to be the only bottle of Bordeaux to be found in the whole of Warsaw. You were on the way to Moscow.

> *S.* Yes, it was very interesting, and even now, those who are still alive, Montel and Denjoy, remember all that. In 1930, there was a congress of Russian mathematicians in Kharkov and the Russians had invited these four mathematicians from France. Goncharov[5] came for us at the border and took us to Kharkov. But we passed through Warsaw. I don't know which one of us told the Polish mathematicians; Sierpinski came to meet us at the station. My father, who was a handsome old man with a long white beard, was very happy, very honoured, to invite French mathematicians. Although they had been invited by Polish mathematicians, they preferred to go to my father's house and my father really gave them a very good reception. I still remember the letters Montel wrote to me about the handsomeness

---

[5]V.L. Goncharov (1896-1955), professor at the Kharkov Institute of Public Education, a specialist in function theory and approximation theory.

of this old man, and Hadamard talked a lot about him to me. We dined at my father's, my sisters were there, your father was there [III.265, p. 27-28].

It was Hadamard who helped Wiener meet Mandelbrojt in Paris. Wiener writes: "I had long wished to meet Mandelbrojt face to face and to discuss with him the relation of our researches. Hadamard told me that he too was eager to talk these matters over with me. This was all the more important as he had taken an active part in the organization of the International Mathematical Congress at Oslo [1936], which I was to attend that summer. Hadamard wrote to Mandelbrojt and made arrangements for me to visit him on my trip to Oslo and possibly put in a few days of joint work with him" [III.422, p. 200]. Their meeting resulted in several joint papers on lacunary Fourier series and quasi-analytic functions.

When Hadamard retired, Mandelbrojt, who was a Professor at the University of Clermond-Ferrand, succeeded him at the *Collège de France* in 1938.

The following letter he wrote to Louise Hadamard from America, on the occasion of her eightieth birthday, is indicative of his warm feelings toward the Hadamards:

> October 21, 1948
>
> Dear Madame,
> Gladys wrote to me that you celebrated quite a birthday. I took part in your celebration by sitting in an armchair and dreaming, thinking about the beauty of your life, yours and that of your companion, my *Maître*.
> I kiss you affectionately,
> S. Mandelbrojt.
> P.S. I send a small packet of chocolate for you, but I give you permission to offer some to Monsieur Hadamard. [IV.21]

# 5.3 L. Schwartz: "He influenced me enormously"

On October 28, 1992, we visited Marie-Hélène and Laurent Schwartz in their flat on the Rue Pierre Nicole in Paris. Schwartz is a grandson of one of Louise Hadamard's sisters, and Marie-Hélène, a daughter of Paul Lévy. L. Schwartz described Hadamard's influence on his own mathematical development as follows:

$$1^{-z} + 2^{-z} + 3^{-z} + \cdots = 0$$

I can say that he influenced me enormously, but not in a direct way, because he was not very pedagogical. He did not understand exactly the level of the pupil and he talked too fast or too slowly. Once, when I was sixteen, he told me about the zeros of the zeta function and I did not understand a word. Then he remarked: 'Oh, you are not very strong in mathematics'. Then he asked me about the second order equation $z^2 + pz + q = 0$ and I gave him the solution. 'Oh, you know many things!' he replied.

So you see, there was some reason not to speak very much with him because he was not well adapted to the intellect of a pupil. I saw myself that he could not be my professor.

$$z = -\frac{p}{2} \pm \sqrt{\frac{p^2}{4} - q}$$

Schwartz continued his reminiscences:

Right after graduating from the Lycée I tried to go to Hadamard's seminar, and attended two or three meetings. He asked me to give a report on some paper of Pólya. I was just nineteen. I read that paper and felt myself unable to speak about it. At that time I found it too difficult to go to the seminar and I stopped, and afterwards I went there occasionally.

Later I studied his work: the radius of convergence, the order of singularities, the theorem on genus and the factorization theorem, the zeros of the zeta function, the calculus of variations, etc. All from his work, not from him directly. I always preferred to read a book.

As for the finite part of a divergent integral, I found it myself (for dimension 1 only) when I was a student, and found some nice applications. Then I spoke to Hadamard about it. He told me that he had introduced it a long time ago for arbitrary dimensions, and referred me to the book on partial differential equations of hyperbolic type (Yale). His aim had been the elementary solutions of these equations, my aim had been analytic continuation. Of course, when I quote it, for instance in my book about distributions, I say 'Hadamard's finite part', since he had found it long before me and it is not the place to talk about my

personal story. But I am now writing my "memoirs", and there
I say what I told you above.[6]

*Laurent Schwartz, 1992*

His definition of a well-posed problem influenced me very
much too, because with distributions I had to say that all the
problems which have a unique solution are well-posed by the
closed graph theorem. So all of that was very important for me.

To our question on Hadamard's reaction to weak works, L. Schwartz
answered: "It was very, very negative. He did a fantastic criticism of some
papers".

## 5.4 Hadamard and young colleagues

The difficulty of direct contact with beginners, which manifested itself during
his years at the *Lycée Buffon* and was mentioned again by Laurent Schwartz,
did not prevent Hadamard from being a perfect teacher in the later stages
of mathematical development. Once, he remarked on his attitude towards

---

[6]Schwartz' reminiscences [III.358] appeared in 1997.

pupils: "If I gladly accept students who come to ask me for thesis subjects, I have a secret preference for those who have their own idea, especially if it seems good according to my scientific taste" [I.377, p. 168].

The success of the seminar was due not only to Hadamard's exceptional mathematical talents, but also to his cordiality, humanitarian culture, and sense of humour. André Weil wrote: "All those who were acquainted with Hadamard know that until the end of his very long life, he retained an extraordinary freshness of mind and character: in many respects, his reactions remained those of a fourteen year-old boy. His kindness knew no bounds" [III.418, p. 29].

Wiener describes the unusually democratic manner of Hadamard's communication with younger mathematicians:

> French mathematics, however, has followed a largely official course, and when the professor has retired to his little office and has signed the daybook which gives a record of the lecture he has just finished, it is customary for him to vanish from the lives of his students and younger colleagues.

*Norbert Wiener working at his MIT desk, 1920s*

To this withdrawn existence, Hadamard forms an exception, for he is genuinely interested in his students and has always been accessible to them. He has considered it an important part of his duty to promote their careers. Under his personal influence, the present generation of French mathematicians, for all the tradition of a barrier between the younger and the older men, has gone far to break down this barrier [III.422].

At the celebration of Hadamard's centenary, Tricomi recalled the following story:

Apart from the important general matters with which he was frequently charged, he liked to entertain himself with young people, to inform himself of their research, without sparing his enthusiasm and even, in the opposite event, his no less useful criticisms.

*Giacomo Tricomi (1897-1978)*

Talking about this, I will always recall an excursion on Lake Zurich, at the time of the 1932 International Congress. We were on the bridge of the boat, sitting around a table. With Hadamard, Élie Cartan and others whom I cannot remember, there was a young (at that time!) Italian mathematician, now dead. This young man explained with great passion, but without the same prudence, how, with certain functionals, he could integrate 'an arbitrary differential system'. Then I saw Hadamard, who was at first a little distracted, prick up his ears and said with great enthusiasm: 'But Monsieur, that's amazing! That's the greatest mathematical discovery of the century.' Then the young man became quite appalled and, with a completely changed tone, added: "Yes, yes, but this is for systems with constant coefficients." And Hadamard, coming down to earth from heaven said: "But my dear young man, your discovery is almost a stupidity! Everybody knows how to integrate systems with constant coefficients!" [II.5, p. 24]

We have already mentioned Hadamard's influence on the young Fréchet, who later called him his "spiritual father" [II.18], as well as on his pupils Paul Lévy (1886-1976) and René Gâteaux (1880-1914). Hadamard noticed P. Lévy as early as 1904, when Lévy took the entrance examinations to both the *École Normale* and the *École Polytechnique*. He was placed first in the *École Normale* with brilliant marks in mathematics given him by Hadamard, but he decided to enter the *École Polytechnique* (see [III.241, p.30]). Lévy recalls the following episode about the academic year 1909-1910 when Hadamard gave a course on the calculus of variations at the *Collège de France*: "To show how he behaved with his students, I shall tell you that one day, after his lecture, I told him that he had omitted talking about an important problem. 'I know', he replied, 'I thought that I would not speak about this problem of integrability before I solved it. But since you have mentioned it, I give it to you'. This was the beginning of my researches in functional analysis. I have always been profoundly grateful to him for having given me this subject of research" [II.37, p. 7].

Paul Lévy is also known for his important contributions to probability theory. "If there is one person who has influenced the establishment and growth of probability theory more than any other, that person must be Paul Lévy", wrote S.J. Taylor [III.392]. However, this part of Lévy's work won recognition in France much later than in other countries and indeed only at the end of his life. Hadamard also seemed to underestimate Lévy's probabilistic results, although in the twenties he himself was actively interested in Markov chains and statistical mechanics. Here is what L. Schwartz said about Hadamard's attitude toward Lévy's work in probability:

Between the two wars, Hadamard, together with Henri Lebesgue and Élie Cartan, dominated the French mathematics, and the weekly *séminaire Hadamard* was a remarkable meeting place for mathematicians from the entire world. Hadamard held in high regard for Paul Lévy's papers in analysis, but was not very interested in his work on probability. This is even more curious because Hadamard was enormously interested in physics, especially in everything having to do with partial differential equations. But he considered that the mathematicians had to be interested in physics and not to become physicists; they had to remain the apostles of rigor. And he once said that he did not forgive Paul Lévy for being devoted almost exclusively to probability and for having left mathematics for physics. It has to be said, that even if the probability of Paul Lévy resembled the physics more than the mathematics of that time. All the same, Hadamard's attitude remained unexplainable. Paul Lévy was always affected by it. Hadamard had been his teacher, and Lévy had an immense (justified!) admiration for him. Here is what Paul Lévy wrote, visibly about Hadamard, in the preface of his book of 1937, on the addition of random variables, 2nd edition (1951), page XII: "I would now like to say a few words of warning to some analysts, and not the weakest ones, who are opposed to the calculus of probabilities, or at least the probabilists, who have no sense of rigor, who could not from this point of view, profit from the progress of the analysis since Cauchy, who even are tempted to forget that a continuous function is not always differentiable... This criticism could probably have been justified in the last century. To repeat it now, one has to ignore the recent development of the probability calculus...". And it is quite true that, really, there was ignorance! Paul Lévy wanted very much to enter the *Académie*, and he was elected only very late, as the successor of Jacques Hadamard, in 1964, at the age of 78, 30 years after his first candidature [III.357, p. 16].

However, in a note to the *Secrétaires perpétuels de l'Académie des Sciences* from November 4, 1965, Lévy wrote "Before succeeding Hadamard at the *Académie des Sciences*, I was his pupil, then his colleague at the *École Polytechnique*, and always his friend" [IV.23].

Gâteaux obtained his main results in 1913-1914. When leaving for the front, he gave his handwritten material to Hadamard. Gâteaux was killed in September 1914. In 1916, the *Académie des Sciences* posthumously awarded him the *Prix Francœur*. After the war, Hadamard asked P. Lévy (who continued his studies in functional analysis on returning from the front) to prepare Gâteaux's work for print. Two papers appeared in the *Bulletin de la Société Mathématique de France*, in 1919 [III.145], [III.146] with prefaces by Hadamard and Lévy.

*Paul Lévy (a drawing made by Michel Mendès-France in 1957-58)*

Wiener's regard for Hadamard was always very warm:

> I myself benefited from Hadamard's largeness of outlook. There was no reason why Hadamard should have paid any particular attention to a young barbarian from across the Atlantic, just at the beginning of his career. That is, there was no reason except Hadamard's good nature and his desire to uncover mathematical ability wherever he could find a hint of it.
>
> Many years later, when I was to meet him again at various mathematical congresses and as a fellow lecturer in China, I was surprised and gratified to find that he still remembered me and had an accurate idea of the entire development of my work. Thus one very positive result of the Strasbourg meeting was to bring me together with the long succession of French mathematicians who owe their recognition and their careers to Hadamard [III.422, p. 67-68].

One feels the same cordiality in the reminiscences of André Weil:

The warmth with which Hadamard received me in 1921 [Weil was fifteen years old] eliminated all distance between us. He seemed to me more like a peer, infinitely more knowledgeable but hardly any older; he needed no effort at all to make himself accessible to me. Soon he had the opportunity to do me a favour that had a decisive effect on my future. Every year the *Lycée Saint-Louis* awarded an endowment prize to the best student in *math élém.* The prize consisted of the equivalent in books of the annual interest in the endowment fund. I was allowed to choose these books myself, and sought Hadamard's advice on the selection. Thus it was that at the annual 'solemn award ceremony' (anyone who attended a French lycée at the time can picture the scene) I received Jordan's three-volume *Cours d'Analyse* and Thomson and Tait's two-volume *Treatise of Natural Philosophy.* Thanks to Hadamard, I learned analysis from Jordan (infinitely better than learning it from Goursat, as most of my classmates did) and I was initiated into differential geometry by Thomson and Tait [III.418, p. 29-30].

Fréchet, Gâteaux, Paul Lévy, Mandelbrojt, Bouligand, André Weil, Wiener, and Laurent Schwartz are only a few in a series of young mathematicians who were influenced or supported by Hadamard. For example, he did his best to help the young physicist and mathematician Miron Mathisson to find a position in the 1930s (see Hadamard's letters to Einstein [III.120, p. 122-126]). Another example of his support is the following letter of recommendation for Jesse Douglas, who at that time was unknown, but became, together with Ahlfors, one of the two first Fields Medallists for his solution of the Plateau problem on minimal surfaces, which was a breakthrough in the calculus of variations:

Rio de Janeiro
May 20, 1930

Dear Professor Wiener,

I understand that Dr. Douglas may apply for nomination at your University. I want to tell you how interested we were in seeing and hearing him at Paris, and what a high opinion I have of him. He has been one of the best collaborators to my Seminar of the *Collège de France.* Overall, we were highly delighted in his beautiful work on Plateau's problem, which he attacks in the most original and successful way. The solution is an unexpectedly elegant one, and one of the most interesting results in the last years, thinking of the difficulty of the subject.

I consider him as one of the most promising mathematical minds in the young American generation. Moreover, I must insist on the remarkable clearness and good order of his *exposés*, so that, from every point of view, it would be regrettable that American Science and teaching would be deprived of his collaboration.

Yours truly,

J. Hadamard. [IV.31]

During the celebration of his seventieth birthday, Hadamard said:

There is no sentiment more comforting for me today than to see around me the young researchers whom I was able, and still hope to encourage in their first steps in research, as well as their elders, these disciples who are my friends and by whose efforts certain aspects of science have evolved to the point of being sometimes unrecognisable to me. Nothing can be more precious for a scientist than to feel himself passed by on the same road he started to build... [II.27, p. 55].

*Jesse Douglas (1897-1965)*

## 5.5　Luzin and Menshov about Hadamard

*Nikolaĭ Luzin (1883-1950)*

Luzin is famous for his work in descriptive function theory, analytic function theory, and differential geometry. A prominent teacher himself, he was one of many who without being Hadamard's pupils, benefited from Hadamard's teaching and mathematical work.

As a twenty-two-year-old student at Moscow University, he went in December 1905 to Paris, where he spent about six months. "I especially recommend to you lectures by Hadamard," his teacher Egorov wrote to him in February 1906, "he gives perfect and very pithy lectures" [III.119, p. 338].[7]

In the spring of 1913, Luzin, already assistant professor at Moscow University, left for a trip to Göttingen and Paris. From the laconic lines of his report about the trip, which lasted about a year, one learns that he attended lectures by Picard, Borel, Bochner, and participated in Hadamard's Seminar at the *Collège de France.*

In 1914-1916 he formed a very active scientific group in Moscow, "the first generation of Luzitania" (cf. [III.260]), but over the years, the many talented scientists from his circle went their own ways in science, and left Luzin's area of research. That is one of the reasons why, during his third trip to France, in the winter and spring of 1926, he was deeply concerned about the future of the Moscow mathematical school, and hence the great attention he paid to mathematical life in Paris. "At present, when I live in contact with several mathematical 'schools' simultaneously (Hungarian, Polish, Serbian, Rumanian, Scandinavian), I see without difficulty, as if in a mirror, what is going on in the womb of each of them," wrote Luzin to O.Yu. Schmidt[8] in February, 1926 [III.254, p. 280]. He was delighted by

---

[7]During the second half of the 1905-1906 academic year, Hadamard gave a course in analysis at the *Sorbonne* twice a week.

[8]Otto Yu. Schmidt (1891-1956), an unusually versatile Russian scientist who worked in mathematics, astronomy, geography, and geophysics. In 1937 he was one of the four

"the modern openness, accessibility for observation of the French school". Looking back on his previous trips to Paris, he reflects:

> One compares, involuntarily, the present with former times, when the magnificent decorum of academic life and the stately inaccessibility were so beautifully stylish that spiritual thirst immediately became of secondary importance.
>
> What changed the former outlook? On reflection, a series of external conditions are apparent, amongst which, a lack of *Normaliens* sufficiently prepared for creative activity: as a consequence of war, the young men have been killed. But the internal factor is the enormous organizing force of Hadamard, who 'democratised' science [III.254, p. 283].

In another part of the same letter, Luzin again stresses the role of Hadamard: "Hadamard's seminar is an epochal event for the French school... in short, I can only say that in Paris I have found a boundless field for thought".

Among Luzin's works is the paper *Sur le problème de M. J.Hadamard d'uniformisation des ensembles* [III.253], communicated to the *Comptes Rendus* by Hadamard on February 10, 1930. In this article, Luzin gave a solution of a problem formulated by Hadamard in his first letter to Borel in the epistolary discussion with Borel, Lebesgue and Baire [I.123] on set theory.

When in 1936 Luzin became a target of public baiting in the Soviet mathematical community, he was charged, among other things, with the unpatriotic worship of French mathematics and publishing his best papers abroad.

A pupil of Luzin, a well-known specialist in the theory of trigonometric series, Menshov recalled that Luzin helped him edit the text of his paper in French and then "sent it with his letter of recommendation to the Parisian Academician J. Hadamard, who presented it to the *Comptes Rendus*" [III.274, p. 321]. Menshov wrote:

> The problems of the theory of functions of one real variable were outside of Hadamard's interests, but he trusted Luzin completely. I remember that in his reply, Hadamard expressed his satisfaction that serious scientific work was being done in Moscow and it was clear that he very much appreciated Luzin's role in its organization. Interestingly, the articles in *Comptes Rendus*

---

Soviet polar explorers on a drifting ice-floe at the North Pole. As a mathematician he contributed to group theory and its applications.

should, strictly speaking, not exceed a certain size, but I got a lot more. However, at that time the attitude toward Russia and Russians in France was especially good – it was at the height of the First World War – and my paper was not shortened...

*Dmitriĭ Menshov (1894-1989)*

I would like to note one more thing. Khinchin arrived at a generalization of the Denjoy integral, which was done independently by Denjoy himself. The difference in time of the presentation and publication of their papers on this subject was very small. Hadamard wrote that, chronologically, Khinchin had priority. I do not know. Probably, it was done out of pure politeness [III.274, p. 322].

About his stay in France, Menshov recounts:

I went to Paris in 1927 after getting a stipend from the Rockefeller Foundation for a year. Such stipends were awarded by a Scientific Committee, to scientists aged no more than thirty-five; I was then thirty-four... I lived in Paris in a small hotel,

*Parisiana*,[9] near the Panthéon, in *rue Tournefort*, 4. Luzin used to stay there, so he recommended that hotel to me. Soon, M.A. Lavrentiev arrived, who came for a shorter period, and in the spring of the same year, Luzin with his wife...

I immediately started attending Hadamard's famous seminar regularly, at the *Collège de France*. Here, reports were given and discussed, in various fields of mathematics and its applications. For example, one French scientist made a report on the works of E. Schrödinger in quantum mechanics. I also made two reports on my works in the theory of conformal mappings and in the theory of orthogonal series. There, it became clear to me that Hadamard was really very poorly acquainted with modern function theory: he asked me to remind him of the definition of the measure of a set. One can not reproach Hadamard for that: he was already sixty-two years of age, and he continued research in classical domains of analysis which he had chosen long before [III.274, p. 332].

---

[9]It remained a hotel until 1994.

# Chapter 6

# In the Thirties

## 6.1   Political commitments

Political and social activity is a characteristic of the French intelligentsia. At various times French scientists even became ministers. Amongst the mathematicians, ministerial positions were occupied by Monge, Laplace, Borel, and Painlevé even became prime minister. Hadamard never took on any state duties other than academic ones, but he was always preoccupied by political events that occurred during his long life, the most important one having such tragic results for his family.

After the First World War, Hadamard continued his activities at the Central Committee of the League of Human Rights. His papers on social and political issues con-

*Hadamard in the thirties*

stitute an interesting work in their own right. During the 1920s he wrote in the *Cahiers des droits de l'homme*, and in one of his articles he gave the definition of an aggressor:

The following will be considered an aggressor:

1. Anyone who is the first to declare war or to commit acts of hostility without having proposed submitting the cause of disagreement to the permanent court of international justice or, if

necessary, to an interim judgement made there.

2. Anyone who is the first to have refused to submit themselves
to this judgement or to have begun hostilities during examination
of the cause of disagreement by the court.

3. Anyone who refused to carry out the judgement. [I.219]

This formulation was used, almost unchanged, in a speech in 1924 by the
prime minister Herriot, on behalf of the French government at the Geneva
conference on disarmament. In this connection, Hadamard wrote to Einstein
on September 16, 1929:

The Geneva protocol of 1924 has formulated precise rules allow-
ing one to define the aggressor. I have perhaps a somewhat paternal indulgence for this definition of aggressor since, two years
before the Herriot declaration, I proposed it in the *Cahiers des
droits de l'homme*, even completing it with two points which, in
my opinion, it would be better to add [III.120, p.118].

LES CAHIERS DES DROITS DE L'HOMME                    789

# LES RESPONSABILITÉS DE LA GUERRE
### Par M. J. HADAMARD, Membre du Comité Central

Une fois de plus (et, à mon sens, il faut plutôt
s'en féliciter) les responsabilités de la guerre
reviennent sur le tapis. (1) On nous en parle beau-
coup: on ne peut pas dire qu'on en parle toujours
dans le même sens — celui des nationalistes allemands et de leur nouvel ami Mgr Andrieu —;mais
peu s'en faut. Il semble même qu'à force de répé-
ter cette thèse ou les arguments invoqués en sa
faveur, on soit arrivé à en faire article de foi pour
nombre de gens. J'avoue que je ne suis pas de
ceux-là, et je voudrais dire à M. Challaye, à
M. Challaye plus qu'à tout autre, pourquoi il ne
m'a pas convaincu, encore qu'à le lire jusqu'au bout
je trouve entre nous un accord inattendu.

critique en ce moment, les responsabilités partagées ou même ,dirais-je, éparses, ne seront jamais
difficiles à trouver: on peut trouver autant de
« responsabilités partagées » qu'on le veut dans les
guerres de Napoléon ou dans les conquêtes romai-
nes. Si vraiment toutes ces responsabilités devaient
s'équivaloir, — et c'est cela qui est en réalité à
examiner dans chaque cas, en particulier dans celui
qui nous occupe, — cela reviendrait à dire que la
guerre ne serait pas considérée comme un crime,
puisque la faute en serait à tout le monde, autre-
ment dit à personne.

Voilà pourquoi cette thèse me paraît crimi-
nelle par ses conséquences; pourquoi j'ai été et suis

*The beginning of one of Hadamard's papers in Cahiers des droits de l'homme*

The following details about the political discussions between Hadamard
and Einstein are taken from the book *Einstein on Peace* [III.291, p. 99-100],
edited by O. Nathan and H. Norden:

Einstein's friend, the French mathematician Jacques S. Hadamard (himself an avowed pacifist), in a letter dated September 16, 1929, took sharp issue with the decision for complete

non-participation in wars which Einstein had announced in his statement for the Czech journal *Die Wahrheit* earlier that year. Hadamard cited a good deal of historical evidence to show that countries refusing to defend themselves against aggression did not thereby prevent aggression, and that aggressors were not deterred either by opposition in their own country or by the pressure of world opinion. And what about the League of Nations? Should it also be restrained from using force? Hadamard considered the Geneva Protocol of 1924, in which an attempt was made to define aggression, a significant step on the road to peace. Reluctant to oppose Einstein on such an important problem, he had withheld from publication an article written even before he had known about Einstein's declaration, pending clarification of their disagreements. On September 24, 1929, Einstein mailed a reply to which he sometimes referred in later years, after the change in his pacifist view caused by the Nazi victory in Germany:

"I was very glad to receive your letter, first because it came from you, and then because it displays the great earnestness with which you are considering the grave problems of Europe. I reply with some hesitation, because I am well aware that, when it comes to human affairs, my emotions are more decisive than my intellect. However, I shall dare to *justify* my position. But let me first make a qualification. I would not dare preach to a native African tribe in this fashion; for the patient there would have died long before the cure could have been of any help to him. But the situation in Europe is, despite Mussolini, quite different ... "

In November 1929 the two men met in Paris and discussed the issues on which they so seriously disagreed. Hadamard then prepared a statement which he submitted to Einstein before publication. He had narrowed down the area of disagreement to some extent. He quoted Einstein as having admitted it might well be that he was ahead of his time, but that there are things that must be said before their time in order to pave the way to the future. On the other hand, Hadamard stuck to his point that the very possibility of a country's gaining a victory without firing a shot would merely serve to advance despotism.

Hadamard published the articles *Une culture qu'il ne faudrait pas détruire* [I.295] and *La physique et la culture générale* [I.288] in the newspaper *L'Œuvre*, as well as writings on moral and juridicial problems: *Les respons-*

*abilités des guerres: Comment déterminer l'agresseur?* [I.219], *La peine de mort et le Code pénal* [I.276], and *Réponse à une enquête sur la révision des traités* [I.306].

In the *Archives Nationales* in Paris, there is the following unsigned message related to the article [I.219]:

Ministry of the Interior

Paris, March 7, 1929
Confidential

I have the honour to enclose herewith, by way of information, a copy of a report originating in the Service of the Prefecture of Police, giving an account of a meeting organized by the *University Republican and Socialist League* during which Mr. Hadamard, Professor at the *Collège de France*, talked about the question of the responsibilities for the war.

For
The Minister of the Interior
Members of the Council of State
General Secretary of the Minister
Director of the *Sûreté Générale* [IV.26]

At the time of the Manchurian war between China and Japan, Hadamard proposed the sending of international peacekeeping forces there, under the auspices of the League of Nations. This idea was ridiculed with great journalistic zeal by Clément Vautel in *Le Journal* of November 17 and 20, 1931, and February 28, 1933:

Mr. Scribbler-Hadamard is the very essence of these terribly dangerous pacifists who would fight from the mountains. But, after all, why does he not form a legion of soldiers of peace himself and take them, in person, to Manchuria, to teach the Japanese and Chinese sense? [November 17, 1931]

To which Hadamard answered:

The idea of sending the French army to Manchuria is perhaps very original. But Mr. Clément Vautel is the sole inventor of this. I spoke of an international guard made up of contingents – with voluntary recruitment, that goes without saying – provided by those countries which make up the society of nations, and not interested parties in the conflict [November 20, 1931].

He was, however, unable to convince Vautel, whose arguments were full of sarcasm:

> Maybe, but it is all fantastic as well. Observe first that the countries which are "not interested parties in the conflict" have only to do one thing: not to get involved. There is no wiser precept in the code of good sense than the following: mind your own business.

> ... Imagine a meeting between Mr. Hadamard's disparate mercenaries and the soldiers of the Mikado. The former shouting in all languages, even Esperanto: "Long live peace!" The latter shouting: "Banzaï!" Oh, Professor! You see here the defeat of your pacifist heroes... Also, I advise you, never go there. But I think that is not what you intend [November 20, 1931].

Vautel's final contribution to this debate was in 1933:

> The reality is this: when a country – whether it be Paraguay or Japan – decides to make war, it does it, and nothing can prevent them from it. This will perhaps change in three or four centuries, but now it is thus.

The beginning of the 1930s was a time of economic and political instability for France, leading to a polarization between the extreme right and left. After a great scandal involving some well-known political figures in financial machinations, the right organized a mass demonstration on February 6, 1934, which then broke out into riots and was quelled by armed police. Hadamard wrote to Volterra: "As for us, we are, as you may imagine, quite saddened by what is happening in Paris at the moment" [IV.44]. The government, led by the Radicals,[1] resigned. After the riots, the *Front populaire* was created, uniting Socialists, Communists, and Radicals, as well as other leftist groups. It was also joined by the League of Human Rights, in which Hadamard, Langevin, and other prominent people were active. In the general election of April-May 1936 following the government's resignation, the Popular Front won by a narrow margin. During the campaign, Hadamard was one of two presidents of the *Bureau du Comité des gauches* in his 14th *arrondissement* [IV.22].

---

[1] The French *Radicaux* are a weaker form of French Socialists and have nothing to do with extremism.

## 6.2   Three letters to Volterra

The relations between Hadamard and Volterra were always cloudless, and grew warmer with the years. On July 10, 1928, Hadamard wrote:

*Vito Volterra*

My Dear Friend, If you feel too hot at Saint Gervais and you use the guide I gave you to get up to Saint Nicolas de Véroce, I want to put right an omission I made when speaking to you: we know a wonderful house there, the most beautiful in the area there. The people who own it are on very good terms with us, ladies of society who have this delightful house where they have a few lodgers. I shall write to them, so that if you wish to go there they will give you that welcome which I would be only too happy to arrange for you. Of course, I will not announce you in any way. The address is the following: Madame Ossent, les

Plans Champs, Saint Nicolas de Véroce. You can moreover, if you wish only to make a trip there, take tea in their meadow.

Madame Hadamard, having been prevented from coming to see Madame Volterra as she would have liked, charges me with her best wishes and profoundest congratulations for the happy event you announce to me.

Please accept, my dear friend, my cordial greetings with all my respects to Madame Volterra.

J. Hadamard. [IV.44]

*Vito Volterra during a dinner at the Hotel Lutetia (Paris, 24 June 1937) in honour of Montel. To the left of Volterra is Montel, to the right is Leray.*

Tricomi recalled Hadamard's frequent trips to Italy "...until the unhappy years during which a protracted political malady covered the face of our country with a mask, of which it was difficult to say whether it was more tragic than ridiculous ... it was not its true face." [II.5, p.23]. When Volterra refused to take an oath of allegiance to the Mussolini regime in 1931, he was dismissed from the University of Rome and, the following year, excluded from all Italian academies. The rise of fascism compelled him to spend as little time as possible in Italy, lecturing at the *Sorbonne* in Paris as well as in Czechoslovakia, Rumania, Spain, Belgium, and Switzerland. Hadamard tried to help him in obtaining positions, as the following letter of November 8, 1933, shows:

My Dear Friend,

In reply to the letter I wrote to Egypt, to suggest to them the idea of calling you, I received the message that an invitation has already been sent to Mr. Chapman, some months ago. I am sorry even more, as I fear that the occasion was unique, with the Egyptian University planning, from next year, to appoint a permanent professor instead of visiting professors.

You must have shared our pain on the death of Painlevé, our old friend, and one of the best people of present-day humanity.

Best regards from my wife and my deepest respects to Madame Volterra, with my best wishes for you, my dear friend.

J. Hadamard. [IV.44]

The last letter from Hadamards to Volterra, kept in the library of the *Accademia dei Lincei*, is dated 1938:

Dear Friends,

We think of you intensely without yet having been able to tell you. We think of you at the threshold of the new year just as we never ceased doing during the present year which finishes full of anguish and pain. It is our deep friendship that I send you.

Louise Hadamard,

J. Hadamard. [IV.44]

Volterra died on October 11, 1940, in Rome.

## 6.3   Hadamard and Lebesgue

It was in 1936 that Hadamard's old friendship with Lebesgue was put to the test. Henri Lebesgue was one of the founders of modern integration theory and theory of functions of a real variable. He is particularly famous for a new notion of integral which he introduced in 1902. The Lebesgue integral, as well as his notions of the measure of a set and of a measurable function, extended the scope of mathematical analysis.

*Henri Lebesgue (1875-1941)*

Lebesgue was ten years younger than Hadamard. Both were former students of the *Lycèe Louis-le-Grand*, they were *archicubes*, or graduates of the *ÉNS*, and they were professors at the *Collège de France* as well as members of the *Académie des Sciences*. However, Lebesgue did not share Hadamard's political views and anxiety at the threat of nazism. Their discussions became so heated that Lebesgue stopped visiting the Hadamards, to avoid a serious quarrel. On December 1, 1936, he wrote the following letter:

My Dear Friend,

I could have replied immediately, yesterday: "thank you, I shall not go to dine with you," but it is better that I explain myself here; without expecting you to understand me and that you will not think me quite insane.

You have known, you have felt for a long time that I am far from agreeing with you in political questions. This does not date from today, nor yesterday, for sure. But today, political matters inflame everybody and you yourself worry about it all the time.

So if I came to you, you would all feel embarrassed; you would endeavour not to speak of current affairs, but every word spoken would take us back to them and it would be a constant effort, which would cause us to feel our disagreement more painfully and make us fervently wish that the hour of departure would finally strike.

This is assuming the best situation. But the more probable would be that we would not be able to avoid the dangerous subject of conversation.

So? You know that I am violent, explosive; what if I were to hurt one of you in my passion? Oh! This would not be what I would wish and against my feelings, but it would be done. Perhaps you are sure about yourself, but I am not about myself. So it is better, is it not, that I avoid you?

There are times when it is no longer a matter of discussions, of argumentation; we are living in the year of these sad times. Let us allow it to pass, and when times have become better we

*The first page of Lebesgue's letter*

shall be more able to understand what was legitimate and noble with the attitude which we, for the moment, do not comprehend, and our friendship, not having been sullied by fighting, will find itself again as clear and whole as it is now. It does not cease to be so for one moment.

Tell Madame Hadamard, tell Jacqueline that I am a fool, but a fool who likes them so much.

Yours,

H. Lebesgue. [IV.16]

Jacqueline Hadamard visited Lebesgue in the winter of 1940-1941, when Paris was occupied by the Germans. She wrote: "We fell into each other's arms. He told me that he had understood his error and that his friendship for the Hadamards had remained intact and even reinforced by this explanation. I later had the pleasure of being able to repeat these words to my father;

which was fortunate, since Lebesgue died during the occupation and my father was never to see him again" [IV.1, p. III(32)].

## 6.4 Hadamard's first trips to the U.S.S.R.

We have already mentioned several Russian mathematicians who were in contact with Hadamard. He was elected Corresponding Member of the Russian Academy of Sciences in 1922 and Foreign Member of the Academy of Sciences of the U.S.S.R.[2] in 1929. That was during the period of the formation of the Soviet mathematical school when the older professors D.E. Egorov, N.M. Krylov, S.N. Bernstein, N.N. Luzin and others were joined by a number of brilliant new names. Liapunov and Steklov were no longer alive when Hadamard made his first visit to the country in 1930. On learning about Steklov's death, Hadamard wrote to Karpinskiĭ, the president of the Academy of Sciences of the U.S.S.R.:

*V. A. Steklov (1864-1926)*

November 3, 1926

Mr. President,

Since I had honour to be a close acquaintance of such a great scientist and man, outstanding in many respects, as Vladimir Steklov was, I am doubly grieved by the loss which Russian science and, as you rightly noted, world science, has suffered.

Would you accept and convey to the Academy my deepest condolences and be assured of my most devoted feelings.

J. Hadamard. [IV.40]

The reason for Hadamard's visit to the U.S.S.R. was an invitation to participate in the first All Union Congress of Mathematicians in Kharkov.

---

[2]The Russian Academy of Sciences acquired this new name after the creation of the Union of Soviet Socialist Republics in 1922.

The French delegation also included Montel, Denjoy, Mandelbrojt, and É. Cartan.  On July 26, 1930, Hadamard was the chairman of a special session dedicated to the fiftieth anniversary of the Kharkov Mathematical Society of which he had been a member since 1903.  The next day he gave a talk entitled *Équations aux dérivées partielles et fonctions de variables réelles* [I.336].  Then, on July 28, he was chairman at the morning plenary session and in the evening he gave a talk entitled *Principe de Huygens et théorie d'Hugoniot* [I.337].

After the congress, Hadamard visited Kiev, where his host was N.M. Krylov, who in 1907-1908 attended Hadamard's lectures in Paris[3] and in 1926-1927 participated in Hadamard's seminar.  He is well known for his work in approximation theory, mathematical physics, variational calculus, and nonlinear mechanics.  Together with his pupil N.N. Bogolyubov he developed a powerful asymptotic method in the theory of nonlinear oscillations.

*In the museum of the Ukrainian poet Shevchenko, Kiev, 1930. Sitting from the left: S. Mandelbrojt, A. Denjoy, N.M. Krylov, J. Hadamard, P. Montel.*

Hadamard's next trip to the Soviet Union was in May 1934, with a delegation of nine French scientists, on the occasion of the French Science

---

[3]His lecture notes are kept in the Archive of the Russian Academy of Sciences.

Week in the U.S.S.R., visiting Moscow, Leningrad, and Kharkov. He was received with honour, being universally regarded as one of the greatest living mathematicians.

The following story was told us by Laurent Schwartz, who heard it from I.G. Petrovskiĭ: "At the first lecture given by Hadamard in Moscow University, Petrovskiĭ introduced him in French, speaking with a strong accent. Hadamard answered: 'I am very grateful to Monsieur Petrovskiĭ for his presentation and I am sure it was very courteous. It is a pity that I, obviously, could not understand anything, because it was in Russian.' Petrovskiĭ finished his account to Schwartz with the words: 'I decided never again to present someone in French'."

*I.G. Petrovskiĭ (1901-1973) obtained results in the general theory of differential equations, algebraic geometry and probability theory which are now classical. From 1951 to 1973 he was the rector of Moscow University.*

Of course, Hadamard was shown the sights of Moscow. At that time the first lines of the Moscow underground were being constructed, and thousands of workers were employed on this project. The streets were dug up, and great heaps of soil were everywhere.

As Solomon Mikhlin, who happened to be among Hadamard's guides on this tour, remembered, Hadamard quipped *"Un peu trop de métros"*, while jumping over the ditches. That was a memorable event for the young Mikhlin who had graduated from Leningrad University five years earlier and whose first mathematical result was an extension of the Cauchy-Hadamard formula for the radius of convergence to double power series [III.278].

*S.G. Mikhlin (1908-1990): his main results are in numerical methods for solving differential equations, the theory of integral equations, and elasticity*

In Leningrad (now Saint Petersburg), Hadamard was welcomed by A.N. Krylov (1863-1945), an applied mathematician and naval architect, whom Hadamard had met before and corresponded with. In the Saint Petersburg section of the Archives of the Russian Academy of Sciences there are four short letters from Hadamard to him. Here is one of these letters:

> December 25, 1926
>
> Dear Sir,
>
> May I ask you to look through a paper of Mr. Milne in the *American Monthly* and to give me your opinion about it? Does it contain a development of Adams' method and is it worthwhile studying? I ask you this favour believing that I will not disturb you seriously, since you are a master in the subject.
>
> I send you my thanks beforehand and assure you of my best feelings.
>
> J. Hadamard.[4] [IV.41]

---

[4] Adams' method is a finite difference method for solving the Cauchy problem for first order systems of ordinary differential equations. Hadamard included "*un excellent exposé de la méthode d'Adams par M. Kriloff*" into the second volume of his *Cours d'Analyse* of 1930.

A.N. Krylov made contributions to the theory of gyroscopes, the stability of ships, ballistics, and structural mechanics. During the period 1927-1934 he was the director of the Physical-Mathematical Institute of the Academy of Sciences. At the reception in Hadamard's honour, A.N. Krylov said:

> It is not a real merit to happen to be the oldest among my colleagues, the Leningrad mathematicians who have entrusted me with the honour of greeting the illustrious *Maître* who gave brilliant flashes to the bright flame of French science. It is not necessary to mention the personal contribution of Mr. Hadamard; it has become classical [IV.42].

*A.N. Krylov (1863-1945)*   *V.I. Smirnov (1887-1974) (a lithograph by G. Vereiskiĭ)*

Hadamard met scientists from Leningrad University and visited the newly founded Research Institute of Mathematics and Mechanics. Its director was V.I. Smirnov, a great authority in function theory and partial differential equations, and the author of a famous course of mathematics in five volumes. Smirnov was a pupil of Liapunov and became a professor in 1915. He was a man of rare nobility in his bearing, and despite a rather elevated position in the Soviet academic hierarchy, he never compromised so as to bring harm to others, even during the hardest years of ideological suppression. On the contrary, he used his influence to help many people.

At the meeting at Leningrad University was the twenty-two-year-old L.V. Kantorovich. He later became famous for his works in various areas of mathematics and was a Nobel Prize winner in mathematical economics. Entering

university at the age of fourteen, Kantorovich became a full professor when
he was only twenty. Shortly before the meeting, he wrote a paper containing
the construction of a conformal mapping of a disc onto a domain close to it.
Many years later he recalled:

> At the time of Jacques Hadamard's visit to Leningrad, in
> 1933 I think, at a meeting with him in the office of Smirnov,
> who was the rector of the University,[5] a short report was made
> on some of the achievements of Leningrad mathematicians, in-
> cluding a discussion of the paper on approximate conformal map-
> pings. The paper interested Hadamard, and he jokingly ex-
> pressed the fear that the same fate would befall Kantorovich
> as Galois.[6] In reply someone said that I did not have such an
> aggressive character [III.208, p. 244].

*L.V. Kantorovich (1912-1986) in 1938     S.L. Sobolev (1908-1988) in 1979*

In Leningrad, Hadamard met S.L. Sobolev (1908-1988), then twenty-six
years old and already a corresponding member of the Academy of Sciences
of the U.S.S.R. In June 1985 Sobolev told us:

> I saw Hadamard for the first time and listened to his talk
> at the Kharkov Congress. I remember that he was present at

---

[5]Smirnov was the director of the Research Institute of Mathematics and Mechanics at
the University, and not the rector. (Auth.)

[6]E. Galois was killed in a duel at the age of twenty.

my talk on a new approach to solving the wave equation with a variable coefficient.

But we were only able to speak later, in Leningrad, at the Physical-Mathematical Institute. Then quite young, I was very proud to see the famous man. Hadamard, of medium height, with a beard and a little moustache, lively and active, made an impression on me of being a jovial man. One felt at ease with him. We talked in French, mostly on the relation between my method and that of his pupil Mathisson on the Cauchy problem. Hadamard was interested to know whether Mathisson's result followed from mine. I remember he asked me to send him a reprint of the paper with a detailed exposition on that subject, which I did, of course, later, when my paper appeared.

The success of further studies in hyperbolic equations with variable coefficients is inspired by Hadamard. His use of the finite part of a divergent integral is just a great foresight. I realized its meaning only later, when I tried to describe distributions satisfying the Cauchy problem.

## Le mouvement scientifique en U.R.S.S.

### Par M. HADAMARD,

*membre de l'Institut,*
*membre de l'Académie des Sciences de l'U.R.S.S.,*
*avec la collaboration de M. PRENANT,*
*professeur à la Sorbonne*

---

Ce rapport devait comprendre trois parties étudiées séparément par MM. les Prof. Hadamard, Prenant et Langevin. Mais ce dernier qui devait traiter des sciences physico-chimiques en a été empêché par son état de santé. C'est la raison pour laquelle le rapport qui suit est incomplet dans son ensemble.

Dans une révolution aussi ample, aussi totale que celle que la nation russe vient de traverser, on pouvait craindre qu'un organisme délicat comme l'est la recherche scientifique, qu'un travail qui vit de calme et de patiente méditation ne fût gravement compromis.

Pour parler d'abord du domaine qui m'appartient en propre et qui m'est familier, les mathé-

*Hadamard's report on his visit to the U.S.S.R. in 1934. It begins with the phrase: "In a revolution as ample and total as the one which the Russian nation just went through, one could fear that an organism, delicate as is the scientific research, that a work which survives on calmness and patient meditation, would be gravely compromised"* [IV.22].

In 1935 Sobolev was the first to give a rigorous definition of distributions [III.369], [III.370] (for a survey see Lützen's book [III.249, part 8]). Initially he called them ideal functions, but soon dropped the term when warned by a colleague that he could be accused of idealism.[7] He studied function spaces which were later named after him, proving imbedding theorems that are widely used in the modern theory of partial differential equations. Towards the end of his life Sobolev worked in numerical analysis, in particular on the theory of cubature formulas.

## 6.5   New apartment. Jubilee

The old house in the *rue Jean Dolent* (it became *rue A. de Humboldt* in 1925), once noisy and gay, reminded the Hadamards of their losses and became too big for them. By 1935, on Hadamard's initiative, new houses for professors had been built near the *Cité Universitaire*, where the old fortifications of Paris had stood. The Hadamards decided to move to a new flat at *rue Émile Faguet* 12. Jacqueline Hadamard recalls:

*The house at rue Faguet 12, where the Hadamards lived from 1935 to 1940 and again from 1945 to 1963*

It was a great effort, because the house in the rue Jean Dolent had three storeys and a great number of cupboards, all filled with things which had been acquired and especially boxes of the kind I had seen at my grandmother's house with a label "Little bits of string no longer useful for anything" (authentic!). The most difficult thing was to get my father to sort his papers, an abominable task for him [IV.1, p. III(29-30)].

---

[7] "Idealism" as opposed to the official philosophy of "materialism" meant ideological infidelity, which could have serious consequences.

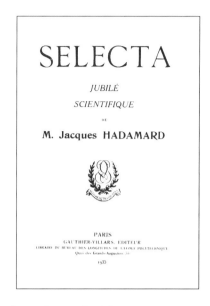

*The title page of Hadamard's Selecta. In 1968, five years after Hadamard's death, the CNRS published four volumes of Œuvres de Jacques Hadamard which contain his main mathematical papers. Unfortunately, all misprints of original publications are reproduced in this collection, and there are no editorial comments.*

In June 1935, a series of international mathematical lectures was organized in Geneva, with Hadamard as president. His own lecture was devoted to well- and ill-posed problems for partial differential equations, in particular for hyperbolic equations, a topic which had fascinated him for more than three decades.

On December 8, 1935, Hadamard became seventy years old. By then, his colleagues and pupils had prepared the volume *Selecta*, containing a number of his papers and a bibliography of 280 of his works. The book begins with the dedication:

> To the illustrious scientist who opened new and sure routes in mathematical research and whose unceasing labour established itself in vast areas,
>
> To the eminent master whose inexhaustible devotion has encouraged and guided so many efforts,
>
> To the man of goodness for whom no human anxiety appeared strange,
>
> His admirers, his friends, his pupils offer this collection of some of his works, a representation of a magnificent and fecund work.

The celebration in honour of Hadamard took place on January 7, 1936, in the assembly hall of the *Collège de France*. His friends, colleagues, and prominent people gave speeches. Among them were Lebesgue, and Fréchet, Picard, Vessiot, and Bédier. At the end of the ceremony, the minister of education, Roustan, announced the award to Hadamard of the rank of *Commandeur de l'Ordre de la Légion d'honneur*.

*Légion d'honneur*
*croix de chevalier*

*Légion d'honneur is a French order (established by Napoléon Bonaparte in 1802) awarded for outstanding military and civil service. It has five ranks; Commandeur is the third. Hadamard had been awarded the first two earlier (Chevalier on December 6, 1910, and Officier on October 1, 1923). After the Second World War, he was made Grand Officier (February 7, 1948) and finally entered the highest rank, Grand-Croix, on his ninetieth birthday.*

When finishing his speech of reply, Hadamard cited the following words of the well-known microbiologist Émile Duclaux. During a discussion, someone had said to Duclaux disdainfully: "In twenty year's time, your treatise on microbiology will have become waste paper", to which Duclaux replied: "Not only do I know that very well, but I also fervently hope that it will be so. A scientific book is a shelter at the wayside. It collects the late arrivals, receives the new recruits, it polishes them a little, gives them their knapsack and then, en route! Outmatch us, you younger people, because your legs are young, but for God's sake, keep moving, because if you stand still and allow us to linger on, then, in truth, neither you nor we will have done our duty" [II.27, p.57].

An account of the ceremony was published in 1937 as a small book [II.27] which also contains numerous greetings signed by the greatest mathematicians and physicists of the world. There were, however, no greetings from Germany.

Papers dedicated to Hadamard on the occasion of his Jubilee were collected in two volumes of the *Journal de Mathématiques Pures et Appliquées*, published in 1937 and 1938.

## 6.6   The trip to China

Hadamard was the Chairman of the International Committee on Mathematical Education at the time of the International Congress of Mathematicians in Oslo in July 1936, but he did not attend the Congress, being in China at that time. He had been invited by the Tsinghua University in Peking (now Beijing) and the Academia Sinica.

The invitation was a good reason for Hadamard to undertake a long trip to the Orient together with Louise. They decided to travel mostly by air. Jacqueline Hadamard recalls:

> We were a little anxious, my sister and me: they were sixty-seven and seventy years old. We contacted our cousin Robert Debré asking him to examine them and decide whether all these flights could be dangerous to them. After doing so, he called me: 'Really, I would not prevent them to leave. They are both in a perfect state'. Before their departure, they asked Langevin, who had already been there, about the rules of Chinese courtesy. 'It does not matter what we do', he told them, 'we cannot match Chinese politeness' [IV.1, p.III(30)].

The Hadamards set off in the early spring of 1936. They visited Indonesia, Indochina, and Japan on their way to Peking. Montel told that Hadamard profited of his stop in Ceylon editing several pages for his *Leçons de Géométrie* [II.5, p.23].

On October 4, 1935, Hadamard wrote to Wiener who lectured at the Tsinghua University during 1935-1936 academic year:

> Dear Professor Wiener,
>
> My best thanks for your kind letter. I am most happy to think that Mrs. Hadamard and I shall meet you and become acquainted with Mrs. Wiener.
>
> Of course, I am also most pleased to hear of the complete safety at Peking, and I am most thankful to Professor Hiong and you to have informed me about that. Indeed, when I saw Mr. Yang, I was reading in the newspapers of accidents experienced by travellers in China – probably exaggerated stories.
>
> Now, your letter gives me a complete security in that line and I am settling everything for my departure. Let us hope that

each of us has not to reckon with the general uneasiness over the whole world, which is a too real fact.

Kindly transmit my best remembrance to Professor Hiong. Thinking most agreably of the pleasure to see you both very soon, I beg you to believe me.

Yours sincerely,

J. Hadamard. [IV.31]

When they arrived in Peking, the University, established in 1911, "...was in the process of a reorganization: from being an ordinary high school, preparing an auxiliary scientific workforce for the U.S.A. (as a compensation for the Boxer Uprising), it became an independent educational establishment" [III.422, p.176]. Hadamard's lectures had to be given in English, with which he had no problems. Hadamard left no account of his stay in China, so we shall cite from Wiener's book *I am a mathematician*:

He was installed near us in the Old South Compound, but he soon moved to an apartment in the city in or near the legation quarter. The Hadamards were much happier in this livelier environment. Professor Hadamard was already well along in years, and the discomforts of the isolated life on the grounds of the university rather terrified him. It was a pleasure to see him again. He was a great mine of reminiscence of the good old days of French mathematics. His wife was also a mine of anecdotes concerning French academic life, and she had known Pasteur when she was a child [III.422, p.199].

*The Hadamards in China*

Wiener writes that he, his wife, and his Chinese colleague Lee often went to visit the Hadamards.

We [...] used to go down into the tangled, squalid streets of the so-called Chinese city (as opposed to the rectangular Tatar city) to rummage in the antique shops. There we would often come across ancestor portraits which showed dignified Chinese gentlemen or ladies, in stiff poses, with hands on the knees, dressed in marvellous silken gowns, which for the men were robes of office, civil or military. For all their pomp and stiffness, it was common for the faces in these pictures to be of a remarkable fineness, humour, and sensitivity.

We found one such ancestor portrait which was so like Professor Hadamard himself, with his somewhat sparse, stringy beard, his hooked nose, and his fine, sensitive features, that it would have been completely adequate to identify him and to pick him  out of a large assembly of people. There was, it is true, a very slight slant to the eyes, and a very slight sallowness of complexion, but not enough to confuse the identification. We bought this picture and gave it to its likeness. He appreciated it very much, but I don't think that Madame Hadamard cared for it [III.422, p. 200].

Wiener mentions also that his rickshaw driver once surprised him by the question "whether it was true that Ha-Ta-Ma Hsien-Sheng [Mister Hadamard] was as great a mathematician as the newspaper said he was" [III.422, p. 193].

In Peking Hadamard also told Wiener "a delightful story about his own youth", which we already mentioned in Section 2.5, when he visited Hermite at the time of the Dreyfus trial. However, Wiener's account of this is not completely accurate: he gives Dreyfus the rank of colonel, and he writes that Hadamard was about to be examined for his doctorate, whereas he had obtained it two years before the Dreyfus trial.

Hadamard's lectures were devoted to the theory of partial differential equations, and became the basis of his last book. "My father is delighted with his pupils", wrote Jacqueline Hadamard [IV.1, p. III(31)]. Four months later, the Hadamards left Peking and returned home by train through Siberia.

*Jacques and Louise Hadamard*
*on the title page of the Peking*
*monthly La politique de Pékin*

In August 1936, the Popular Front government issued a decree in which the age of retirement was set at seventy for professors. Although still full of energy, Hadamard was obliged to leave his positions at the *Collège de France* and *École Polytechnique*. On October 1, 1937, he was granted a pension after "fifty-two years, eleven months and twenty-six days of service" [IV.29].

## 6.7   Before the storm

The 1930s were also a time of tension, and the danger of war in Europe increased inexorably. Mussolini's regime became more aggressive, invading Abyssinia in 1935; in 1936 a civil war began in Spain. Deep anxiety was caused by the ominous actions of Nazi Germany: the remilitarization of the Rhineland in 1936, the Nazi coup in Austria in 1938, and its subsequent annexation, the complete militarization of the country.

When many Jewish scientists had to escape from Germany, Hadamard did his best to find positions for them in France and other countries. Wiener, who was a member of the Emergency Committee in the Aid of Displaced German Scholars, tried to get philanthropists to help place the refugees in

America. In 1935 he included the following letter from Hadamard in one of his articles [III.423, p. 928], in order to convince American readers that French help to the refugees had been exhausted:

> Dear Professor Wiener,
> The task of affording help to German intellectual refugees – a heavy one, for as you know refugees of every kind have come to us in great number – has been very difficult for France.
> Our country is small in comparison with yours, and we have a limited number of universities. At present, the number of young French scholars, I mean men of very great distinction, is great, so that the way to a career is very difficult. Therefore, not one refugee has been appointed in our public educational system and we cannot think of making any such appointments.
> For about one year we had been able to support financially those who were on our soil. However, this has ceased since last summer as the funds were exhausted.
> Happily, at least as far as mathematicians are concerned, practically all those who had come to France have found, or I hope are soon to find, employment abroad. I regret this for my own country. Some of them have already gone to America, – in my opinion, the better for you and the worse for us. It will be a great thing for your country and for civilization if you have full success in that direction. We can do nothing more as we are overcrowded by a mass of refugees of all sorts, not merely of intellectuals.
> On the other hand, I have just written to Geneva to Dr. Kotschnig, who will inform you, if you do not know already, what has been done on the part of the High Commissioner appointed by the League of Nations.
> Wishing you complete success, I beg you to believe me.
> Yours truly,
> J. Hadamard.

The same topic was the subject of the following Hadamard's letter of January 27, 1935:

> Dear Professor Wiener,
> The situation as concerns France is slightly better than I depicted in my previous letter. I just hear that instead of fifty-eight German scholars who could be supported last year, it has been possible, hitherto, to help – rather poorly – thirty. But it is feared that even that last effort must cease very soon.

I wanted, of course, to tell you accurate information, and this is why I write to you again.

Believe me.

Yours truly,

Hadamard.

As a member of the *Comité du foyer des enfants juifs allemands*, Hadamard tried to organize the reception of Jewish children from Germany whose parents had been arrested. On July 6, 1933, he wrote to Einstein:

My Dear Friend,

We are trying at this moment to organize here the reception of Jewish German children, whom their parents would like to get away from the terrible conditions of life made for them. The project seems to me quite worthwhile, I am active in it: it is a matter of bringing up the children not only in good material circumstances (that goes without saying), but also in moral conditions so that they later do not find themselves disoriented in free countries.

Amongst all the efforts at the moment being tried for the benefit of the exiles, this seems to me the one most worthy of interest. [III.120, p. 123]

In April 1938, the Popular Front government collapsed, and a centrist coalition formed the new cabinet under the premiership of the Radical Édouard Daladier. On September 29, 1938, he signed the Munich agreement, together with Hitler, Mussolini, and Chamberlain. It was a betrayal of Czechoslovakia, a close ally of France. After this act, the Hadamards addressed themselves to Czechoslovak mathematicians in the following letter.

Dear Friends,

In these sorrowful days, it is necessary to say how close we are to you. You have, at least, no cause for shame and you can say with pride that you have kept your honour. The stance of your President, the dignity with which he never wavered for a moment, aroused general admiration, and will make history, and we have every right to hope that justice will remember this. Over the heads of the western governments which have betrayed us, we shake hands with you.

J. Hadamard, L. Hadamard. [IV.22]

The letter is undated, but the envelope is stamped October 7, 1938.[8] On March 15, 1939, Hitler invaded Czechoslovakia, and on September 1, a week after the Soviet Union and Germany signed their pact of non-agression, he invaded Poland. Two days later the Daladier government announced that it considered itself obliged to fulfill the terms of its treaty with Poland.

---

[8]In the *Archives de l'Académie des Sciences* in Paris there is a letter dated June 10, 1966, from Professor V. Kořínek of the Charles University of Prague, to Szolem Mandelbrojt, which begins with the following words: "Dear Colleague, as you wished, I send you the letter which Professor Jacques Hadamard sent to the Czechoslovak mathematicians after Munich, through the intermediary of Professor Budžovsky" [IV.22].

# Chapter 7

# World War II

## 7.1 War again

From September 1939 to May 1940 there was no military action between France and Germany: this was the period of the *drôle de guerre*, the phoney war. Hadamard, with his wife and grandsons, Étienne and

*Jacques and Louise Hadamard*

Francis, spent this time at Rambouillet (22 rue Pasteur), fifty kilometres southwest of Paris. "The inhabitants of the little town, who saw him walking in the street or through the forest, were generally unaware that this man with the frail silhouette, who was friendly to everybody he met, was a member of the *Institut* and, even more, a scientist with a world reputation, who was interested in the plants which grew in the woods as a scientist and not only looking for edible mushrooms" [II.27, p.312-313].

The boys were in the second and fifth classes at the *lycée*, and their grandfather, whom they called *Pare*, was greatly interested in their lessons and homework. "One day he talked to me, not without humour, of the last Latin translation I had given in the class in which his grandson, Étienne Picard, was my pupil," wrote A. Rossat-Mignod [II.58, p.312]. Francis Picard recalled that his dominant memory of his grandparents was of extraordinary kindness.

However, the clouds of war were advancing: after the breakthrough at Sedan and the battle of Dunkirk, the German army advanced rapidly on Paris. Jacqueline Hadamard vividly describes the Hadamards' flight from Paris. On June 9, 1940, the members of Langevin's laboratory, where she worked, received instructions to evacuate to Toulouse. She rushed to her parents, who were in Rambouillet, but on arrival she discovered that it was not possible to take her family immediately. Her father, characteristically obliging, had gone to Paris by train, to act as chairman of some committee at the *Institut*, and turned out to be the only one present. The railway station from which Hadamard was to catch his train back to Rambouillet was under bombardment, and so it was very late by the time he managed to return home.

Hadamard and his wife, with their daughters Cécile and Jacqueline, drove to Brittany where Cécile's children were staying with relatives. The roads were jammed with streams of fleeing Parisians. On June 13, Paris was declared an open city so as to avoid its destruction. Jacqueline Hadamard wrote:

> We reached Concarneau where we found the children who were staying with cousins. In their garden we buried, carefully wrapped in oilskin, the letters of my two brothers, killed in 1916, which my parents did not want to fall into the hands of the Germans. Then, with a heavy heart, we listened to the radio, and it was the request for an armistice, made by Pétain![1] Unnecessary to speak of our thoughts.
>
> But in the middle of the night, on June 17, my cousin came to wake us: he had heard the BBC, which had explained that the conditions of the armistice were absolutely unacceptable. So we had to flee from this *cul-de-sac* which Brittany was. Hurriedly, we repacked our suitcases. That was when my mother fell down the stone staircase of the house and hurt herself a lot. Fortunately, she had not broken her hip! [IV.1, p. IV(2)]

The family left Brittany, which the Germans were approaching. Cécile and Jacqueline attached mattresses to the roofs of the two cars, as this was believed to give protection during bombardments. They barely had time to cross the bridge over the Loire before it was blown up. The gendarmes tried to stop the stream of refugees, but the Hadamards were able to continue,

---

[1] Marshal Philippe Pétain (1856-1951), hero of the World War I, was in command of the defence of Verdun. Head of the collaborationist government from 1940, he was condemned to death in 1945 which was replaced by life imprisonment.

as Jacqueline had a pass, which she had obtained recently in order to take platinum out of Paris on the instructions of her laboratory.

## 7.2   In Toulouse

After two hard days on the road and one night spent in the open air, the family arrived in Toulouse on the 19th of June 1940. Soon afterwards they were joined by Mathieu Hadamard with his wife and children. Jacqueline Hadamard recalls:

> We were directed to the *cité universitaire* where we were sheltered. I found my colleagues from the laboratory and Langevin. We went to where the news was. Toulouse was full of refugees. The news centre was a large wall on which everyone stuck their message, saying where they could be found and asking for news about their relatives.
>
> We addressed the question of food, the main preoccupation of the time. The little meat and fat which one got meant that we consumed incredible quantities of vegetables: I still remember the eight kilogrammes of Jerusalem artichokes we peeled every day! It was fortunate that we were occupied with that, because the time would have seemed very long without anything to do and brooding over such sad thoughts [IV.1, p. IV(3)].

On June 22, France signed an armistice accepting the German occupation of large areas of the country. The French government moved to Vichy, a small city in Central France. Toulouse, at the south of France, remained in the unoccupied zone. The following story from that time was told us by Laurent Schwartz, whose family also lived in Toulouse:

At the beginning of the war, when Hadamard was still in
France, he and my brother went to buy eggs in a shop which
had just got a delivery. Since the dealer, the civicism incarnate,
only gave half a dozen per family, they decided not to speak to
each other so as to avoid any suspicions. Soon he forgot completely about the agreement and, standing for half an hour in the
queue, he and my brother became involved in a loud conversation. When they approached the salesman, Hadamard suddenly
remembered. He stretched out his hand to my brother and said:
'My dear Sir, I am very happy to make your acquaintance'. So
they got a dozen. On leaving the shop, Hadamard exclaimed:
"See you at home in a moment". [Interview, October 28, 1992]

After learning that Langevin had returned to German-occupied Paris,
Jacqueline followed his example, but on arriving at the *Collège de France*,
she discovered that she, being Jewish, had been fired. She removed her
father's mathematical library to the *École de Physique et de Chimie*, and
his collection of ferns and fungi to the *Muséum d'Histoire Naturelle*. She
refused to obey the edict that all Jews should register at the Commissariat
and returned to her parents in order to avoid arrest.[2] It was not without
great difficulty that she did this, and she arrived in Toulouse with only
her personal belongings and the family picture album commenting: "How
many people have been deported, unable to give up their furniture!" [IV.1,
p. IV(12)] When, after the first months of the occupation, the underground
Resistance movement began to emerge, Jacqueline managed to get in touch
with a group from the Resistance, and became their courier in the southern zone. Meanwhile, Hadamard tried to use his international contacts to
help Jewish scientists from Germany, Austria, Czechoslovakia, Poland, and
France. Writing to Einstein on this matter in a letter dated January 16,
1941, Hadamard included this personal request:

To this general request, I add a word on my own behalf. My
case is, moreover, completely separate from those about whom I
also inform you, because I have been retired for three years. But
it would be useful and valuable for me to come, if possible (for
a temporary stay) to the United States, by the only way I could
think of doing this, that is by invitation to some conferences.

If you learn of some possibilities of this kind (even relatively
few and not specially advantageous from the pecuniary point of
view, which is not what I attach myself to particularly), would

---

[2]For the account of the anti-Semitism in France under the Vichy Regime see the book
by R. Poznanski [III.330].

you cable them to me, at the same time as informing me by letter. It is the only way of obtaining, for me, the authorizations, French as well as American.

Once again warm thanks from your devoted friend

J. Hadamard. [III.120, p. 127]

*Louis Rapkine (1904-1948)*

Help came from Louis Rapkine, a young man who played an extraordinary role in rescuing many French scientists. Rapkine was born in Byelorussia in 1904. His family moved to Paris when he was seven years old, and then went on to Canada. After three years as a student at McGill University, he returned to France in 1924. Despite many hardships, Rapkine was able to perform important work in cell biology. However, he did not limit himself to academic research. In 1934, he organized the Welcome Committee for Foreign Scientists which helped those who had fled from the Nazis. At the beginning of the war, he moved to the U.S.A. and started organizing the departure of scientists from France. He did all he could to find private means to support them on arrival and to finance research and academic positions for them in the U.S.A. In this way, he was able to help about thirty French scientists, among whom were Hadamard, Jean Perrin and André Weil. Weil wrote:

Louis Rapkine, a brilliant young biochemist from Canada [...] happened to be in New York at the crucial moment. He had drawn up and presented to the [Rockefeller] Foundation a list of French scientists, including both Jews and non-Jews, whom he thought it advisable to rescue from France as rapidly as possible. As soon as he found out – how, I do not know – that I was back in France in October of 1940, he added me to his list, and it was thus that I received the offer from the New School. I used to call him Saint Louis Rapkine. To this day I am touched when I remember the kindness that he showed me. After the Liberation, he managed with some difficulty to find a position at the *Institut Pasteur*, only to die soon after of lung cancer. I was deeply saddened by his death [III.418, pp. 175-176].

In May 1941, Hadamard received the following telegram from New York:

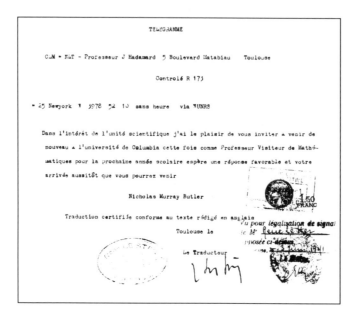

Professor J Hadamard 5 boulevard Matabiau Toulouse

In the interest of scientific unity I have the pleasure of inviting
you to come again to the University of Columbia this time as
Visiting Professor of Mathematics for the coming academic year
Hope a favourable reply and your arrival as soon as you can come

Nicholas Murray Butler.[3] [IV.17]

Jacqueline Hadamard wrote: "We were extremely happy for this invitation, me and my sister, because my father was one of those who risked most,
and to place our parents in safety fulfilled our desires, because it would have
been very difficult to hide my father" [IV.1, p. IV(14)].

The family waited for their visas from the U.S. consulate in Marseilles,
and their rapid issue was aided by Joint, an American Jewish organization.
Finally, in August the visas were ready and it was then possible to leave for
Lisbon, in order to sail on to the U.S.A. Cécile remained in Toulouse.

So, Jacques, Louise, and Jacqueline Hadamard, together with a group
of refugees, arrived in Spain. Now it was possible to sit in a café with a

---

[3]N.M. Butler (1862-1947) was president of Columbia University from 1902. In 1931 he
became a cowinner of the Nobel peace prize.

cup of coffee and a cake. A long-forgotten pleasure! "Don't forget your briefcase, the American visas are inside", Jacqueline said, when they were leaving the café. "But they should be with you", Hadamard replied. The visas were not to be found in the luggage: they were left in Toulouse. When the Hadamards applied for permission to return for their visas, the Spanish police told them that they could leave Spain but not enter again. A telegram was dispatched to Cécile, asking her to send the visas to the offices of Joint in Marseilles. From there, they would be sent to Lisbon with another group of refugees.

Arriving in Lisbon, the Hadamards found no visas and they were unable to join their ship to the U.S.A. The only way out was to leave on board an Italian ship, chartered by Joint, for a large group of refugees from Germany without passports. On arrival the Hadamards would have to wait in jail until the visas were delivered by airmail. "An American prison? Compared with what we fear, it's not a very grave prospect. Agreed," wrote Jacqueline Hadamard [IV.1, p. IV(16)-IV(17)].

## 7.3  Life in America

On arrival in New York, the Hadamards were placed in a taxi, together with an armed policeman, and sent to Ellis Island (an island in Upper New York Bay which was a U.S. immigrant station from 1891 to 1943). Jacqueline Hadamard described the prison:

It seemed a model prison, with a large day-room having a library where we spend the whole day (except for the "walks" between the walls of wire netting). One can telephone from there to the outside world. The atmosphere there is quite tense, suspicion reigns, everyone coming to you in turn to denounce the others: "Don't speak to him, he is a nazi spy" (or a fascist spy or whatever). The rooms are impeccably clean, but my mother categorically refuses to go to the showers with the others (she is seventy-three years old!).

After ten days, we are called in front of the judge. Careful, we were told, we have to continue pretending not to know where our visas went astray; otherwise Joint or the shipping company will have to pay large fines for having taken us on board without visas. When they question my father I become afraid that he will make a gaffe; I offer my services, explaining that he can not hear very well. Phew! It's accepted, because the judge is convinced that all these immigrants are vermin contaminated with microbes and parasites, and he forbids my father to approach. So I explain – from a distance – the situation as we agreed to tell it: visas handed in on arrival in Lisbon. Who mislaid them? Nobody knows. And I go back to my place [IV.1, p. IV(17)-IV(18)].

*In the 1940s Ellis Island was used as a detention centre for unwanted immigrants. This is the registration room, located in the main building, where they were waiting for the decision of the immigration officers.*

At that moment, Louis Rapkine appeared in the court room. He showed the judge a telegram from a representative of Joint in Lisbon, saying that the visas were being forwarded by airplane, and the visa numbers were given. Jacqueline Hadamard wrote: "He added that if people got to know that Mr. Hadamard, a scientist with a world reputation, and, moreover, a Yale man (that is, a former professor at Yale University), was in prison because his visa had got lost on their way, it would cause a scandal in the university world. And he added: 'I don't see why I shouldn't tell the newspapers' " [IV.1, p. IV(18)].

The next day, the Hadamards were released on Rapkine's recognisance. Jacqueline Hadamard continues: "Perfect!... No! My father refuses to

leave the prison: he had begun reading a
book on the history of the United States
and didn't want to leave until he had finished it! Finally, after having promised
him that we would buy the book as soon as
we had left, he agreed (between ourselves,
this promise was never kept, I forgot it ...
and so did my father)" [IV.1, p. IV(19)].

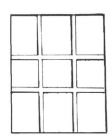

Hadamard got a visiting position at
Columbia University and the family settled in a flat belonging to the University. He prepared lectures and published
some papers on problems of mathematical physics. As before, he was interested
in the question of well-posedness for non-
classical boundary value problems, in particular for the Dirichlet problem for the

linear hyperbolic equation with two independent variables. He also wrote
two obituaries about Émile Picard on learning of the death of his teacher.

*The Library and Main Approach to Columbia University in 1930s*

In 1943 he gave a course on the psychology of invention at the newly-
founded *École Libre des Hautes Études* in New York. He had long been
interested in the role of intuition in mathematics, and he tried to study it

from the psychological point of view, appealing to the notion of the subconscious. This material can be found in his book *The psychology of invention in the mathematical field* which he wrote for Princeton University Press. Hadamard sent the manuscript to the publishers in August 1944, just before leaving America.

VOLUME I    OCTOBRE-DÉCEMBRE    FASCICULE 4

# RENAISSANCE

REVUE TRIMESTRIELLE PUBLIÉE PAR
L'ÉCOLE LIBRE DES HAUTES ÉTUDES

SOMMAIRE

*ARTICLES PAR*

JACQUES HADAMARD, LIONELLO VENTURI, JULIEN BONFANTE,
FRANÇOISE DONY, RENÉ ETIEMBLE, HENRI SEYRIG,
M<small>me</small> DE WYZEWA, JEAN-SYLVAIN WEILLER, AUGUSTE VIATTE.

*COMPTES RENDUS PAR*

HENRI GRÉGOIRE, PAUL JACOB, LÉON KOCHNITZKY, ALEXANDRE
KOYRÉ, CLAUDE LEVI-STRAUSS, M<small>me</small> THÉRÈSE MARIX-SPIRE,
ANDRÉ SPIRE.

NEW YORK, 1943

LA SCIENCE ET LE MONDE
MODERNE

I

Il y a quelque mélancolie à parler d'un élément de la civilisation alors que cette civilisation est menacée — menacée pour un temps nous en sommes sûrs — dans ce qu'elle a de précieux. Peut-être au contraire est-ce le moment de réfléchir sur le rôle de cet élément, de la Science, puisque c'est d'elle que je veux parler.

Il s'en faut que ce rôle ait toujours été bien compris. La Science a connu dans l'opinion des faveurs et des disgrâces excessives.

En présence de ses succès retentissants, ce fut, pendant presque tout le XIX<small>e</small> siècle, l'enthousiasme: celui qui s'exprime dans les belles pages de *L'Avenir de la Science*[1] où Renan, publiant en 1890 ce qu'il pensait dès 1848, écrit: "La Science seule peut résoudre à l'homme les éternels problèmes dont sa nature exige impérieusement les solutions" et où dans sa préface il précise: "Tout en continuant de croire que la Science seule peut améliorer la malheureuse situation de l'Homme ici-bas, je ne crois plus la solution du problème aussi près de nous que je le croyais alors."

La réaction est venue plus tard et fut sans mesure. Ma génération a vu une fameuse campagne de Brunetière sur la "banqueroute de la Science". On lui reprochait, à la Science, de n'être en état de résoudre aucun problème de métaphysique ou de morale. Nous dirons tout à l'heure pourquoi telle n'est pas sa mission et pourquoi on ne saurait le lui demander. Comme on l'a dit avec raison, "la Science n'a pas fait

Page 108. On doit être reconnaissant à M. Georges Sarton de n'avoir pas laissé oublier, à l'heure actuelle, ce livre et de lui avoir consacré l'intéressante étude qu'on vient de lire dans *Renaissance*, vol. I, fasc. 2.

*The title page and the beginning of Hadamard's paper in the journal Renaissance published by the École Libre des Hautes Études, 1943*

In the archive of Hadamard's family there is letter of Einstein to Hadamard giving his reaction to an invitation to participate in the "Philosophy – Science" week at the Mount Holyoke College to be held during the period July 25 – 30, 1944:

April 14, 1944

Professor J. Hadamard
54 Morningside Drive
New York City

Dear Mr. Hadamard:

I was quite happy to receive a letter directly from you, which gave me the feeling that you are getting along well in this strange world. I feel quite embarrassed, however, that I cannot do what you so kindly have proposed to me. The reason is that the only means to save my soul and my time under the prevailing circumstances is to abstain from all activities in public. If I would make an exception I would get in the devil's kitchen without reasonable hope to escape. There is, furthermore, the fact that I have not to say anything of special interest or originality.

I hope that your life in this country is satisfactory and I am expressing the sincere wish that you may see again soon your beloved France.

Cordially yours,
Albert Einstein. [IV.19]

While in the United States, Hadamard looked for a position. Laurent Schwartz recalls the following story told by Hadamard: "He came to a small university and was received by the chairman of the department of mathematics. He explained who he was and gave his Curriculum Vitae. The chairman said: 'Our means are very limited and I can not promise that we shall take you'.

Then Hadamard noticed that among the portraits on the wall was his own. 'That's me', he said. 'Well, come again in a week, we shall think about this'. On his next visit, the answer was negative and his portrait had been removed" [Interview, October 28, 1992].

Jacqueline Hadamard writes: "During all these years, our great support was the French community which shared our anguish and our hopes" [IV.1, p. V(8)]. The Hadamards often participated in the meetings of *Français Libres*, the patriotic mouvement inspired by General de Gaulle's appeal to fight for the liberation of France. Jacqueline Hadamard describes this as follows: "My father naturally took an active part in the discussions, even when he appeared to be sleeping deeply, which gave me and my mother terrible frights; for when someone suddenly asked him to say something, he asked, to our great surprise, a pertinent question, even though he seemed to have slept throughout all of the talk. Sleep had always been his enemy: I still hear him say: 'I am never tired, but I always want to sleep' " [IV.1, p. V(8)].

*In the U.S.A. during World War II*

In December 1943 two American newspapers published a declaration of the *Reichstag Fire Trial Anniversary Committee*, which started with the words:

> On a wintry evening early in the year 1933 the Nazis set fire to the Berlin Reichstag. Herr Hitler, the new Chancellor, rushed to the scene and declared loud enough for foreign journalists to hear: "This is a signal from Providence. No one will prevent us now from dealing with Communists with an iron hand."
>
> The world knows now that it was not for the Communists alone that Hitler prepared his iron hand. It was for all the German people: the Jews and Catholics, the Freemasons and trade unionists, the workers and students... The iron hand for Austria, Czechoslovakia, Norway, the Low Countries, France, the Balkans — all of Europe... The iron hand to enslave the Soviet Union and Asia... The iron hand closing on the United States and Latin America... The iron hand for all the world! [III.395]

THE FIRE
THAT HAS BEEN BURNING
FOR TEN YEARS

*The title of an advertisement in the* New York Times *with a portrait of
G. Dimitrov*

The declaration was signed by about two hundred persons among whom
was Einstein. The following letter from the black singer Paul Robeson,
chairman of the Committee, shows that Hadamard was somehow involved
in this episode of anti-Nazi activity:

> January 4, 1944
> Dear Professor Hadamard,
> We are enclosing herewith a reproduction of the advertise-
> ment entitled "The Fire That Has Been Burning for Ten Years",
> as it appeared in the New York Herald Tribune on December 28.
>
> Its appearence in the Herald Tribune, you will be interested
> to know, was made possible by the warm response we received
> from the original advertisement in the New York Times on De-
> cember 22.
>
> I think you will be pleased to learn also that there has been
> considerable evidence of favorable public sentiment toward the
> Declaration. We feel that its publication in these two leading

newspapers has indeed helped to further national unity in this critical period when we are called upon to exert our utmost efforts for the speedy defeat of the Axis and the establishment of a just and lasting peace.

I wish to extend my personal thanks to you as well as the appreciation of the Committee for your cooperation in making this important project a success.


Sincerely yours,
Paul Robeson. [IV.18]


The suffering of the victims of the genocide perpetrated by the Germans on the other side of the ocean, impelled Hadamard to action. In New York he addressed the American Jewish Joint Distribution Committee in an open letter of July 24, 1944:

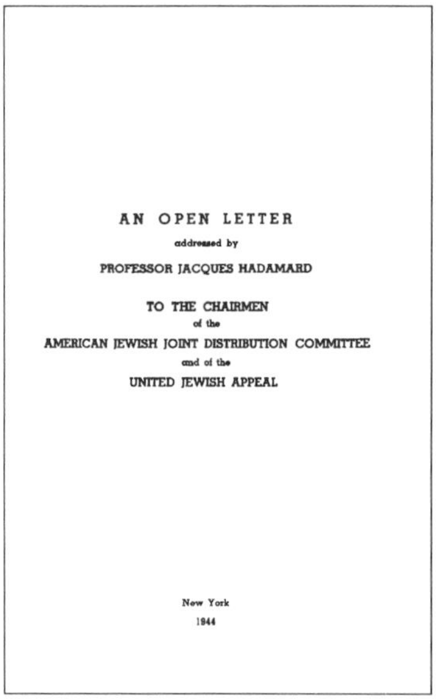

> AN OPEN LETTER
>
> addressed by
>
> PROFESSOR JACQUES HADAMARD
>
> TO THE CHAIRMEN
> of the
> AMERICAN JEWISH JOINT DISTRIBUTION COMMITTEE
> and of the
> UNITED JEWISH APPEAL
>
> New York
> 1944

The tragedy, unprecedented in history, of the European Jews, of whom two to three millions are estimated to have been assassinated by the Germans, and who are almost all totally ruined, famished and distressed, demands an energetic collaboration of all the humanitarian forces, in order to save those who are menaced of death, to alleviate the misery of the living, and to establish healthy and strong bases for the reconstruction of their social and economic life [I.367, p. 3].


Then he proposed some concrete measures to increase the help given to European Jews.

The Hadamards could not adjust to this new country: they were home-sick and had only a few American friends.  "A notable excep-

tion, however: an American mathematician whom my father already
knew: Norbert Wiener, who de-
lights us: a small tobacco jar, his
laugh is almost as booming as Einstein's!" as Jacqueline Hadamard
noted [IV.1, p. V(18)].

They were anxious about their
family in France, having had no
news from them for many months.
Cécile and her husband René Picard were active in the Resistance.
After several months of waiting, a
letter from Mathieu, the last son
of the Hadamards, arrived: he had
managed to escape from France to
Spain. Then, after a long silence,
he wrote from Algeria, saying that
he had joined de Gaulle's Free
French Forces. He died on July 1,
1944, in Tripolitaine. In Jacqueline Hadamard's reminiscences one
reads:

*Norbert Wiener (1894-1964)*

> The first morning, at roll-call, they call out Lieutenant Hada-
> mard and, to his amazement, two came forth from the ranks!
> That's how he got to know a distant cousin from Avignon, whom
> we had never met and whose wife was in New York.

> Some days afterwards, my friend Janine Bernheim came to
> see me and, alone with me, told me that this other Jacqueline
> Hadamard had received a letter from her husband saying that
> my brother had been killed! The only one we thought saved, my
> parent's last son... How should we announce this terrible news?
> It was a new tragedy. How does one endure such blows, I still
> ask myself.

> As the news given to me by Janine was uncertain, there was
> no question of telling my parents. So I decided to say nothing, to
> hide my pain and for ten days I did not even allow myself to cry
> alone, my mother might hear me. Fortunately, for years we had
> been accustomed, or even trained, not to think of things difficult
> to bear during the night as they are ten times worse then.

> So I hold out. But alas, the news is confirmed... I call my

cousins for help and ask them to make this terrible revelation,
and I collapse. The only one we thought unharmed is no longer,
and he has left a wife and two small children in France! [IV.1,
p. V(26)-V(27)]

Mathieu's death made the Hadamards' longing to be nearer to France
even stronger. Moreover, although Hadamard wanted to study problems of
a military nature, his attempts to become involved in the war effort were
futile. In 1943 Rapkine went to Britain and organized the *Mission scientifique française en Grande Bretagne*, which was intended to attract French
scientists to the British military programme "Operational Research". With
the help of the *Mission*, Rapkine arranged for a group of French scientists
to move from the United States to London. When it became apparent that
the conditions for his work were more favourable there, Hadamard decided
to move to England. On August 21, 1944, the Hadamards left America, as
Jacqueline Hadamard recalls:

> We are only ten civilians who board the liner taking 18,000
> American soldiers, a liner which was part of one of the last con-
> voys to cross the Atlantic. The most important members of the
> Mission – Louis Rapkine, Pierre Auger, Francis Perrin [Jean Per-
> rin's son] and, I think, one or two others – had taken the plane
> and we took their luggage with ours.
>
> It was in the middle of the Atlantic that we learned of the
> liberation of Paris over the radio on board.
>
> Relaxed, radiant, we enjoy this incredible spectacle which a
> convoy journey is: around us, instead of a great empty ocean,
> dozens and dozens of ships (I think there were 140) with, around
> them, corvettes whose duty it was to find and sink submarines.
> In the very middle of the Atlantic, these corvettes (which do not
> have enough fuel to make the journey at the speed of the convoy)
> come to suck their mother, that is to say to fill up from a tanker
> through a long pipe, all this without stopping the pace of the
> convoy for one second. We had, moreover, been warned that if
> we were torpedoed, there was no question of the convoy stopping
> to fish us up, we would have to wait, with our life belts, for other
> boats to pick us up (which was not very probable as the routes
> followed by the ships were very variable, to avoid submarines).
> In short, I understood then the old refusal of Breton mariners to
> learn to swim! Better to die quickly in this case...
>
> We know that we are going to England, but where shall we
> arrive? It's only when we disembark that we do learn that we
> are in Liverpool.

On the quay, we, the ten civilians, ask the for help of porters
– no, female porters, it's women who deal with this kind of job.
At that moment, my mother turns to me and says: "But there
is soul in the eyes of these women!" And it is true that we have
the impression of returning to our own world, of dealing with
people who have the same preoccupations as us, the same fears,
the same hopes. So enthusing!

We are taken to the station, and there we have a problem: we
have, quite simply, one and a half tons of excess luggage! And
each of us has five pounds sterling. How are we to pay the way
to London?

The station authorities know quite well what is wrong and
ask us politely: "How much are you able to pay?" We are very
embarrassed. Then after laboriously looking in their books they
charge us the tariff for ploughing equipment (which already emp-
ties our pockets). But they warn us that they can not put all
these cases in our train, as the carriage is too small. One of us
has to stay behind and wait for the next train with the rest [IV.1,
p. V(29)-V(30)].

## 7.4   A year in London

In London the Hadamards rented a furnished apartment, together with Mandelbrojt, in Pall Mall, near Carl-
ton Gardens, where the French scientific mission was located.

Hadamard soon became involved in Operational Research, a group of scientists working together with the
Coastal Command and the Anti-Aircraft Command in
order to make the British radar network more effective.
Judging from an account by Jacqueline Hadamard, he
dealt with problems concerning the defence of naval con-
voys from submarines. She wrote the following reminiscences of their life and work in England:

It was agreed that we would be shown everything classified
'secret' but nothing classified as 'top secret'! The English, their
fingers burned by their difficulties with de Gaulle, were watching
us. They saw after a short time that we scrupulously observed
this rule and finally they gave us their confidence.

Our life became organized: at first, each of us spent the day with the group to which we were assigned: Fighter Command, Bomber Command, Coastal Command (where I was placed). These groups were located outside of London and therefore we normally came home at night, except when we came at lunch time, to work on our reports in the afternoon. On those days, we had lunch – as any self-respecting Englishman – at our Club, the Club of Visiting Scientists, a club which was founded to take in the foreigners. In the evenings, we had dinner at home, served by our admirable butler.

*Hadamard in London, 1944*

Our liking for the English increased each day: they received our admiration for their courage and their sense of civic duty. They needed their courage during the Blitz, that is, when London was bombed every night to such an extent that they had to evacuate the city. A large part of the East End (nearest to the coast) was destroyed. Then there were the V1s (pilotless planes carrying bombs, arriving at any time, which one could hear passing overhead and which then, running out of fuel, would turn into a spin and finally crash, so that one was not sure whether they would fall on you when they had passed over you).[4]

It was at the end of that period that we arrived, that is at the beginning of the V2s, which were rockets with an enormous explosive charge, travelling faster than sound so that they fell before one heard them. If you heard them, then they had already fallen elsewhere.

Many of the people we met had already been in bombed houses, and their nerves were quite naturally more tense than ours. But they reacted in an admirable way. I remember one day passing by a shop gutted by a V1 which had fallen in the vicinity during the night. Now the London shops generally had their windows strengthened with strips of paper and a sign in the middle saying "Open as usual". The owner was busy with his shop and the first thing he did was to change the sign by writing on it: "More open than usual"!

Another time, seeing a demolished house and the owner looking at his furniture strewn all over the pavement, I told him how sorry I was, and he said: "Oh, it could have been worse, it could be raining".

We blossomed in this atmosphere where they begin by having confidence in you, counting on your honesty and on your sense of duty. So I was sent to the Farnborough Centre to help locate by radar the position of the V2 firing ramps, by triangulation. I took the train to this large suburb. On arrival, I couldn't find my ticket to give to the inspector. He asked me: "Do you come every day?" "Yes, as from today". "Good, then you can give it to me tomorrow". That would surprise me from an inspector at Orsay, but in England it seemed normal; and the following day, I gave him two tickets.

Evidently, the day you betray this confidence, you are finished in the eyes of an Englishman, and this seems to me to be reasonable [IV.1, p. VI(3)-VI(5)].

---

[4]From June 13, 1944, to March 29, 1945, about 2500 jet flying bombs V1 reached the London area. (Note of the authors.)

Hadamard gave lectures at universities, and in January 1945 he spoke to the London Mathematical Society. Introducing him, Hardy called the eighty-year old Hadamard as "the living legend of mathematics" [II.4].

In one of the 1965 issues of the journal *Nature*, there is an article in memory of Hadamard, in which it is written that in 1945 he " gave a lecture on his personal reminiscences, and which is still remembered with pleasure by many of his British colleagues" [II.17, p.937]. Mary Cartwright writes: "I remember him as a little figure lecturing in an overcoat because of the cold, and when he tripped, instead of falling, he ran lightly down the steps" [II.4, p. 84].

*G. H. Hardy (1877-1947)*

*Louise Hadamard*

Mandelbrojt, at the centenary celebration in honour of Hadamard, in 1966, told the following story from that time: "I take advantage of this opportunity given to me, to keep a promise. One day, during the war, in a London street which had probably just been bombed by the V1s, Hadamard told me abruptly: 'Mandelbrojt, if anything happens to me, don't forget to say everything that I owe to my wife.' This has now been done" [II.5, p. 27].

## 7.5 The return home

The Hadamards returned to Paris in the spring of 1945. Their large flat in the *rue Émile Faguet* was empty, their possessions having been taken by the Germans. There were only five or six pieces of furniture left, some dishes and some silver which had been kept by the neighbours. The books and the botanic collections, preserved during the war at the *École de Physique et Chimie* and at the *Muséum National d'Histoire Naturelle*, were returned. (After Hadamard's death his botanic collections were donated by Jacqueline Hadamard to the *Muséum National d'Histoire Naturelle* and placed at the *Laboratoire Maritime du Muséum* at Dinard in Brittany.)

Jacqueline Hadamard describes their life just after the war:

> We have recovered our flat, and we take care of the children. We are all on top of each other, my parents, my nephews, my niece and myself, in this flat, where life is not very easy. In fact there's no question of central heating, because we only have one fireplace, and, fortunately, some wood given to members of the *Institut*. So, only one room heated for six people (it happened to be my room which had a fireplace). My father, who walks up and down, dictates mathematics to my mother; my oldest nephew is preparing for entering the *École Supérieure d'Electricité*; the two younger ones do their homework. It wasn't possible for them to do this at the same time, and so the older boy waited for the others to go to bed, in order to work for his examination, while I went to sleep [IV.1, p. VII(3)].

In 1945, on his eightieth birthday, Hadamard gave a lecture entitled *Subconscient, intuition et logique* at the *Palais de la Découverte*. In the same year he published the paper *Problèmes d'apparence difficile* [I.368] as well as an obituary of George David Birkhoff [I.370].

Hadamard and Birkhoff had known each other for a long time, and Birkhoff had been a guest speaker at Hadamard's Seminar. When congratulating Hadamard on his seventieth birthday, Birkhoff wrote:

> My own particular indebtedness to Professor Hadamard is very large indeed. A number of his papers have been most stimulating to me and have proved to be the starting point for my own researches. I have felt inspired not only by his scientific contributions but also by his unfailing kindness toward me and interest in what I have undertaken. He has always seemed to me as being the true successor to Henri Poincaré by virtue of his breadth and profundity. In the centuries that are to come it is certain that his name will take place in the splendid and unsurpassed line of celebrated mathematicians of which France can be so justly proud.

*G. D. Birkhoff (1884-1944) was a leading American mathematician in the first half of the twentieth century. He contributed decisively to ergodic theorems in statistical mechanics, introduced new ideas in the theory of dynamical systems and gave the first proof of Poincaré's last theorem on fixed points in a ring. Birkhoff did important work on difference and ordinary differential equations.*

As Birkhoff's son Garrett said in 1996 in a letter to us: "I had been aware of Hadamard as a scientific friend of my father for thirty years, largely because of their mutual interest in Poincaré's creative ideas about dynamical systems".

Hadamard's book *The psychology of invention in the mathematical field* was published by Princeton University in 1945 with the dedication "To the companion of my life and my work". It summed up many years of reflections on the mechanism of scientific thinking and on the psychology of creative activity.

The extended and revised French edition of Hadamard's book *The psychology of invention in the mathematical field*, translated by Jacqueline Hadamard, was published in 1959

# Chapter 8

# After Eighty

## 8.1   Once more in the U.S.S.R. in 1945

 The last visit of Hadamard to the U.S.S.R., in June 1945, was in connection with the jubilee sessions of the Academy of Sciences held on occasion of its 220th anniversary in Moscow and Leningrad, where he was a member of the French delegation. This event was celebrated on a large scale.

Only a short time previously, victory over Nazi Germany had been celebrated, and despite the difficult times, the public was elated. In the words of the report printed in the *Vestnik Akademii Nauk* [III.133, p. 149]:

> The doors of academic institutes, laboratories, and museums were open for guests from various countries [...] They were acquainted with the scientific life of Moscow and Leningrad [...] Many foreign scientists saw an exibition of portraits of academicians at the Bolshoi Theatre. Foreign delegates attended Glinka's opera 'Ivan Susanin' at the Bolshoi. At the newly repaired Mariinskii theatre, which had been severely damaged by bombing during the siege of Leningrad, they saw the wonderful ballet 'Swan Lake' by Tchaikovsky. For participants of the sessions, excursions were organized to the Kremlin, the Tretiakov gallery, along the Volga-Moscow canal, to Yasnaya Poliana,[1] to the exhibition 'The defense of Leningrad', to Pushkin, and Petrodvorets.

---

[1]The former country estate of the writer Leo Tolstoi, where he lived about sixty years and wrote *War and Peace*, *Anna Karenina*, etc. (Note of the authors.)

*Hadamard being interviewed in Pushkin, June 1945.   The
beautiful suburbs of Leningrad, Pushkin and Petrodvorets,
used to be places of summer residence of the Tsar's family.*

Laurent Schwartz told us the following story which he heard from F. Joliot-Curie, who was also a member of the French delegation:

> In one of the large museums the scientists were led through
> was the exhibition of 'official painting'. When he stopped in front
> of one of the canvasses, Hadamard recognized Lenin, Stalin, and
> somebody else. "And that is, of course, Trotsky," he said. Every-
> body became silent. "No," answered the guide very quickly, "it
> is so-and-so." At the next canvas, Hadamard, unwilling to give
> up with his idea, asked: "And where is Trotsky?" "Not here,"
> the guide answered. Again and again he posed the same ques-
> tion until he heard: "Trotsky was the people's enemy". "What
> is this, Joliot?" - asked Hadamard. - "Are these pictures for
> politics or for history? If they want to make politics here, then
> this is not a museum" [Interview, October 28, 1992].

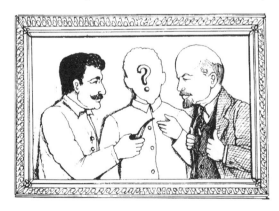

V. Arnold retells Kolmogorov's reminiscences of the following episode:

Hadamard was a passionate collector of ferns. When he came to Moscow, Andreĭ Nikolayevich [Kolmogorov] together with Pavel Sergeyevich [Alexandrov] took him boating [apparently on the Obraztsov pond near Klyazma[2]- V.A.].

*A.N. Kolmogorov (1903-1987) was a mathematician of phenomenal depth and productivity who made fundamental contributions to the theory of trigonometric series, approximation theory, probability theory, topology, and hydrodynamics.*

Suddenly, Hadamard saw something on the bank and asked to pull in immediately. He went to the prow of the boat and, as it approached the bank, he was so excited and eager to reach the shore that he fell into the water. It turned out that an unusual variety of fern grew there, which he had been looking for everywhere for many years. Hadamard was absolutely thrilled. However they had to take him immediately to a reception with the president [of the Academy of Sciences] in the presidium of the Academy of Sciences of the USSR [the president at that time was apparently Komarov[3]- V.A.].

We had to dress Hadamard in Pavel Sergeyevich's suit. However, it was very obvious that the suit was borrowed (Hadamard was much taller). At the reception, everybody asked Hadamard "What happened to you, Professor? This is not your suit – did you fall into the water?" To which Hadamard answered proudly "Why do you assume that a professor of mathematics can not have other adventures?" [III.363]

---

[2]Klyazma is a small river near Moscow. (Note of the authors.)

[3]V.L. Komarov, botanist and geographer, was the president of the Soviet Academy of Sciences from 1936 to 1945. (Note of the authors.)

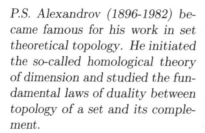

*P.S. Alexandrov (1896-1982) be-came famous for his work in set theoretical topology. He initiated the so-called homological theory of dimension and studied the fundamental laws of duality between topology of a set and its complement.*

At the same reception somebody asked Hadamard: "Professor, being an academician, you have to communicate a lot of mathematical papers for publication. How do you distinguish between good ones and bad ones?" Hadamard answered with the following oriental parable:

> There was once a Sultan who, in a bad mood, ordered his eunuch to bring him a new girl. The beauty chosen by the eunuch had the eyes of a gazelle, but the next morning the Sultan was not pleased. "Bring another one", he demanded. The eunuch did his best to select a houri who seemed perfect to him. However, the Sultan was not satisfied at all, and said: "Try again and either I like her, or you will lose your head."

The poor eunuch went out of the palace in despair. "Why are you crying?" asked a ragged beggar who sat by the road. When the eunuch told him the reason for his sadness the beggar said: "Don't worry", and brought him a girl, who did not impress the eunuch very much. However, he took her to the palace and this time the Sultan was extremely happy. "Why were you so sure she was good?" the eunuch later asked the beggar. "How can I explain it to you? After all, you are a eunuch!" came the answer."[4]

It is not possible to confirm whether Hadamard actually told this anecdote at the reception, but *se non è vero, è ben trovato.*

At one of the jubilee sessions, the thirty-year old Yu.V. Linnik was introduced to Hadamard, and he presented him with reprints of his papers on the asymptotics of the density of zeros of Dirichlet series. After that, when meeting Linnik at the sessions, Hadamard greeted him with the exclamation: *"Densité de zéros!"*

*Yu.V. Linnik (1915-1972) made outstanding contributions to number theory, probability theory, and mathematical statistics.*

On June 22, in Moscow, Hadamard gave a talk on the psychology of mathematical creation and he repeated it in Leningrad on June 27, where he met his old acquaintance Smirnov. At that time Smirnov was continuing with his research into special solutions of hyperbolic equations. Hadamard spent a pleasant evening at Smirnov's apartment in Röntgen Street, talking

---

[4]This story was retold us by Ya. G. Sinaĭ, who heard it from S.V. Fomin (1917-1975).

about hyperbolic equations and drinking tea with cloudberry jam. (This northern berry proved to be completely unfamiliar to Hadamard.)

On his return home, Hadamard shared his impressions with the readers of the *Revue France–U.R.S.S.* in a short article, which he ended with the sentence: "Because it will be, we are sure, one of the unforgettable results of the journey we have just made, to strengthen the intellectual ties which exist between our two countries and to enrich French logic with the élan and powerful originality of the Russian temperament" [IV.22]. Was that a feeling that the Russian hosts remained closer to the romantic vision of mathematics of Hadamard's day than did Bourbaki?

In the Great Soviet Encyclopedia of 1949 (volume 1, p. 388) there is a short unsigned article about Hadamard, written quite adequately from the mathematical point of view. However, it contains the following curious passage (which gives an indication of the atmosphere in Soviet science at that time):

> In his methodological statements, Hadamard usually opposes any restrictions on the selection of topic or method of mathematical research (for example, he is in favour of the unlimited use of the axiom of choice) and agnosticism, stemming from the conviction that every mathematical problem is solvable, which is natural for an eminent mathematician. However, Hadamard's philosophical foundation of this positive view of the boundless possibilities of scientific progress is obviously unsatisfactory. It represents a synthesis of objective idealism and narrow empiricism.

This article is included in Kolmogorov's list of publications in [III.216, p. 538].

## 8.2   The 1947 trip to India

In January 1947, Hadamard and his wife went to India: he was given an honorary doctorate by the University of Delhi and invited to participate in the Indian Scientific Congress. There were participants from the United States, England, the Soviet Union, Canada and

China, with Hadamard being the only delegate from France, as Joliot-Curie, who had also been invited, was unable to come.

"Splendid reception. Interesting congress", wrote Hadamard in his notes about the visit to India (possibly a draft of a report to the *Académie*) [IV.3]. In fact, the congress was a very important event because India was to become an independent country six months later, after two centuries under British rule. Pandit Nehru,[5] the future Prime Minister, gave an address at the opening ceremony, speaking about the future development of the country and its peaceful role in the world.

Among the Indian scientists at the congress was the physicist C.V. Raman (1888-1970), famous for his work in optics, acoustics, and molecular physics. (He was awarded the Nobel Prize in physics in 1930.) Two years after Hadamard's trip to India, Raman wrote the following letter to him:

April 27, 1949

My dear friend,

It was extremely kind of you to have informed me of my election as Correspondent of the Academy on the very day that the election took place. I have since received the official intimation of the same and have written to the Permanent Secretaries expressing my high appreciation of this honour. I am indeed immensely pleased to be associated in this way with your illustrious Academy. France has literally showered me with honours of which this is the latest. I am deeply touched by the very kind sentiments expressed in your letter.

Thanking you,

I am

Yours sincerely,

C.V. Raman. [IV.7]

At the time of Hadamard's visit to India, France was still a colonial power, occupying, among other countries, Vietnam and Cambodia. The first Indo-China war was then in progress and it provoked strong anti-French sentiments in India, as Hadamard wrote:

I must note great sympathy for France shown by Pandit, who always expresses himself with a deliberate moderation on the events in Indo-China. These events give rise, there, to great indignation and do incalculable harm to the moral situation of France.

---

[5]Pandit is a learned man, scholar; used as a title of respect.

During my stay in Bombay, I learned that there had been
a hostile demonstration at the Consulate by students shouting
"out from here" [IV.3].

However, Hadamard was welcomed very warmly:

Photograph taken during the tea party given by Pandit Jawaharlal
Nehru to meet foreign scientists on January 10.    Among the
distinguished gathering were members of the Government, Mr.G.V.
Mavlankar and Speakers of Provincial Legislatures,    Sir Terence
and Lady Shone and Mr. George Merrell.

( Picture issued January 1, 1947)

The present animosity towards our country has not extended to
university circles...I should note the particularly friendly attitude
of Pandit towards me...The English authorities have also shown
us consideration and a great friendliness [IV.3].

Hadamard's visit was remarked upon by the press, as the following newspaper articles witness:

Professor Jacques Hadamard:  His profound researches in
pure mathematics extending over a long period have placed him
in the forefront of the mathematicians of the world and he has

added to his fame by his studies in the psychology of mathematical invention. He has maintained the thesis that mathematical thought is possible without even the vehicle of language [Times of India, January 11, 1947].

Under the auspices of the Tata Institute of Fundamental research three eminent scientists, Prof. P.M.S. Blackett of the Manchester University, Prof. S.S. Chern of the Academia Sinica of China and Prof. J. Hadamard of the *Académie des Sciences*, Paris, who are now in India will deliver a series of lectures during their visit to Bombay in the course of the next few days. Some of these lectures will be of a popular and others of a technical nature. The lectures will be open to the public [Free Press Journal, January 14, 1947].

Prof. Jacques S. Hadamard is unquestionably the doyen of all present-day mathematicians. There is no branch of Mathematics upon which he has not left an indelible mark. Though in his eighties he still retains mental powers which many of his younger colleagues envy [Morning Standard, January 14, 1947].

As usual, the Hadamards travelled a lot: as well as Delhi and Bombay, they visited Benares, Mysore, Lucknow, Aligarh, Bangalore, and wherever they went, Hadamard gave lectures. "I have received requests for lectures for next year" he wrote [IV.3]. In Bangalore they were invited to a women's college where the girls performed scenes from *Cyrano de Bergerac*, and Madame Hadamard was asked to greet them. There were also many official receptions, as Hadamard reported:

The Consul in Bombay has largely done what he thought should be done: besides the help he gave me as regards our life in Bombay, he gave a reception in our honour which the Governor of Bombay and Lady Colville attended... The Governor of Bombay invited us, Madame Hadamard and myself, to a dinner where he insisted on having the Marseillaise played, which we listened to standing. It was the first time it was heard in India since the war.

On his part, the Governor of French India, advised of our presence by the Consul in Bombay, invited us to come to Pondicherry where we were his guests during our short stay [IV.3].

After two months in India, the Hadamards returned home.

## 8.3   Mathematics as always

*Hadamard working outdoors*

Hadamard continued to work for a further ten years, his publications appearing regularly. The papers from these years dealt mainly with problems of mathematical physics, the psychology of mathematical thought and the history of mathematics.

He was interested in the history of mathematics during all his life. In contrast to his joke that "the triumph for a historian of science is to prove that nobody ever discovered anything" [I.375, p. 35], Hadamard payed homage to many of his great predecessors and contemporaries. Besides his articles analysing Poincaré's work, he published papers on the life and work of Maurice Lévy, Painlevé, Duhem, Picard, and George Birkhoff.

In 1946, when the third centenary of Newton's birth was celebrated in Cambridge and London, Hadamard was one of the main speakers and he gave a brilliant lecture on *Newton and the infinitesimal calculus* [I.375]. In 1950 his paper *Célébration du deuxième centenaire de la naissance de P.S. Laplace* appeared. In 1954 he published the paper *Sur des questions d'histoire des sciences: la naissance du calcul infinitésimal*, where he underlined the dangers of too free an interpretation of facts in the history of mathematics.

There were, however, works on other themes, among them the book *Non-Euclidean geometry in the theory of automorphic functions* [I.383],

*Hadamard working indoors*

which came out in 1951, and only in Russian. It contained mainly a detailed survey of Poincaré's results in this area. The story of that work is the following. During the 1930s a group of Soviet mathematicians, headed by V.F. Kagan, started the preparation for publication of the complete works of N.I. Lobachevskiĭ. As a supplement they initiated a series of monographs "Lobachevskiĭ's geometry and the development of its ideas". Hadamard responded to this undertaking by a manuscript which was translated into Russian and published in 1951.

Hadamard renewed contacts with Chinese scientists in 1946, three years before the establishment of the People's Republic of China, as is witnessed by the following letter from the Chu Chia-hua, the Minister of Education in the Kuomintang[6] governement and president of the Academia Sinica:

Chiaoyupu
(Ministry of Education)

Chungking, China

April 10, 1946

Dear Professor Hadamard,

I have just received your letter of January 30 in reply to my telegram of December 13, 1945. It gave me real pleasure, and I thank you for it.

You reminded me of the stay you made in my country before the Second World War, during which stay you left such a deep impression on us. I hope that you will continue the task of contributing to the intellectual and scientific co-operation between our two countries. You can rest assured that, for my part, I shall not cease to work, in the fullest measure possible, for the development of this co-operation which is the best guarantee of friendship between our two peoples.

Please accept, dear Professor Hadamard, my cordial regards.
Dr. Chu Chia-hua. [IV.6]

It appears that it was then that Hadamard returned to the project of a new, large book on partial differential equations, which was intended to include all the most important results in this area. Already in 1936, while in China, he had promised to publish his lectures but, as Mandelbrojt put it, "things happened, there was the war of China against Japan, and then Japan against [...] and then the real war began. Hadamard was unable to do his book. And [...] he had this on his conscience, he carried on with his book after the war. He had promised something he thought he could do in twenty-two chapters" [III.265, p. 19].

---

[6]Nationalist political party of China.

The following letter by A. Vasilesco, replying to a question by Hadamard, shows that by 1947 he was working on chapter two of the book, the section on regular and irregular points in the Wiener sense:

Paris, September 24, 1947

Dear Sir,

I have just received your letter. The two notions which you mention are the same. The notion of conducting potential, due to Wiener and being the oldest, was the immediate generalization of the analogous classic notion obtained from the generalization of the classic Dirichlet problem. On the other hand, the notion of capacitary potential, due to Mr. de la Vallée-Poussin, comes from the *direct proof* of the fact that for a given closed set, there can be one and only one mass distribution whose potential is equal to 1 quasi-everywhere on the set (that is to say, with the exception of a set of zero capacity). Although obtained independently of the generalized Dirichlet problem, this proof shows the *identity* between the two notions of conducting and capacitary potential. It is capacitary because it is due to a *unique* mass distribution equal to the capacity of the set. In sum, the progress obtained in comparison with the first notion of Wiener was the proof of this uniqueness.

I remain entirely at your disposition.

Yours sincerely, A. Vasilesco. [IV.30]

*Hua Loo-Keng (1910-1985)*

In the beginning of 1954, Hadamard received an official invitation to publish the book in China. It was signed by Hua Loo-Keng, an outstanding number theorist who had returned to Mao Tse-Tung's China from the United States in 1950. He became Director of the Mathematics Department of the Tsinghua University and two years later was nominated the head of the Institute of Mathematics of the Academia Sinica. Hua Loo-Keng could have been one of Hadamard's audience in the spring of 1936, since he worked at the Mathematics Department of the Tsinghua University at that time (in the summer of that year he went to Cambridge University) [III.354].

On January 5, 1954, Hua Loo-Keng wrote to Hadamard:

> Professor,
>
> We are very happy to learn that, after conscientious efforts during many years and a considerable increase, the work which follows from your lectures on the theory of partial differential equations at the Tsinghua University in 1936 is finally about to be finished. Considering the value of this work in the scientific world, the Academia Sinica would very much like to take care of having it published in French in Peking. Professor, if you would kindly accept this idea, we would ask you to send the definitive text to the Institute of Mathematics through our legation in Berne.
>
> And to facilitate the reading of your work, we would like, of course with your permission, to make an edition in Chinese which would be considered as a publication of the Institute of Mathematics.
>
> Our Institute will then pay more attention to the development in China of research in the domain of partial differential equations. Our young researchers will begin incessantly the study of your important contributions in this area. We would very much like to have your precious advice which will certainly aid us in this aim.
>
> Please accept my deepest respects.
>
> Hua Loo-Keng
> Director of the Institute of Mathematics
> Academia Sinica
> Peking, China. [IV.8]

In 1994 Benoît Mandelbrot sent us the following story ("which I have often told and seems to impress and amuse everyone") about Hadamard from that time.

> During the year 1949-50, I had a moving and amusing encounter with Jacques Hadamard. At the time, he was about 85 years old. His daughter Jacqueline was active in far-left political causes and was at that time especially involved with Mao's China. She arranged for her father to be asked to contribute a token of esteem to the New China, by offering them an especially written book. Hadamard chose to write on partial differential equations.
>
> He tried to bring himself up to date by attending the very demanding lectures of Jean Leray at the *Collège de France*, but

this did not work out, so he asked my uncle Szolem (his successor at the *Collège*) to find a set of notes that he could use.

My uncle asked for my notes, but, like Hadamard, I had already dropped out. However, I knew that one of my friends, Léon Trilling, had been a very diligent attendee and note taker. I asked Léon, if he could copy a fresh set, and present his notes to Hadamard in person. Léon could not resist the offer, so in due time we went to Hadamard's apartment near Porte d'Orléans.

Hadamard asked us to wait while he looked through the notes. He scanned them page after page, not taking time to actually read anything but constantly nodding "yes, yes." Then he froze and looked at Léon Trilling. "What is this?" Léon acknowledged he was not sure. Hadamard's reply was, more or less, "never mind, never mind, Leray is applying the Fourier method to the wave equation; I know both and will straighten out the link."

He went on, reading page after page, until he froze again. "What is this?" My friend acknowledged again that he was not sure. He had missed one class and taken notes from another friend's notebook. "Never mind, never mind," replied Hadamard and finished turning pages with no further comment.

The book ultimately appeared. According to reports, it did not have much effect on the progress of the field. However, while the elevator was going down, Trilling and I could only marvel that this frail and extremely old-looking man could jump in the middle of notes of Leray's very heavy argument and know at a glance that something was out of kilter.

S. Mandelbrojt, who often visited the Hadamards, also recollects:

Hadamard was ninety years old at the time, or almost that, couldn't see very well, didn't read much of what was being done, but he received the *Matematicheskiĭ Sbornik*. I remember when I used to visit him.... "Oh, Mandelbrojt, he discusses partial differential equations here, tell me what he says." And so, by chance, without thinking, I would tell him. He put it into his book if it was interesting... But he worked all the time. I can still see him working at home, walking up and down in his study, thinking... Madame Hadamard knitting, typing what Hadamard was dictating. She would interrupt her knitting all the time [III.265, p. 19-20].

Pékin, le 16 décembre 1957

Monsieur le Professeur,

Nous sommes bien heureux de recevoir les manuscrits de votre Ouvrage sur la théorie des équations aux dérivées partielles. Nous avons maintenant en tous 350 pages, et quelques feuilles de suppléments. Nous espérons toujours que votre santé va mieux et que votre Ouvrage peut être achevé dans un delai pas très longue.

Bientôt ce sera votre 92ième Anniversaire, nous vous envoyons nos félicitations les plus chaleureuses et vous souhaitons des plus grands succès scientifiques pour l'année prochaine.

Veuillez agréer, Monsieur le Professeur, l'expression de mes sentiments les plus respectueux.

L. K. Hua

*Letter of Hua Loo-Keng*

Ou-Sing-Mo, Hadamard's student at the Tsinghua University, and whom Hadamard met later in Paris, dealt with the preparation of the book in Peking, and periodically contacted his old teacher. On January 27, 1959, he wrote: "Our colleagues are happy and grateful for what you have done for the scientific progress of our people. And it is very gratifying to inform you that your work has grown to 420 pages. For your age, it is an immense effort, and it would be very significant if you were to finish your work before the Tenth Anniversary of our People's Republic which will be celebrated solemnly this year on the first of October" [IV.9].

Hadamard sent each finished chapter in turn to the Chinese embassy in Berne, since there was no embassy in Paris at that time. By the spring of 1959 eight chapters had been forwarded to China. Jacqueline Hadamard describes the last period of work on the book:

> His capacity for work lessened, and we saw, my mother and I, that this burden was becoming heavy. Furthermore, we had another worry: my father was more than ninety years old; was what he was writing worthy of Jacques Hadamard? So we asked for help from my cousin Laurent Schwartz, a young mathematician, and gave him the copies of the manuscript which was sent off. He assured us it was impeccable. So, no worry on that side.

But his tiredness was painful for us to see. Finally, we were able
to persuade him to stop. We wrote to the Chinese to tell them
that his great age prevented him from continuing and that he
was prepared to repay Beijing a part of the little amount they
had payed in advance. But again there was no reply. It was
several years after the death of my father that I was called to
the Chinese Embassy, where they gave me copies of the book in
question...[IV.1, p. VII(21)].

J. 阿 达 玛

# 偏 微 分 方 程 論

JACQUES HADAMARD

## LA THEORIE DES EQUATIONS

### AUX

## DERIVEES PARTIELLES

科 学 出 版 社
EDITIONS SCIENTIFIQUES
PEKIN, 1964

*The title page of Hadamard's last
book which appeared in 1964, a year
after his death*

Hadamard did not lose interest in current developments in mathematics, especially in the theory of partial differential equations. He attended
seminars, and the depth of his understanding of quite new results surprised
his colleagues. On February 10, 1952, he wrote the following letter to F.
Bureau in connection with a recent note of Bureau in the *Comptes Rendus*
on the Fourier method for hyperbolic equations:

My Dear Friend,
It is needless to tell you that your note stimulates a great
interest in me. When I previously proposed this subject to the
Academy for the *Prix Bordin* competition [in 1933] I had long
seen in it a curious mystery and I shall be very glad to see the
details of your solution.

I shall have the galley proof sent to you for correction. In the absence of [mathematical] proofs which you are obliged to leave until later (and in which I have every confidence), I mention the first lines of your page two where the system "assumed incompatible" would perhaps call for a word of explanation.

Yours sincerely,

J. Hadamard. [III.64, p. 8]

Laurent Schwartz recalls that Hadamard listened to talks until he was over ninety:

Several times he attended the Bourbaki seminar. It was full of young people. A very old man among only the young people! He was called *le petit père Hadamard*. And very respected! Everybody respected this very old man.

How different the language must have seemed to him from that of his own time! It is already difficult enough to understand what a mathematician five or ten years younger is doing. Hadamard recognized in new notions the old things he knew, and at the end would ask: "Well, what you say here, does it mean what I call...?" and so on. And it turned out to be so, because he understood the principle without following the details [Interview, October 28, 1992].

Even when he was very old, Hadamard was active in his contacts with colleagues all over the world as is illustrated by the following letter from Garret Birkhoff who, starting with 1950, worked on *A source book in classical analysis* [III.39] and got a list of suggestions of topics for inclusion from Hadamard:

August 22, 1956

Dear Mr. Hadamard,

Thank you so much for your nine suggestions for the collections which I am trying to compile. They will be very valuable to me, and I shall look forward to reading your reprints when I return to Cambridge in September. — I have been in the West of our country all summer.

If you find time to write me about other items on your return to Paris, I should also appreciate this *vivement et sensiblement*.

I hope you had an agreable summer, and that this letter finds you in excellent health.

Sincerely yours,

Garrett Birkhoff. [IV.10]

During the last years of his life Hadamard was still a very active member of the *Académie des Sciences*. Jacqueline Hadamard describes how her father

> [...] estimating that the average age of the Academicians was too high and that some young people deserved a place [in the *Académie*], made the following proposal: that the members over seventy years old (of whom there were obviously many) be made honorary members. Needless to say, he was supported by only a few of his colleagues and his proposal was not accepted. And yet he was certainly the "youngest" of these "old men", at least in his mentality. I am not sure that any other of them would follow him in his favourite pleasure which consisted in buying a paper cone of chips after the sessions, and eating them while walking [IV.1, p. VII(3)].

Hadamard never stopped attending the meetings of the *Académie*. He would place his chair as near as possible to the speaker, sit down and turn his ear in the direction of the speaker, so as not to miss a word. After the talk, he would get up and animatedly join in the general discussion. This unquenchable interest in everything around him dispelled peoples' thoughts about his frailty, and reassured them that the next meeting would be the same.

Wiener has similar memories of his visit to Paris in 1951: "I saw a lot of dear old Hadamard and his wife, both of whom seemed to us almost ageless, although they were in their eighties" [III.422, p. 334].

*Jacques and Louise Hadamard in Paris, March 13, 1947*

## 8.4   At the Harvard Congress

On August 30, 1950, the first postwar International Congress of Mathematicians in Cambridge, Massachusetts, was opened.  Hadamard was elected one of the honorary presidents together with Castelnuovo and de la Vallée-Poussin. At this congress Fields Medals were awarded to Laurent Schwartz for his theory of distributions and to Atle Selberg for important results on zeros of the Riemann zeta function and for his contribution to number theory, in particular, for an elementary proof of the prime number theorem.

However, up to the last moment, it was unclear whether Hadamard and Schwartz would participate in the Congress. The political sky was covered by clouds of cold war. Moreover, in June the Korean war had begun. In Halmos' book *I want to be a mathematician* we read: "The U.S. was between the euphoria of postwar patriotism (we won!) and the hysteria of McCarthyism (they're winning!). I am sure that the mathematical powers, such as Lefschetz, wanted a free Congress, but the U.S. State Department, scared possibly of the House of Representatives, or of public opinion, or of the Red Menace, was making difficulties that were not easy for the AMS to overcome" [III.163].

All the members of the French delegation got American entrance visas except for Schwartz and Hadamard. Schwartz explained:

"We were refused because I was a former Trotskyite and Hadamard had great sympathy for the Communist Party" [Interview, October 28, 1992].

The American Mathematical Society tried to persuade the authorities to change their decision, and Schwartz obtained his visa a few months before the Congress thanks to President Truman's personal intervention. The fate of Hadamard's visa was still uncertain. Indignant, sixteen of the twenty-eight French delegates decided to boycott the Congress, provided the visa not being granted. The American mathematicians went on

*At the Congress in Harvard, 1950*

to press Washington. Finally, Truman took the decision once again. Hadamard's visa was issued five days before the departure.

Schwartz told us the following story:

When we, the French delegates, came up to the ship 'Queen Mary', Hadamard could not find his ticket. So all of us stood there, wondering where it could be. Suddenly he said: 'Oh, I remember very well. When I passed through the customs, the officer asked me for my ticket. I showed it, and, I am sure, he took it'. We asked for the authorization to postpone the departure, which was uncommon, and went to the man in the customs.

Hadamard said: 'I remember very well that you got my ticket'. 'No, Sir', was the answer, 'I did not keep your ticket, but I saw that, when I returned it, you just put it with the change in your pocket, here'. Hadamard slipped his hand into his pocket and took out the ticket [Interview, October 28, 1992].

Just before the opening of the Congress the following cablegram was received from the President of the Soviet Academy of Sciences:

The U.S.S.R. Academy of Sciences appreciates receiving kind invitation for Soviet scientists take part in International Congress of Mathematicians to be held in Cambridge. Soviet Mathematicians being very much occupied with their regular work unable attend congress. Hope that impending congress will be significant event in mathematical science. Wish success in congress activities.

S. Vavilov, President, U.S.S.R. Academy of Sciences [III.332, p. 122].

This cablegram was read at the opening plenary session of the Congress which was held on Wednesday afternoon, August 30, in the Sanders Theatre of Harvard University. Denjoy recalls that, when Hadamard's fragile figure appeared on the stage, all 2000 participants stood up and applauded him [II.56, p. 737].

Here we reproduce a portrait of Hadamard made during the Congress, together with a letter from Alexander Ebin (the President of the American Institute of Man) written three weeks after the death of the addressee.

The following is the beginning of the letter of Ebin, which clarifies the origin of the portrait:

8 November 1963

My Dear Professor Hadamard:

The attached copy of a pen sketch of yourself, which you inscribed to the artist, will remind you of a pleasant incident at the 1950 International Congress of Mathematicians at Harvard University. A young woman introduced herself to you as an artist and asked if you would sit for a sketch. You sat for two: one front view and one profile. You preferred the front view and the artist inscribed it to you. You then inscribed the profile sketch to her.

I will now tell you the background for this incident. The artist was my wife, a professional fashion illustrator and teacher of art. I attended the Congress as a member of the American Mathematical Society, and my wife, who knows nothing about mathematics or mathematicians, came along with me. I expected her to find interest in the art museums and sights of Cambridge. Instead she came to me in high excitement and said she had seen a "wonderfully interesting" man and that he had agreed to let her sketch him. I asked her who the man was, and she could not tell me his name. When she returned to the hotel triumphantly with her inscribed sketch, I exclaimed: "My God, you have sketched our guest of honor: the honorary chairman of this Congress!"

*Hadamard with A. Ghaffari (in the middle) and Th. Motzkin at the Congress in Harvard, 1950*

She was not impressed by this news; she was only impressed by you as a person. I find it always interesting to be married to an artist because one never knows what will result from her enthusiasms.

I think of you, as we all did at the Congress, as a personal embodiment of the mathematical intelligence of France. Now, as the result of mathematical and historical investigations carried on for many years, I have to report certain unexpected facts to the mathematical intelligences of France. The facts are very personal to mathematics and to France, and so it is appropriate to introduce them in a personal way. I regret that my command

of the French language, which should be perfect for this trans-
action, is not equal to the occasion. I must write in English, my
native tongue for I was born in England; but to give my English
words a contemporary Gallic association I am taking advantage
of the admiration which my wife and I have for you, for different
reasons, to address myself to the mathematical intelligence of
France through you. I believe, and hope, that what I have to
say will justify for you and others this personal arrangement.

And now to our mathematical muttons.[7] [...]

With admiration and respect,

Alexander Ebin,

President. [IV.11]

Hadamard did not give a mathematical lecture at the Congress. However, he gave a talk on the subject *Are we lacking words?* in the logic and
philosophy section. He was concerned by the problem of finding proper terms
for new mathematical notions, which had interested him for some time. As
he put it three years earlier in his talk about Newton: "The creation of a
word or a notation for a class of ideas may be, and often is, a scientific fact
of great importance, because it means connecting these ideas together in
our subsequent thoughts" [I.375, p. 38]. In concluding remarks of his talk at
the Congress he said:

> A useful thing, in order to avoid lack of words, would be not to
> waste them; I mean, not to use several words where only one
> is needed. Why speak of the "roots" of an equation or of a
> polynomial while an equation has solutions and the polynomial
> has zeros. There is also no reason to denote the solutions of a
> differential system as "integrals", an expression which, moreover,
> is needed in a different meaning... [I.386]

After the Congress, it took much effort on Mandelbrojt's and Schwartz'
part to dissuade Hadamard from a new trip: he
intended to travel to the Mexican forests
on horseback to enrich his botanical collection. Instead, he was taken to a nearby forest with interesting ferns and became completely satisfied [III.358, p. 48].

---

[7] Then follow several pages of mathematical history.

## 8.5   Social activity

Hadamard's involvement in social issues did not diminish neither in the post-war years, nor in the 1950s, a period of political instability and dangerous confrontation. He could often be seen on the rostrum or in assembly halls, at meetings organized by the opponents of the wars in Indo-China and Algeria, of the remilitarisation of Germany, always ready to defend what seemed to him human and just, and to try to convince others. He was a member of the Honorary Committee of the *Mouvement contre le Racisme, l'Antisémitisme et pour la Paix* and continued with his work on the Central Committee of the *Ligue des Droits de l'Homme*.

Malgrange wrote in the article *Jacques Hadamard: un grand savant, un homme de progrès* [II.24]:

> I also had the opportunity of becoming acquainted with him at a meeting of the Peace Movement. There also I had the same feeling of being in the presence of this man who, from his younger days (especially after the Dreyfus affair) had taken part in the struggle for progress and peace, with all the weight of his prestige. A man born during the Second Empire, and five years older than Lenin!

It has sometimes been said that Hadamard was a communist. In fact, he was never a member of the Communist Party, although he did sympathize with the left. While participating in the Resistance, Hadamard's daughter Cécile and her husband René Picard became communists. As for Jacqueline, at first she did not want to join the party because of the show-trials in the Soviet Union. However, after the war, this seemed not to be so important any longer, and, on returning from the United States, she saw the main danger as being the penetration of the American life style in France. So she joined the Communist Party, which

*Hadamard with his daughter Jacqueline*

seemed perfectly natural to Hadamard and his wife. The atmosphere in the family is colourfully described by Jacqueline:

The cell would meet at our home, because we had the biggest flat, and my parents were invited as honorary members, being deeply interested in our discussions. One evening my father, hearing us organizing a team of poster stickers, asked us if he could join us in the hope, he explained, of being arrested and being able to show the Police Officer his card as a Member of the *Institut* with its tricolour. He was disappointed by our refusal, because we were not going to take him on such an adventure!

My father, always in the abstract, did not always follow our reasoning; but my mother, my *marquise révolutionnaire* as I called her, had an innate political sense and showed her opinions clearly. One day, boarding a bus, I saw the dumbfounded looks of the people turn to this old distinguished lady, with a hat, veil and gloves as well, opening *L'Humanité* [IV.1, p. VII(4)].

Once, in January 1957, Hadamard's call to vote for a communist candidate was printed under his photograph in *L'Humanité* [IV.22]. Daniel Mayer, the president of the *Ligue des Droits de l'Homme* commented on Hadamard's political views: "He entered the struggle of the Resistance because it was one against fascism. The same spirit moved him when he denounced the forms taken, here and there, by McCarthyism, that is to say, by intolerance. His activity was to bring him closer to those whom McCarthyism persecuted with its systematic suspicion, and his sympathies for the extreme left were not only the product of a family atmosphere, but of an *élan du coeur*" [II.58, p. 356-357].

Among Einstein's letters to Hadamard in the book [III.120] there are five answering some of Hadamard's proposals connected with his participation in the Peace Movement in the years 1948-1952. Einstein was also deeply involved in social activity, but his peace program and political views (in particular, the creation of a world government and supranational security forces, and his call for civil disobedience) were rather far from those of Hadamard at that time, and his attitude to Hadamard's initiatives was one of scepticism. For instance, when Hadamard asked him to write a message for the World Peace Congress, held on April 20, 1949, in Paris, Einstein refused:

I must confess frankly that my experience of the first congress of this kind held in Wroclaw last August, and what I have observed about the recent New York congress, gives me the impression that this type of procedure does not really serve the cause of international understanding. The reason is simple: it is more or less a question of a Soviet initiative and all the organisation

comes from there. This in itself would not be a bad thing if the
Russians and the people from countries affiliated to Russia were
really free to express their personal opinions and did not have
to present the official Russian point of view, as happens at the
moment. The impression of the majority of people is thus characterized by the words "Soviet propaganda". Those who speak
in the name of western countries are chosen according to the criterion of not questioning the global model. From this results an
intensification of stupid controversies and polemics which characterize the present international situation ([III.120, p. 131]).

On a different occasion, during the Korean war, Hadamard wrote an open
letter to Einstein accusing the Americans of developing biological weapons:

What are we to think of the
terrible news which has reached
us, of the possibility of spread-
ing the most dangerous diseases
by American aeroplanes? Are
we to think that the propagation of such plagues can be considered as being a part of military operation? I have said: the
possibility, because the information, according to which such
horrors are already taking place,
has been the subject of a de-
nial.

But what is undeniable, is
the work – already in progress
for a number of years – on the
preparation of this war... [I.387]

On March 26, 1952, Einstein responded with a personal letter:

*Hadamard's letter to Einstein*
*in* Action, *March 1952*

Monsieur,

I am really the last person who would excuse these abominable
weapons, whether they are atomic bombs or biological means

of destruction. Also, there is nothing surprising if those who, by their own admission, systematically prepare such abominable things are suspected of making use of them.

But the evil into which we are sinking so deeply cannot be imputed to one side only. I think that one cannot make it better by simple explanations of the times we live in. A constructive supranational action of security would be the only outcome. The European peoples could contribute to this end if they did not believe themselves obliged to seek support from the United States, for their miserable colonial politics. One must unfortunately say that nowhere, except perhaps in India, does there exist an influential public opinion which resists all these unhealthy compromises.

Cordially yours, A.E. [III.120, p. 132]

*Einstein in Princeton*

Hadamard asked for permission to publish this last letter, but Einstein disagreed, replying: "This political propaganda and counter-propaganda, which can not *de facto* be based on any known fact, will only result in provoking hatred and hostility" [III.120, p. 132].

In the United States, in 1950, Julius and Ethel Rosenberg were arrested and accused of providing atomic secrets to the Soviet Union. They denied the charges, but although the proof of their guilt was weak, they were sentenced to death and executed, and Hadamard, who had "two passions: mathematics and justice", as Jacqueline Hadamard put it, felt very keenly about them, as their trial reminded him of the Dreyfus affair. In December 1952 he wrote letters to some American colleagues, expressing his uneasiness. To Wiener he sent the following letter:[8]

December 31, 1952

Dear Professor Wiener,

Before writing to you, I temporized for a long while. Was a foreigner like me qualified to say anything to you about a question relative to your country? That was the reason of this long delay, while this question weighs on my conscience for a very long time.

Of course, I do not know what you think of the Rosenberg case; I can not but tell you of our uneasiness on this issue, which is very great.

The number is great of those who, if the Rosenbergs are executed, will consider that as a murder: a murder disguised under legal formalities remains a murder.

The honour of American Justice, its reputation over the world are at stake.

This is, you must know it, the opinion of many scientists, not only in France, but in many countries – and there is, as you will agree with me, something which we can call a solidarity among scientists all over the world. They feel about many things and ideas in a similar way.

Forgive me if I write this letter to you and to three of our colleagues. Perhaps, I should not have done so if I was not a survivor of the Dreyfus case, still keeping in mind the struggles of that epoch and the energetic intervention of intellectual men. I could not help – else I should have even felt unfair – speaking of such a remembrance to colleagues whom I highly appreciate.

Yours sincerely,

---

[8]Hadamard wrote this (and the next) letter to Wiener in English.

J. Hadamard.

Let me use that opportunity to send you our best and heartiest New Year wishes, from Mrs. Hadamard and myself, for you and Mrs. Wiener. [IV.31]

Wiener replied immediately, but confidentially, and three weeks after the first letter Hadamard sent a second:

January 19, 1953

Dear Professor Wiener,

Thank you for your letter. Of course, there is no possible question of speaking of it to anybody. It remains strictly between you and me. But you can use mine if you wish.

I agree with you on the dreadful atmosphere of present times and, as you, I do not think it reasonable or fair to pronounce an irrevocable sentence in such an atmosphere. That was the position taken by the Court in the Dreyfus trial, and they did not condemn to death.

Of course, there is no possibility of proving the innocence of the Rosenbergs, as innocence is a negative thing, and as such is very often impossible to prove. But a condemnation, especially to death, is a crime as long as the prosecution has not proved the guiltiness without any possible doubt – until then, the accused must be assumed to be innocent... [IV.31]

In Jacqueline Hadamard's papers there is a draft of her father's letter to Pierre Mendès-France (1907-1982), who was *Président du Conseil* (Prime Minister) for seven months during 1954-1955. Mendès-France ended the war in Indo-China and negotiated a plan for German rearmament which was acceptable to the French National Assembly (French Parliament). Hadamard addressed him with a plea to prevent German militarization:

December 18, 1954

*Monsieur le Président*,

As you see, it took me quite a long time to decide to write to you on the subject which grips the heart of all of France and, I am certain, yours. But I could no longer be silent.

In the course of my long life I have lost all my sons in wars begun by Germany: all three volunteers, the two older ones during the first and the last in the one of 1939. It is in thinking of my grandchildren and great-grandson that I say to you: in the

light of History, do not take this responsibility of having given
– what am I saying, almost imposed on – arms to a people who
is still not completely cured of militarism, to a people which is
quite alone in the world, I believe, in having *Kriegsstrasse*.

Noboby dares pretend any longer that nazism is dead in West
Germany: there have been too many declarations by some of Mr.
Adenauer's Ministers to be able to say that the Germany of Bonn
is a pacifist country. And when we are told that rearming her,
it is a matter of assuring peace, I remember too well – you were
perhaps too young at that time to have noticed – that Hitler
declared already in 1933 that German rearmament would assure
peace in Europe!

*Monsieur le Président*, the people of our country do not agree
with a German rearmament. The proof of this is, moreover,
that no one has dared, under all the successive governments
favourable to this rearmament, to ask their opinion by referen-
dum, which would otherwise have been quite natural and would
still be so – on a question so vital to the future of our country.

Do not allow the weight of your moral authority, which the
peace in Indo-China so justly endows you with, in this weighing-
scale tray, to support something that would remain the most
sombre page of French history.

Dare I add that it is inconceivable for me that it is a Jew
who puts all his authority, all his political career in the service
of Nazi rearmament? It is something I would not remind anyone
else of; but, be sure, others will say it, will find a reason for
antisemitism in it. Another responsibility to History, allow one
of the rare surviving founders of the *Ligue des Droits de l'Homme*
to tell this to you.

To dare speak to you with this frankness has required the
anguish of seeing the irreparable being prepared.

*Monsieur le Président*, I want to keep all my hope with you.
[IV.12]

Even in his very old age, Hadamard did not allow himself indulgencies
when his civil duties were involved. Mandelbrojt, in [II.42, p. 18], recalls
that once, when he was over eighty, Hadamard and his wife were hit by
a car, but the injuries did not prevent Hadamard from appearing at the
presidium of a crowded meeting.

J.-P. Kahane recalls:

I saw him in different meetings. He was always late,
always tiptoed to the floor,
asked for a chair, and let
his fingers drum until he
was invited to say a few
words. What I remember
best is the expression of
his face. From the pho-
tographs you can appre-
ciate its sharp and bibli-
cal beauty, but more striking was the acuteness of
his look and the constant
motion of his eyes [II.29,
p. 26].

*Jean-Pierre Kahane*

## 8.6   Reminiscences of Ernest Kahane

For many years, Hadamard was a member of the *Union Rationaliste*. This
association, created in 1930, is open to everybody, its aim being "the diffusion of the rationalist spirit and scientific method". The union publishes the
journal *Cahiers Rationalistes* and regularly organizes conferences at the *Sorbonne*. At the end of his speech given at one of these conferences, the eighty-
seven-year-old Hadamard said: "Historically, scientific truth and moral truth
have gone hand in hand since the sixteenth century. This moral progress,
which we must define, all our moral evolution, evoked by Renan and which
Fascism wants to destroy, all this has gone hand in hand with scientific
progress, and should never be separated from it" [I.389]. For Hadamard,
these were not mere words. In his own life, moral and scientific values were
really indissolubly connected. When Hadamard was ninety, he was elected
Honorary President of the *Union Rationaliste*.

A former Secretary General (from 1954 to 1967) and then President
(from 1968 to 1970) of the *Union Rationaliste*, Ernest Kahane kindly gave
his permission for us to include his unpublished reminiscences of Hadamard,
whom he first met after the Second World War when serving as the secretary
of the *Syndicat National de l'Enseignement Supérieur*. He was a biochemist

and the head of the Laboratory of Organic Microanalysis of the CNRS (*Centre National de la Recherche Scientifique*) from 1946 to 1955, then going on to be a professor at the Faculty of Sciences at Montpellier University from 1955 until his retirement in 1973. Always interested in the history of science, Ernest Kahane wrote about Pasteur, Teilhard de Chardin, Claude Bernard, Lavoisier, and many others. About Hadamard, he recounted the following:

> Trade union life is not rich in amusing memories. Here is one I recall with pleasure. It concerns someone I had the pleasure of calling our venerable colleague Hadamard, and whom we all affectionately called *le petit père Hadamard* when we talked about him.

*Ernest Kahane, J.-P. Kahane's father*

> He had been one of the first members of the *Syndicat National de l'Enseignement Supérieur*, when unions for civil servants were still illegal, in the thirties. Then he rejoined the union on its reestablishment in 1945. He was a constant participant of demonstrations which demanded something (or demonstrations of another kind), and when he arrived, whether invited or not, he went without hesitation to the rostrum, sure that he would

be in his proper place, that he would be welcomed – the good man, the good comrade! He was able to give personally and also, because false and stupid modesty were foreign to him, he, the prince of mathematicians, had his share of glory to offer in service to the union movement.

The first great demonstration after the war, towards the end of 46, was organized by the Paris regional *Syndicat de l'Enseignement*, a departmental section of the FEN,[9] then a member of the CGT.[10] The assembly was to take place in the *avenue de l'Opéra*, at the corner of *rue Louis-le-Grand*, so that the procession would go to the Ministry of Finances, at Louvre palace. The journalist from the *Huma*[11] collared me so that I could point out the personalities present. They were numerous and there were some illustrious ones. I had a job to gather my flock, and I hurriedly pointed out Joliot, Nobel Prize, Teissier, Director of the CNRS, Chapelon, professor at the *École Polytechnique*, Orcel, professor at the *Muséum*, a bevy of professors at the *Sorbonne*, and the *père* Hadamard, without much detail, as is usual, speaking of the greatest.

The journalist from *L'Humanité* had faithfully taken notes, and the next day one could read in the list of participants *le père Hadamard* amongst the other notables. I was dismayed, but delightful Hadamard laughed about it until tears came to his eyes.

$$* * *$$

How we respected him! How we cherished him! Also we complained about the way he had received us, when it was a question of all or nothing and he had exasperated us.

It's just that he knew what he wanted, and did not let himself be manoeuvred (in which he was damned right, and from the depth of my heart I pay him tribute). He wasn't one of those people who, open-mouthed, accept whatever is told them, and who sign without discussion the papers one is given. He retained this noble critical mind which had been an undisputed constant all through his long life and he was hardly disposed to cede in the name of some *raison d'état*.

---

[9] *Fédération de l'Éducation Nationale.*
[10] *Confédération Générale du Travail.*
[11] The communist newspaper *L'Humanité.*

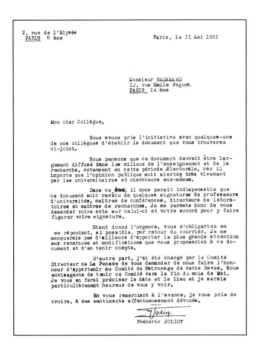

*F. Joliot-Curie's letter to Hadamard with a request to sign a certain document*

We mentioned him one day when some of us were discussing our work as activists, and Joliot, who loved to tell revelatory stories and told them well, with a curious mixture of loquaciousness and conviction, described how Hadamard received him when he went to appeal to him. I was not at the scene with Joliot, but how many times have I myself been the actor facing Hadamard, and have I not noticed just like Joliot the mastery with which our recalcitrant intended to run the scenario in his own manner!

"He's a sadist, that man," Joliot said comically, "he doesn't let me read my paper, he puts on his spectacles and wants to see the text with his own eyes. He dissects it with a vengeance, in content and form, he makes shreds of it and demands reasons for everything, as a whole and in details. He calls on heaven and earth as witness so excessively because one wants him to sign and then leaves the sollicitor in a *tête à tête* with his wife, the exquisite Madame Hadamard, who would like to please everybody, and whose role then becomes that of being the confidante of all the bad things one thinks of the great man.

He, who is not quite as deaf as a post, and plays on his infirmity with virtuosity, has gone into the adjoining room... and has found a way of hearing everything, if he doesn't guess it all. After a short while, he jogs in and asks: 'What else do you say about me?' He mischievously awaits the explanations, his hand to his ear. Of course, they don't come, but the visitor, overstrained, discouraged and somewhat disconcerted all the same, finishes by rising to leave. That's the moment *le père Hadamard* chooses to ask: 'And this paper, where is it? What have you done with it? You're not going just like that without me signing it!' "

$$* * *$$

Dear Hadamard was not at all deaf when I was running after him in one of these poles of his universe, between the *rue Pierre-Curie*, the *rue d'Ulm*, the *rue Gay-Lussac* and the *rue Saint-Jacques*, because I had seen him from a distance and I wanted to offer him my car at his service. He refused to tell me where he was going before he got to know if it fitted in with my itinerary. Too bad for me if, assuming he was going back home, I invented something to do at the *Porte d'Orléans*, when he had business at the *Institut*, or vice-versa, he accepted no second thoughts and set off again at a steady pace which brooked no reply.

I ought to accuse myself of having abused the candour of this artless man in excusing myself with two errands, and hesitating on the order in which I was to do them, one on the quays, the other in the southern suburbs. Like this, he did not put any other conditions to have himself driven to the *Institut* or the *rue Émile Faguet*. Artless? I said artless? My eye! as Zazie[12] would say! Wasn't I the artless one, believing myself to have abused someone like Jacques Hadamard!

$$* * *$$

He had made me the honour of choosing me as confidant for his misgivings about union ethical matters, whether it was about subscriptions, financial declarations or about personnel. His confidence was complete and touched me enormously, with the veritable veneration I had for him. The word is not too strong. I was brought up instructed by a father who flattered himself in having "a nose for respect", and for whom Jacques Hadamard was the most respectable thing humanity could produce, as an eminent mathematician and exemplary citizen. I was quite little when paternal influence had me place Jacques Hadamard in first place in my personal Pantheon. And then there he is, asking my advice, the world turned upside down. However I had to put up

---

[12]Zazie is the heroine of the novel *Zazie dans le métro* (1959) by the French writer Raymond Queneau.

a good show, in such glorious circumstances! There was no effort
of documentation, of reflection, of elaboration which measured
up to the dignity with which I, in my own eyes, found myself
invested! And when the time came to give a reply – because it
was usually a small note, brief and precise, which reached me – I
arrived with a dossier fuller than an apothecary's memory. But
I hadn't time even to open it before this man so gently put me
at my ease – if I may say so, because I was more overwhelmed by
his attitude and more convinced of my insignificance. "I asked
you a question because I couldn't solve this little problem myself.

A leaflet of the *Union Rationaliste*          *Hadamard's paper in* Les Cahiers
                                               Rationalistes

It comes under your union prerogatives, we put our faith in your
competence and judgement, and I do not need reasons, I shall
do what you tell me to do, and in the way you tell me." Doesn't
one need to be great to have such thoughts and to express them
with such simplicity?

Happy as I am, I had Hadamard's confidence for many years
after giving up union duties, and it was quite often that he would
send me one of his little questioning notes. Once when I told him
that I would send his question on to the present union secretary,
he thought for a moment, as if to digest his surprise, and insisted:

"I am an old man, it is time I changed mentor." And when I became Secretary general of the *Union Rationaliste*, he added to the palette of competences he ascribed to me, the ability to answer the most varied points of reason and folly.

The meeting of the *Union Rationaliste* at the Sorbonne in commemoration of A. Einstein, 1955. Presiding: Mlle J. Lévy, J.-P. Vigier, J. Hadamard, A. Châtelet, E. Schatzman (standing), Mme P. Langevin, M. Prenant

\* \* \*

It was on this account that I asked him for a meeting and that I went to see him one fine day. It was me who had something to ask him. The *Union Rationaliste* had a Committee of Honour, but no Honorary President. "Would you accept, for your ninetieth birthday, becoming our Honorary President?" He didn't accept immediately, he had to formulate an objection. "You should wait for my hundredth birthday, ninety is nothing." It was not coquetry, but innocent mischievousness. We were right in not giving in to his argument, since he left us before we could celebrate his hundredth birthday [IV.13].

## 8.7   Awards

During his life, Hadamard received many marks of recognition. In January 1936, at the celebration of his seventieth birthday, Lebesgue said: "I would like, in conclusion, to give you an idea of all his honours, but Hadamard was unable to help me compile a complete list. This was the only time I found a gap in his knowledge" [II.27, p. 14].

Without aiming to give a complete list, we mention the following: he was a member of the *Académie des Sciences* and a foreign member of the *Accademia dei Lincei*, a fellow of the Royal Society of London and the Royal Society of Edinburgh, a member of the Academy of Sciences of the U.S.S.R., the National Academy of Sciences of the U.S.A., the American Academy of Arts and Sciences, the National Academy of Exact Sciences of Argentina, the Royal Lombard Institute of Science and Letters of Milan, the National Academy of Exact, Physical and Natural Sciences of Lima, the Academy of Sciences of Saragossa, as well as of the Belgian, Brazilian, Irish, Indian, Egyptian, Dutch, Rumanian, Polish, and Swedish Academies of Sciences. He was awarded honorary doctorates of Göttingen, Yale, Bruxelles, Liège, Oslo Universities, and of the Hebrew University of Jerusalem. He was an honorary member of the Royal Scientific Society of Uppsala, the London Mathematical Society, the Mathematical and Physical Society of Erlangen, the Mathematical Society of Copenhagen, the Kharkov Mathematical Society, the Benares Mathematical Society, and the Calcutta Mathematical Society. He was on the committee of the *Circolo Matematico di Palermo*, and of the Swiss Society of Natural Sciences, and he was a Corresponding Member of the Royal Scientific Society of Liège.

It is well known that there is no Nobel Prize in mathematics. However, there exist other international awards for mathematicians, the most famous of these being the Fields Medal, created in 1936, and awarded at each International Congress of Mathematicians. Another prestigious honour is the Feltrinelli Prize, founded by the *Accademia dei Lincei* in 1955, the first laureate of which was Hadamard. He was awarded another prize, the *Albert I$^{er}$ de Monaco* Prize, in 1948 for his contributions to science [II.53, p. 1302].

*Jacques and Louise Hadamard with the French President René Coty after Hadamard was awarded the Grand-Croix de la Légion d'honneur, 1957*

The French President René Coty bestowed on him the *Grand-Croix de la Légion d'honneur* at a ceremony at the *Élysée* Palace on May 2, 1957. In the same year he was awarded the gold medal of the CNRS. From 1917 Hadamard was an Officer of the Order of the Crown of Italy.

*The gold medal of the CNRS*

Hadamard's last award was a specially prepared gold medal presented to him by the *Académie des Sciences* in 1962, on the occasion of the fiftieth anniversary of his election to the *Académie*. He received it at home, from

Denjoy, Julia, and the two *Secrétaires Perpétuels* of the *Académie*, Robert Courrier and Louis de Broglie. The latter said:

> When you were elected to the *Académie* fifty years ago, you were given, as all new members are, a small medal. It is a very modest small medal made of silver which has on one side the head of Minerva because this helmeted goddess was always considered as personifying Wisdom and Science, and on the reverse side is the name of the newly elected member and the date of his election. This modest medal gives those who possess it, the advantage, also quite modest, of being able to enter without payment the museums and castles belonging to the *Institut*. We

*The gold medal of the Académie des Sciences*

> thought that, to remember your election, it was natural to offer you another medal like the one you once received, but larger and made of a more precious metal. It therefore had to be in gold because it would serve to celebrate your "golden wedding" with the *Académie des Sciences*. I do not need to say that the proposal, made by the Bureau, to offer you this medal, when it was put to the *Académie*, was received with such lively and unanimous applause that any vote on it was unnecessary [II.56, p. 735].

## 8.8   The end of life

On May 15, 1954, in the great amphitheatre of the *Sorbonne*, during the official ceremony of the centenary of Henri Poincaré, Hadamard talked about

Poincaré's contribution to mathematics.

The following year, Hadamard's ninetieth birthday was reported in many French newspapers. Here is a letter from Montel written the day before the birthday:

> Dear Monsieur Hadamard,
>
> Your family will tomorrow be with you. I do not wish to disturb by my presence such precious intimacy. But I would like you to feel the presence of my grateful and affectionate thoughts. Because I owe you so much for the past half century since I approached you. To the mathematician, my teacher, first of all, but in this I am only one in a crowd. Then to the man who showed me a friendship which touches me intimately and of which I feel some pride.
>
> I have experienced this friendship at the *École Polytechnique* and at the *Académie des Sciences* (where it earned me a song as well), during our travels in Russia and in the Alps.
>
> Your life of honesty, of simplicity, of work with an unquenchable curiosity, through the painful trials which it inflicted on you, serves me as a model.
>
> Accept the wishes I offer, that your health remains robust, that it conserves the integrity of a magnificent mind, that it be the joy of your friends and the happiness of your admirable wife, who is not the least of your discoveries.
>
> My wife joins me in these wishes, in the respect and in the affection.
>
> Paul Montel. [IV.14]

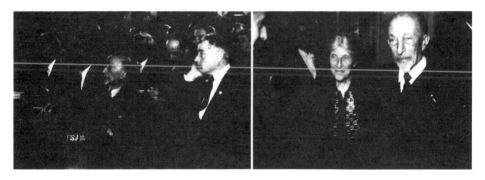

*At the Fourth Congress of Rumanian Mathematicians in Bucharest, 1956. To the right on the left photograph is I.N. Vekua*

In May 1956, Hadamard attended the Fourth Congress of Rumanian Mathematicians in Bucharest. As on many previous occasions, Louise accompanied him. Montel wrote: "Fortunately, an admirable companion took care of him. Madame Hadamard realized immediately the role she could play beside a man of genius. With constant generosity, mathematics acted on her as if by induction. She rang me one morning: "Today it is exactly seventy years ago... that I was eighteen", thus introducing feminine delicacy into arithmetic" [II.5, p. 22].

But the years took their toll. Hadamard's hearing became weaker and he found difficulty in walking. It became more and more arduous for his wife to take care of him. She had a renal colic when she was eighty years old, and she died on July 6, 1960, two days after the second attack. They had lived together for sixty-eight years. "It is from that day that my father ceased having a desire to live," wrote Jacqueline Hadamard.

*Louise Hadamard (1868-1960)*

When a Conference on partial differential equations took place at the *Collège de France* in June 1962, the following greeting to Hadamard was agreed on.

COLLÈGE
de
FRANCE
—

*Paris, le* 25 Juin 1962

A sa séance d'ouverture au Collège de France,

Les participants au Colloque International organisé
par le Centre National de la Recherche scientifique
sur les Équations aux dérivées partielles,

se souvenant que Monsieur Jacques Hadamard renouvela
profondément cette théorie,

se souvenant qu'au cours de nombreuses années son
Séminaire réunit dans cette Maison des Savants du
monde entier,

lui expriment leur profonde et filiale gratitude,
leur respectueuse et affectueuse amitié.

*The participants of the International Colloquium on partial differential equations, organized by the Centre National de la Recherche Scientifique, recalling that Mr. Jacques Hadamard profoundly transformed this theory, and recalling that during many years his Seminar brought together in this building scientists from the entire world, express to him their deep and filial gratitude, their respect and affectionate friendship [IV.15].*

It was decided that three people be chosen to visit Hadamard and present him with this greeting. Olga Oleinik recounts the following story:

> I had the good fortune to be one of the three, together with Leray and Schwartz. We went upstairs to Hadamard's flat and waited for him in a large sitting room. He came in, accompanied by his daughter Jacqueline, who helped him sit down in the armchair. Hadamard, ninety-six years old, moved with difficulty. His strength was running low after the death of his wife. However, he was friendly and smartly dressed, and the buttonhole of his dark suit had the scarlet rosette of the *Légion d'honneur* in it. Hadamard's hearing had become very weak by that time, and Schwartz bent over him to tell him about the conference. When I was introduced to him, he said something about the Wiener criterion. I supposed that he remembered my first paper in *Matematicheskiĭ Sbornik*, where the criterion for the regularity of a boundary point for the Laplace operator, was extended to general elliptic equations of the second order. Several years later, I found the reference to it in Hadamard's last book, published in Peking after his death.
>
> The visit did not last long. When we left, Jacqueline told me at the door to the lift: "You must be so happy to live in such a remarkable country!" [Interview, April 1988]

The same year Hadamard received his final blow: his grandson Étienne Picard died during a mountain climb. After that he stopped going out. Jacqueline and a nurse Mlle Hameau stayed with him. Hadamard died peacefully on October 17, 1963. He was buried in the family grave in the *Père Lachaise* cemetery.

*The grave of the Hadamard family, where Jacques Hadamard was buried, in the Père Lachaise cemetery. (His name is not engraved on the tombstone.)*

Many French newspapers reported Hadamard's death [II.22] – [II.26], [II.35], [II.45] – [II.48]. The Moscow Mathematical Society payed hommage to Hadamard at a special meeting on March 10, 1964 [II.60]. At its session on November 13, 1965, the *Accademia dei Lincei* marked Hadamard's 100th anniversary. The chairman B. Segre shared his reminiscences about Hadamard, whom he had first met in 1924 in Paris, and then he read an obituary of Hadamard written by F. Tricomi. On January 13, 1966, Hadamard's centenary was celebrated at the *École Polytechnique*. Members of Hadamard's family were present, and there were many foreign guests. Louis de Broglie presided over the ceremony which was opened by General Mahieux. P. Lévy and L. Schwartz gave surveys of Hadamard's mathematical results, S. Mandelbrojt described Hadamard's work at the *Collège de France*, F. Tricomi spoke on behalf of foreign mathematical societies and the *Accademia dei Lincei*. Other speakers were P. Montel, M. Roy, then *Président de l'Académie des Sciences*, and L. Armand, then *Président du Conseil de Perfectionnement de l'École Polytechnique*. The first few sentences of Montel's talk, *Jacques Hadamard, l'homme et le savant* give a fitting epitaph to Hadamard:

> The brilliance of the great mathematical discoveries of Jacques Hadamard sometimes dazzles his admirers and prevents them from gauging the extent of his intellectual richness and of his moral grandeur. Teeming with magnificent gifts, he reminds one of those high mountains one has to climb by all their faces, in order to know them well. If he took pleasure in what Anatole France called the silent orgies of thought, Jacques Hadamard was passionately interested in many creations of the human mind and has contributed to their realization [II.5, p. 21].

*French newspapers on the death of Hadamard*

# Part II

# Hadamard's Mathematics

# Chapter 9

# Analytic Function Theory

## 9.1   Singularities

*Augustin Louis Cauchy (1789-1857)*

**Introduction**. Hadamard's first significant publications are related to the theory of analytic functions of a complex variable. The principal features of this domain took shape in the works of Cauchy (integral and residue theorems), Riemann (conformal mappings, theory of many-valued functions), and Weierstrass (representation of analytic functions by power series and analytic continuation). There were important overlaps between their approaches and important differences. For example, Riemann's theory was the most geometrical, Cauchy and Weierstrass prefered the use of power series and convergence criteria. But Weierstrass distrusted the use of integration and could not accept Riemann's geometric constructions and his free use of the variational principle.

Although there was no consensus among the founding fathers, the interaction of their ideas led to the creation of a subtle and deep theory which, at the time when Hadamard was a *Normalien*, was paid tribute by most good mathematicians. In Hadamard's words, the attention of geometers was entirely attracted to the beautiful theory of analytic functions, the unexpected and splendid results which seemed to cover the whole domain of mathematical science. It had revealed to geometers that "the shortest way between two truths in the real domain very often goes through the complex one" [I.220, p. 111].

*Bernhard Riemann (1826-1866)*

Young Hadamard, under the influence of his teachers Hermite, Darboux, and Picard, followed the Cauchy-Weierstrass tradition. In his early significant works of the 1880s and 1890s he obtained fundamental results for two main problems in the theory of Taylor series. The first is the problem of the relation between the singularities of a function and its Taylor coefficients; the second, concerning entire functions, deals with the growth of a function, the distribution of its zeroes and the decay of its Taylor coefficients. His work initiated a new phase in the study of analytic functions and influenced the development of vast domains in the analytic function theory for many decades to come.

**Disc of convergence**. We begin our description of Hadamard's research with a theorem he proved when a student of *École Normale*. We recall that a function $f(z)$, defined in a domain $\Omega$ of the complex plane, is said to be analytic (regular, holomorphic) in $\Omega$ if its derivative exists at each point $z \in \Omega$. According to Cauchy, $f$ can be represented by the Taylor series

$$c_0 + c_1(z - a) + c_2(z - a)^2 + \dots \qquad (9.1)$$

in some disc $D_\rho(a) = \{z : |z - a| < \rho\} \subset \Omega$. This property is the base of Weierstrass' construction of analytic continuation, where the values of a function are extended to a larger domain along chains of small overlapping discs.

In his treatise *Analyse algébrique* [III.73], published in 1821, Cauchy obtained the following result: the series (9.1) converges for $|z - a| < 1/l$ and diverges for $|z - a| > 1/l$, where $l$ is the limit or "the largest of limits" of the sequence $\{\sqrt[n]{|c_n|}\}$. Although he did not explain the term "the largest of limits", one can see that it is the upper limit of a sequence which is referred to here, a concept introduced later by du Bois-Reymond (see [III.110, p. 267]). A precise formulation of the above result, together with its proof, appeared for the first time in Hadamard's 1888 paper [I.10]. Judging from the text of that article, it seems that he was unaware of Cauchy's theorem and of du Bois-Reymond's definition.[1]

*Karl Weierstrass (1815-1897)*

The Cauchy-Hadamard theorem, which can be found in any textbook, gives the formula

$$r = l^{-1}$$

for the radius of the disc of convergence of the series (9.1), that is, for the number $r$ such that (9.1) converges for $|z-a| < r$ and diverges for $|z-a| > r$.

In the same article [I.10], Hadamard is interested in the following question: what information can be obtained from the coefficients of (9.1) about the singular points of the function at the boundary of the disc of convergence?

A singular point of a function is a point where the function in some sense behaves badly, and so provides an obstruction to the analytic extension of the function. A function given by the power series with a finite radius of

---

[1]References to these works are also absent in Hadamard's thesis [I.13], 1892, but he gives them in [I.87, p. 24], 1901.

convergence always has a singular point on the circle of convergence. The simplest class of singularities are poles of finite order: a point $a$ is a pole of order $p$ of the function $f(z)$ if $a$ is a zero of order $p$ of $1/f(z)$. In a neighbourhood of $a$ one has the representation

$$f(z) = \sum_{n=1}^{p} \frac{d_n}{(z-a)^n} + \sum_{n=0}^{\infty} c_n(z-a)^n.$$

The first sum is called the principal part of $f$ at the point $a$. If $p = 1$ the pole is said to be simple.

An analytic function is characterized by its singular points, as well as by its domain of definition (the set of regular points). The former can have several manifestations: they can be isolated (a pole, an essential singularity, a branch point of a multi-valued function as, for example, $z = 0$ for $\cot z$, $\sin z^{-1}$, $\sqrt{z}$, respectively), or limiting points, and even form a continuum. Despite the great successes of the theory of analytic functions, the question of the distribution of their singularities remained open until Hadamard's work on these problems appeared at the end of the last century. Discussing the difficulties involved, Goursat wrote, in his *Cours d'Analyse* (1923-1924): "Only in recent years, however, has this problem been the object of thorough investigations, which have led to some important results. The fact that these results are so recent must not be attributed entirely to the difficulty of the question, however great it may be. The functions that have actually been studied successively by mathematicians have not been chosen by them arbitrarily; rather the study of these functions was forced upon them by the very nature of the problems which they encountered"[III.154, p. 206-207]. The functions were given either as particular series (for example, the hypergeometric Gauss series), or as certain integrals (Euler's gamma function), or as solutions of differential equations (mostly linear). One can say that almost all the great achievements of analytic function theory, as they were at the beginning of this century, were connected with specific functions. Very little was known about the general relations between the behaviour of the coefficients and the nature of the singularities.

An immediate stimulus for Hadamard's study [I.10] was a short article by Lecornu [III.230], which had appeared a year previously, where it was claimed that the existence of the limit

$$z_0 = \lim_{n \to \infty} \frac{c_n}{c_{n+1}} \tag{9.2}$$

implies that the Taylor series

$$f(z) = c_0 + c_1 z + c_2 z^2 + \dots \tag{9.3}$$

has only one singular point $z_0$ at the boundary of the disc of convergence. A sort of converse to this theorem was proved by Kőnig [III.217] in 1876 and by Darboux [III.92] in 1878, who showed that the existence of a unique singular point $z_0$, which is a simple pole at the circle $|z| = r$, implies (9.2). However, Lecornu's proof was wrong, and, what is more, the condition in his formulation turned out to be too weak. In [I.10], Hadamard showed that Lecornu's statement is true, provided that (9.2) is replaced by the more stringent assumption

$$\limsup_{n \to \infty} \left| \frac{c_n}{c_{n+1}} - z_0 \right|^{1/n} < 1.$$

The following simple counterexample to Lecornu's formulation, given by Hadamard, is the series

$$\sum_{n \geq 0} \left( 1 + \frac{(-1)^n}{n} \right) z^n,$$

which represents the function

$$(1 - z)^{-1} + \log(1 + z).$$

Here (9.2) is valid with $z_0 = 1$ but the function has two singular points $z = \pm 1$ on the circle $|z| = 1$. Hadamard's new condition is clearly violated.

Later, in [I.13], Hadamard remarked that his new condition is also necessary if the unique singular point at the boundary of the disc of convergence is a simple pole. In connection with this, we mention a more general theorem due to Pellegrino [III.304], proved in 1942: the conditions

$$\lim_{n \to \infty} \frac{c_n}{c_{n+1}} = 1$$

and

$$\limsup_{n \to \infty} \left| 1 + \sum_{k=0}^{p} (-1)^k \binom{p}{k} \frac{c_{n-k}}{c_n} \right|^{1/n} < 1$$

are necessary and sufficient in order that the series (9.3) has a pole of order $p$ at the point $z = 1$.

In the first part of his thesis [I.13], presented in 1892 to the *Faculté des Sciences* at the *Sorbonne*, Hadamard found a necessary and sufficient condition for $z = 1$ to be a singular point of $f(z)$. It takes the form of the inequality

$$\frac{|f^{(n)}(z)|}{n!} > \left(\frac{1-\epsilon}{1-t}\right)^n,$$

where $\epsilon$ is an arbitrarily small number, $t$ is some number in the interval $(0,1)$ and $n$ takes on infinitely many values [I.13, p. 17]. The proof of this is based solely on the formula for the radius of convergence of the Taylor series. Using this criterion, one can easily obtain the following simply formulated result, known as Pringsheim's theorem ([III.331, p. 1033], [III.268, p. 24]): if the coefficients of the series (9.3), with unit disc of convergence, are non-negative, then $z = 1$ is a singular point for the sum of the series.

Mandelbrojt writes, in connection with this: "The condition on the co-efficients for singularities gave Hadamard other interesting results, but we can not omit mentioning Hadamard's refusal to consider as his own the theorem, which by the way follows immediately from his criterion, according to which the positive real point of the circle of convergence is a singular point if all the coefficients are non-negative. Several authors have, in good faith, attributed this result to Hadamard, but he has always considered it, and quite categorically, as being due to Pringsheim (who, by the way, has not failed to claim its paternity!)"[II.37, p. 27]. Hadamard, in his comments [I.372, p. 51], considers the fact that he did not notice the above result as being one of his failures.

**Gap series**. In studying singularities of functions given by power series in Chapter 1 of his thesis [I.13], Hadamard pays special attention to functions for which all points on the boundary of the disc of convergence are singular. The problem goes back to Weierstrass who as early as in 1880 [III.416], arrived at this phenomen when studying series of rational functions, as, for example, the following one

$$\sum_{n\geq 0} \frac{1}{z^n + z^{-n}}. \tag{9.4}$$

Since

$$\left|\frac{1}{z^n + z^{-n}}\right| \leq \frac{|z|^n}{1 - |z|^{2n}} < \frac{|z|^n}{1 - |z|}$$

for $|z| < 1$, the series (9.4) is convergent in the disc $|z| < 1$. However, all points of the unit circle are singular for this series (note that the poles of its partial sums

$$e^{\pi i/2n}, \ e^{3\pi i/2n}, \ e^{5\pi i/2n,\ldots}$$

are dense on the unit circle). In other words, (9.4) can not be analytically extended through the unit circle. In cases similar to this, when the domain where an analytic function is defined, coincides with its domain of analyticity, the boundary of this domain is called the natural boundary of the function. In accordance with this definition, the natural boundary of the series (9.4) is the unit circle.

Hermite, intrigued by Weierstrass' study, obtained the same results by a different method ([III.174], 1881) and drew the attention of Poincaré to the functions with non-isolated singularities. Poincaré, who had already encountered the phenomenon in his theory of fuchsian functions (1881), extended Weierstrass' and Hermite's work in the memoir of 1883 [III.316], where he proposed the following construction for functions with natural boundary. Let $a_n = e^{2\pi i \varphi_n}$, $n = 1, 2, \ldots$, where $\{\varphi_n\}$ is the sequence of all rational numbers. Consider the function

$$f(z) = \sum_{n \geq 1} A_n (z - a_n)^{-1}, \tag{9.5}$$

where $\sum_{n \geq 1} |A_n| < \infty$. The series in the right-hand side of (9.5) converges and the limit function has no singularities for $|z| < 1$. Since the points $a_n$ are everywhere dense on the unit circle, $f$ can not be analytically extended outside the unit disc.

But (9.4) and (9.5) are series of rational functions, not power series, and it was also Weierstrass who gave the following example of a power series the circle of convergence of which is a natural boundary

$$f(z) = \sum_{n \geq 1} a^n z^{b^n},$$

where $a > 0$ and $b$ is an integer greater then one ([III.416], 1880). In fact, the radius of convergence for this series is

$$r = \lim_{n \to \infty} a^{-n b^{-n}} = 1,$$

which gives at least one singular point $z_0$ on the unit circle. The replacement of $z$ by $z \exp(-2ki\pi b^{-m})$, where $k$ and $m$ are arbitrary integers, changes only a finite number of terms of the series. Hence, along with $z_0$, every point $z_0 \exp(-2ki\pi b^{-m})$ is singular. This means that the set of singular points is everywhere dense on the unit circle, which makes analytic continuation outside the unit disc impossible.

The coincidence of the circle of convergence with the natural boundary in this Weierstrass' example proves to be connected with the presence of large gaps between terms of the sequence $\{b^n\}$. In the period between Weierstrass' and Hadamard's work, numerous other examples of series with circle of convergence as natural boundary were proposed by Darboux, Lerch, Méray, J. Tannery, and Fredholm.

The first general theorem on power series

$$\sum_{n \geq 1} c_n z^{\lambda_n} \tag{9.6}$$

with gaps in the sequence $\{\lambda_n\}$ was proved in Hadamard's thesis [I.13]. It states that the series (9.6) cannot be extended beyond its disc of convergence if

$$\lambda_{n+1} > (1 + \theta)\lambda_n, \tag{9.7}$$

where $\theta$ is a positive constant. Hadamard himself was not quite satisfied with this result, because it covered all known examples but a single one, due to Fredholm (1890). The fly in the ointment was the series

$$\sum_{n \geq 1} a^n z^{n^2} \ , \ \ |a| < 1,$$

which, as well as all its derivatives, is an absolutely convergent series on the disc $|z| \leq 1$, and yet it is not analytically extendable outside the disc. Being unable to explain the phenomenon with his method, Hadamard mentioned the problem to Borel. The latter reacted with an extension of the sufficient condition (9.7) which embraced Fredholm's example (1896). The even less restrictive requirement $n/\lambda_n \to 0$ was given the same year by Fabry. There exist many other generalizations and refinements of this theorem (see, for instance, [III.36]).

**Poles of meromorphic functions**. Some of Hadamard's early results concern the general properties of meromorphic functions, i.e. single-valued analytic functions having no singularities other than poles.

The simplest class of meromorphic functions is formed by rational functions $P(z)/Q(z)$, where $P$ and $Q$ are polynomials. The possibility of expressing a rational function as a sum of partial fractions suggests the problem of a similar representation for meromorphic functions with infinitely many poles. A simple example is provided by the Euler decomposition formula

$$\pi \cotan \pi z = \frac{1}{z} + \sum_{n \geq 1} \left( \frac{1}{z - n} + \frac{1}{z + n} \right).$$

This was proven to be true by Mittag-Leffler, who, inspired by Weierstrass' lectures, proved his celebrated theorem on the existence of meromorphic functions with *a priori* given poles and principal parts of these poles (1876).

In the 1880s and 1890s general properties of meromorphic functions were studied by Picard, Poincaré, Runge, Hadamard, Mittag-Leffler, and Borel.

In the second part of his thesis Hadamard considered the following question. Let the series (9.3) have a disc of convergence $D_r = \{z : |z| < r\}$. What are necessary and sufficient conditions for the function $f(z)$ to have $p$ poles, and no other singularities, on the boundary $C_r = \{z : |z| = r\}$ of the disc $D_r$? Hadamard gives an answer in terms of the determinants

$$
\mathcal{D}_{n,p} = \begin{vmatrix} c_n & c_{n+1} & \cdots & c_{n+p} \\ c_{n+1} & c_{n+2} & \cdots & c_{n+p+1} \\ \cdot & \cdot & \cdots & \cdot \\ c_{n+p} & c_{n+p+1} & \cdots & c_{n+2p} \end{vmatrix},
$$

which were introduced by Kronecker in 1881 [III.219]. Kronecker's condition for the necessity and sufficiency of (9.3) to represent a rational function takes the form

$$
\mathcal{D}_{0,p} = 0
$$

starting from some $p$.

Since one has, by the Cauchy-Hadamard formula,

$$
\limsup_{n \to \infty} |c_n^{1/n}| = \frac{1}{r}
$$

and because the order $(p+1)$ of the determinant does not depend on $n$, one obtains

$$
\limsup_{n \to \infty} |\mathcal{D}_{n,p}^{1/n}| \leq \frac{1}{r^{p+1}}.
$$

Hadamard shows that a necessary and sufficient condition for a function $f(z)$ given by (9.3) to have no more than $p$ poles and no other singularities on the circle $C_r$, is

$$
\limsup_{n \to \infty} |\mathcal{D}_{n,k}^{1/n}| = \frac{1}{r^{k+1}}
$$

for $k = 0, \ldots, p-1$ and

$$
\limsup_{n \to \infty} |\mathcal{D}_{n,p}^{1/n}| < \frac{1}{r^{p+1}}.
$$

Many years passed before these results of Hadamard's were extended by other authors. Of the long list of papers which one could cite, we mention only the memoir by Pólya [III.322], in which he simplified the difficult proofs of Hadamard, and the paper by Piranian [III.313], in which Hadamard's methods were extended to functions having algebraic and logarithmic singularities of the form

$$(z-a)^{-s}[\log(z-a)]^k,$$

where $s$ is a complex number and $k$ is a non-negative integer.

Applications of Hadamard's theory of polar singularities to Padé approximants, i.e. best rational approximations to power series, are discussed in [III.155].

**Fractional derivatives and singularities.** An analytic function always has at least one singular point on its circle of convergence. In general its behaviour on the circle can be quite varied. The singular points can be everywhere dense on some arc, or on the whole of the circumference, in which case it is the natural boundary of the function. These points can also be of different types. Indeed, the function can have finite values at these points (for instance, if it is continuous on a closed disc, but can not be extended beyond it), or it can have an infinite value if the point is a pole. There is thus the problem of characterizing the behaviour of a function on its circle of convergence, or, to be more exact, of constructing a numerical scale characterizing all the possible cases of singularities.

This is the problem which Hadamard formulated in the third part of his thesis [I.13]. The subtle results he obtained here are so general and complete, that we could call this "Hadamard's theory of the singularities of a power series on the circle of convergence". In the following we assume that the function $f$ is given by the series (9.3) and that $C = \{z : |z| = 1\}$ is its circle of convergence.

In order to proceed we need the notion of the Mittag-Leffler star of the series (9.3) with respect to the origin introduced by Mittag-Leffler in 1898 [III.282]. The star is constructed as follows. Mark the poles $a_k$ on the plane, draw the rays starting at $a_k$ in the direction $Oa_k$, and cut the plane along these rays. The remaining domain is the Mittag-Leffler star with vertices at $a_k$. In other words, this is the largest domain which is star-shaped with respect to the origin[2] in which (9.3) can be analytically extended along rays starting from $z = 0$ (see Figure 1).

---

[2] A domain is called star-shaped with respect to $z = 0$ if any ray with origin at $z = 0$ has a unique intersection point with the boundary of the domain.

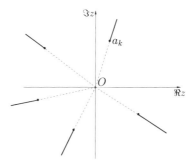

**Figure 1.** *Mittag-Leffler star*

We first remark that, if $f(z)$ is regular at a point $z_0$, then the derivative $f'(z)$ and the primitive $\int_0^z f(t)dt$ are also regular at $z_0$. If $f(z)$ has a pole at $z_0$, then the nature of the singularities of the derivative and the primitive are readily seen. One can easily characterize the effect of these operations on isolated singularities of other types.

The integral can be considered as a derivative $D^{-1}$ of order $-1$, and a further integration gives the derivative $D^{-2}$ of order $-2$ and so on. The derivative of negative integer order $-m$ is given by

$$D^{-m}f(z) = \int_0^z \frac{(z-t)^{m-1}}{(m-1)!} f(t)dt. \tag{9.8}$$

Clearly, the differentiation of order $-m$ does not lead to the appearance of new singularities.

At the beginning of the third part of his thesis, Hadamard considers a general operation having the same property. Namely, he shows that for a function $v(t)$, summable on the interval $(0,1)$, the Mittag-Leffler star of the function

$$g(z) = \int_0^1 v(t)f(zt)dt \tag{9.9}$$

contains the Mittag-Leffler star of $f(z)$.

Hadamard further proposes a certain classification of singular points of analytic functions by using differentiation of negative order (not necessarily an integer), which is a particular case of operation (9.9).

By analogy with (9.8), one can define the Riemann-Liouville derivative of arbitrary negative order $\alpha$ as the integral operator[3]

---

[3]For the history of this and similar operators see Ch. 8 of Lützen's book [III.250].

$$D^\alpha f(z) = \int_0^z \frac{(z-t)^{-\alpha-1}}{\Gamma(-\alpha)} f(t)dt,$$

where $\Gamma$ is Euler's Gamma function. Clearly, the function $z^\alpha D^\alpha f(z)$ can be written in the form (9.9) with

$$v(t) = \frac{1}{\Gamma(-\alpha)} (1-t)^{-\alpha-1}.$$

One has

$$z^\alpha D^\alpha z^m = \frac{\Gamma(m+1)}{\Gamma(m+1-\alpha)} z^m.$$

Hence, for the series (9.3) we obtain

$$z^\alpha D^\alpha f(z) = \sum_{m=0}^{\infty} c_m \frac{\Gamma(m+1)}{\Gamma(m+1-\alpha)} z^m.$$

In order to consider a simpler series, Hadamard modified the Riemann-Liouville derivative by taking

$$v(t) = \frac{1}{\Gamma(-\alpha)} \frac{1}{t} \left(\log \frac{1}{t}\right)^{-\alpha-1}.$$

This leads to the operator

$$H^\alpha f(z) = \frac{1}{\Gamma(-\alpha)} \int_0^1 \frac{1}{t} \left(\log \frac{1}{t}\right)^{-\alpha-1} f(zt)dt,$$

which can also be expressed by the series

$$\sum_{m=0}^{\infty} c_m z^m m^\alpha.$$

Since

$$\lim_{m\to\infty} m^{-\alpha} \frac{\Gamma(m+1)}{\Gamma(m+1-\alpha)} = 1,$$

the operators $H^\alpha$ and $z^\alpha D^\alpha$ have the same basic properties, but $H^\alpha$ is more convenient for the study of power series. Although $H^\alpha$ can improve or even weaken the convergence of the series (9.3), depending on the sign of $\alpha$, it does not alter the radius of convergence. The operator $H^\alpha$ is called Hadamard's fractional integral [III.355].

Before introducing the notion of order of a function $f(z)$ given by the series (9.3) with radius of convergence equal to 1 (which is a sort of numerical characteristics of its singularities on the unit circle $C$), Hadamard defines the functions of finite span (*écart fini*). By this he means continuous functions $u(\phi)$ defined on an arc $(a, b)$ of the circle $C$ and satisfying

$$\left| n \int_{a_1}^{b_1} u(\phi) \cos n\phi \, d\phi \right| < M, \qquad \left| n \int_{a_1}^{b_1} u(\phi) \sin n\phi \, d\phi \right| < M$$

for $n = 1, 2 \ldots$ and for an arbitrary subarc $(a_1, b_1)$ of $(a, b)$. It is a simple exercise to show that functions of bounded variation are of finite span.

The number $\omega = \omega(a, b)$ is called the order of the function $f(z)$ on the arc $(a, b)$ if for any $\epsilon > 0$ the function $H^{\omega+\epsilon} f(z)$ is continuous and of finite span on $(a, b)$, but at least one of these properties fails to hold for $H^{\omega-\epsilon} f(z)$. If such an $\omega$ does not exist, then the order is $+\infty$.

Hadamard shows that if the sequence $|c_n| n \log n$ is bounded and if

$$\sum_n |c_n| < \infty$$

then $f(z)$ is continuous and has finite span on the unit circle. He also proves that the order $\omega(C)$ on the whole of the circumference $C$ is given by the formula

$$\omega(C) = 1 + \limsup \frac{\log |c_n|}{\log n}.$$

Mandelbrojt and Schwartz write: "It appears that Hadamard's idea of order has not been fully exploited by the younger generations of mathematicians. Much deeper results should certainly be obtained from this notion, so rich in content..." [II.43, p. 108].

**Multiplication of singularities.** In Volume 22 of *Acta Mathematica*, 1899, Hadamard published his celebrated theorem on the multiplication of singularities. Two power series

$$f(z) = \sum_{n \geq 0} a_n z^n, \qquad g(z) = \sum_{n \geq 0} b_n z^n$$

are considered, with radii of convergence $r_f, r_g$ respectively. It follows directly from the Cauchy-Hadamard formula, that the radius of convergence $r_h$ of

$$h(z) = \sum_{n \geq 0} a_n b_n z^n \qquad (9.10)$$

does not exceed $r_f r_g$. Hadamard showed that one can say more, namely that the singularities of $h(z)$ are contained in the set $\{\alpha\beta\}$, where $\alpha$ and $\beta$ are the singularities of $f$ and $g$ respectively. (Although this statement is not completely rigorous, we shall accept it for the sake of simplicity.[4]) Note that the function $h$, as opposed to $f$ and $g$, may have no singularities whatsoever. For example, if $f$ and $g$ are arbitrary functions of the form

$$f(z) = \sum_{n \geq 0} a_n z^{2n}, \qquad g(z) = \sum_{n \geq 0} b_n z^{2n+1}$$

then $h(z)$ is identically zero. Hadamard's proof is based on the observation that with an appropriate choice of the closed contour $C$, the expression

$$\frac{1}{2\pi i} \int_C f(\zeta) g\left(\frac{z}{\zeta}\right) \frac{d\zeta}{\zeta} \qquad (9.11)$$

is an analytic extension of the function $h(z)$ outside its disc of convergence. This expression is often called the Hadamard composition.

During the two years that followed the publication of Hadamard's theorem, several papers by Borel, Hurwitz, and Dell'Agnola appeared, developing the result. In particular, Hurwitz [III.185] proved the following theorem on the summation of singularities.

Let the series

$$f(z) = \sum_{n \geq 0} a_n z^{-n-1}, \qquad g(z) = \sum_{n \geq 0} b_n z^{-n-1}$$

converge for $|z| > r_f$ and $|z| > r_g$ respectively. Then the series

$$\Psi(z) = \sum_{n \geq 0} c_n z^{-n-1}$$

with

$$c_n = \sum_{k=0}^{n} \binom{n}{k} a_k b_{n-k},$$

called the Hurwitz composition, converges for $|z| > r_f + r_g$.

---

[4]An analysis of Hadamard's theorem can be found in Bieberbach's book [III.36].

The Hurwitz theorem states that for functions $f$ and $g$ having finite sets of singularities $\{\alpha_i\}$ and $\{\beta_i\}$ respectively, the function $\Psi$ has its singularities in the set $\{\alpha_i + \beta_i\}$.

In the same year, 1899, Dell'Agnola generalized Hurwitz' theorem by using, in analogy with Hadamard's composition, the representation

$$\Psi(z) = \frac{1}{2\pi i} \int_C f(\zeta)g(z-\zeta)\,d\zeta,$$

where $C$ is a contour in the complex $\zeta$-plane [III.96].

Of Hadamard's work on the singularities of analytic functions it has been said by Mandelbrojt and Schwartz that "...the delightful little monograph *La série de Taylor et son prolongement analytique*, written in 1901, inspired many well-known mathematicians, and it is not an exaggeration to say that almost all of the 350 publications of 150 authors, referred to in the recent monograph by Bieberbach on analytic extension, were directly or indirectly influenced by Hadamard's work "[II.43, p. 109].

## 9.2   Entire functions

**Background.** Up to now, we have spoken about Hadamard's work on analytic functions having singularities in the finite plane. Just as important is his work on entire functions, i.e. functions which are given by power series that are convergent for all values of $z$. The study of the general properties of such functions was pioneered by Weierstrass, Laguerre, Picard, and Poincaré. Their results were well known to Hadamard and influenced him.

Any polynomial $p(z)$ is obviously an entire function. If $p(0) \neq 0$, the polynomial can be written in the form

$$p(z) = p(0)\big(1 - \frac{z}{z_1}\big)\ldots\big(1 - \frac{z}{z_n}\big),$$

where $z_1, \ldots, z_n$ are the zeros of $p(z)$. By analogy, one can consider the problem of factorization of entire functions: decomposing functions into factors and reconstructing a function from its zeros. It may happen that an entire function has no zeros (for instance, $e^z$) or a finite number (for instance, $p(z)e^z$), or an infinite number of zeros. The latter case is the most

interesting, since one may expect the function to be represented in the form
of the so-called infinite product, i.e., by the expression

$$\prod_{n=1}^{\infty}(1 + a_n) = \lim_{N \to \infty} \prod_{n=1}^{N}(1 + a_n).$$

After taking the logarithm the analysis of the convergence of infinite prod-
ucts is reduced to that of infinite series.

The first infinite product for an entire function was found by Euler (1742)
(see the representation for the sine in Section 1.10). Another classical example is Gauss' formula for the Gamma function

$$\frac{1}{\Gamma(z)} = ze^{Cz} \prod_{n=1}^{\infty}(1 + \frac{z}{n})e^{-\frac{z}{n}},$$

where $C$ is the so-called Euler's constant. Note that the exponentials in
the infinite product ensure its convergence. Starting with this factorization
formula Weierstrass arrived at the following result.[5]

**Theorem** ([III.415], 1876). *Let $f(z)$ be an entire function, with a zero
of order $m$ at the origin, and let the other zeros be ordered with respect to
their absolute values $|z_1| \le |z_2| \le \dots$. Then $f(z)$ can be represented in the
form of the absolutely convergent product (1.3), where*

$$P_n(\zeta) = \zeta + \cdots + \frac{\zeta^{k_n}}{k_n}.$$

As in Gauss' formula these polynomials are chosen to compensate for the
first $k_n$ terms in the Taylor expansion of $\log(1 - \zeta)$. Elementary estimates
show that the choice $k_n = n$ is always possible and that $k_n$ can be even
made equal to a fixed integer $k$ such that the series $\sum |z_n|^{-k-1}$ converges
(the smallest value of $k$ is convenient).

When $Q$ in (1.3) is not a polynomial or when the degrees of $P_n$ can not
be chosen independently of $n$, one says that $f$ is of finite genus *(genre)*.
Otherwise, by the genus of $f$ one means $p = \max\{k, q\}$, where $k$ is the
common degree of $P_n$ and $q$ is the degree of $Q$.

For example, by Euler's formula, $\sin z/z$, considered as a function of
$z^2$, has genus zero. The notion of genus was introduced by Laguerre in his
studies of the distribution of zeros of entire functions and their derivatives
(1882).

---

[5]For the history of the factorization formula, see Bottazzini [III.53].

In 1882-1883 Poincaré studied the growth of an entire function of genus $p$ as well as the rate of decrease of its coefficients. He showed, for example, that

$$c_n(n!)^{1/(p+1)} \to 0,$$

where $c_n$ is the coefficient of $z^n$ in (9.3).

*Edmond Laguerre (1834-1886)*

Another important result in the theory of entire functions was obtained by Picard in 1879. It describes the set of values of an arbitrary entire function. Take again polynomials as a starting point. Clearly, for an arbitrary polynomial $p(z)$ which is not constant and for an arbitrary complex number $A$ there exists at least one $z$ for which $p(z) = A$. Picard's theorem is an extension of this elementary fact to entire functions: an entire function which is not constant takes every value an infinite number of times, with one possible exception. The result is precise as the entire function $e^z \neq 0$ shows. A crucial role in Picard's proof was played by one of the so-called modular functions, introduced and studied by Hermite in his work on elliptic functions. But the use of a modular function, a notion taken from a different domain, seemed unnatural and the problem to avoid it in the proof of Picard's theorem attracted general attention.

**Hadamard's memoir**. In 1892 Hadamard was awarded a *Grand Prix de l'Académie des Sciences* for his work *Étude sur les propriétes des fonctions entières et en particulier d'une fonction considérée par Riemann* [I.15], which is closely connected with both Weierstrass factorization formula and the just formulated theorems of Poincaré and Picard.[6]

The memoir is divided into three parts. The first one is mostly devoted to the relations between the rate of growth of the maximum modulus $M(r)$ of the entire function $f(z)$ on the circle $|z| = r$ and the law of decrease of the coefficients $c_n$. At the beginning Hadamard found a majorant for $M(r)$ described in terms of the sequence $\{c_n\}$. He noted, for example, that if

$$|c_n| < (n!)^{-1/\alpha} \ , \ \ \alpha > 0, \tag{9.12}$$

then

$$M(r) < e^{Hr^\alpha} \tag{9.13}$$

for some constant $H$. Then, he considered the inverse problem to find the law of decrease of the coefficients departing from the law of growth of the function. The problem was treated by Poincaré for functions satisfying (9.13), and Hadamard modified Poincaré's method in order to include functions satisfying $M(r) < e^{V(r)}$, where $V$ is an arbitrary positive increasing unbounded function.

At the end of the first part of [I.15], Hadamard turned to the above-mentioned theorem of Picard and proved it without the use of modular functions. However, he was able to do this only for a class of entire functions satisfying condition (9.13). In 1896 Borel gave the first proof of Picard's theorem, independent of modular functions and valid for all entire functions.

Picard's theorem is concerned with the problem of the distribution of values of entire functions. The theory of the distribution of values of meromorphic functions began to develop only in the 1920s, but its starting point was the formula due to Jensen

$$\log \frac{r^n |f(0)|}{|z_1 \ldots z_n|} = \frac{1}{2\pi} \int_0^{2\pi} \log |f(re^{i\theta})| d\theta,$$

where $z_1, \ldots z_n, \ldots$ are the zeros of $f(z)$, $|z_1| \leq |z_2| \leq \ldots$, and $|z_n| \leq r \leq |z_{n+1}|$ ([III.196], 1897). In his book [I.372, pp. 50-51], Hadamard writes the following about a formula he obtained at the very beginning of his research: "I decided not to publish it and to wait till I could deduce some significant consequences from it. At that time, all my thoughts, like many other analysts', were concentrated on one question, the proof of the celebrated

---

[6]A story of this contest is given in the section *The first mathematical triumph*.

'Picard's theorem'. Now, my formula most obviously gave one of the chief results which I found four years later by a much more complicated way: a thing which I was never aware of until years after, when Jensen published that formula and noted, as an evident consequence, the results which, happily for my self esteem, I had obtained in the meanwhile".

*Ludvig Jensen (1859-1925)*

In the second part of the memoir Hadamard considers the question converse to that treated by Poincaré: what information on the distribution of zeros of an entire function can be derived from the law of decrease of its coefficients? He shows in particular, that the genus of the entire function is equal to the integer part $[\lambda]$ of $\lambda$ provided $|c_n|(n!)^{-1/\lambda}$ tends to zero and $\lambda$ is not integer. From this statement one concludes, in particular, that a function has zero genus if (9.13) holds with $\alpha < 1.$[7] Hadamard's theorem is less precise for the case when $\lambda$ is integer: the function may be either of genus $\lambda$ or $\lambda + 1$.

Hadamard's result was improved by Borel [III.49], who used two important characteristics of an entire function: the order (he called it *ordre*

---

[7]This remark was used by Poincaré in the section "Application of Hadamard's theorem" of his treatise on celestial mechanics [III.318, v.2].

*apparent*) and the exponent of convergence of zeros (*ordre* in his terminology). The order is the upper lower bound $\rho$ of the numbers $\alpha$ such that inequality (9.13) is valid. Its explicit representation is given by

$$\rho = \limsup \frac{\log\log M(r)}{\log r}.$$

For a polynomial we have $\rho = 0$, and for $e^z$, $\sin z$, $e^{e^z}$ the order is $1, 1, \infty$ respectively. The exponent of convergence of the zeros $\mu$ is defined as the upper lower bound of those $\nu$ for which the series $\sum |z_n|^{-\nu}$ converges. While the order characterizes the maximal possible growth of the function, the exponent of convergence $\mu$ is an indicator of the density of the distribution of the zeros of $f(z)$. One can check that the exponent of convergence is given by

$$\mu = \limsup \frac{\log n}{\log|z_n|}.$$

We formulate Hadamard's refinement of Weierstrass' formula (1.3) by using the notion of order.

**Theorem** (Hadamard's factorization theorem). *If $f$ is an entire function of finite order $\rho$, then the entire function $Q(z)$ in (1.3) is a polynomial of degree not higher than $[\rho]$.*

Borel obtained a sort of converse to this result by showing how the order can be found from the factorization formula. His theorem runs as follows. *Let $\mu < \infty$ and let $Q$ be a polynomial of degree $q$ in (1.3), then $f(z)$ is a function of order $p = \max\{\mu, q\}$.*

In the final, third part of his memoir [I.15], Hadamard applied his result on the genus of an entire function, proved in the first part of his memoir to the famous Riemann zeta function. This function plays a fundamental role in number theory. It is defined and regular on the whole of the complex plane, except at the point $z = 1$, where it has a pole of order one. In the half-plane $\Re z > 1$ the function $\zeta$ is given by the series

$$\sum_{n\geq 1} n^{-z}.$$

The problem of the distribution of the prime numbers is connected with the zeros of the function $\zeta(z)$, more precisely, with those zeros which are complex and are in the "critical strip" $0 \leq \Re z \leq 1$ (outside this strip, $\zeta(z)$ has only real zeros $-2, -4, \ldots$ but they are of no particular importance in the theory). All this will be discussed in Chapter 10 on number theory, and for the present we limit ourselves to explain how Hadamard used the above results to fill in a gap in Riemann's paper [III.342].

Riemann reduced the study of the zeta function to that of the even entire function $\xi$ defined by

$$\xi(z) = \frac{z(z-1)}{2\pi^{z/2}} \Gamma\left(\frac{z}{2}\right) \zeta(z). \tag{9.14}$$

Writing $\xi$ as the series

$$\xi(z) = b_0 + b_2 z^2 + b_4 z^4 + \dots$$

Hadamard proved the inequality

$$|b_m| < (m!)^{-1/2-\epsilon}, \ \epsilon > 0,$$

thus verifying (9.12) with $1/\alpha = 1/2 + \epsilon$. So the genus of $\xi$ as a function of $z^2$ is equal to zero and

$$\xi(z) = \xi(0) \prod_{k=1}^{\infty} \left(1 - \frac{z^2}{\alpha_k^2}\right),$$

where the $\alpha_k$ are the zeros of $\xi$. This property was given in Riemann's paper [III.342] without a rigorous proof.

## 9.3 Other results on analytic functions

**Dirichlet series.** The study of the Riemann zeta function led Hadamard to the study of series of the form

$$\sum_{n \geq 1} c_n n^{-z}$$

with complex coefficients $c_n$, as well as their generalizations

$$\sum_{n \geq 1} c_n e^{-\lambda_n z}, \tag{9.15}$$

where $\{\lambda_n\}$ is an increasing sequence of positive numbers satisfying the condition

$$\limsup \frac{\log n}{\lambda_n} < \infty.$$

The domain of convergence of any series of the form (9.15) is a certain half-plane $\Re z > a$. However, generally speaking, these half-planes are different for uniform, pointwise and absolute convergence. This constitutes one of the difficulties in the study of such series.

In his paper [I.146] on Dirichlet series, Hadamard investigated necessary and sufficient conditions for the representation of a function $f(z)$, which is regular in some half-plane, by a series of the form (9.15). These conditions were originally formulated in an article by Cahen in 1896, but Landau, a great master in finding errors in the arguments of his fellow mathematicians, showed in [III.222] that Cahen's proof contained a mistake. In [I.146], Hadamard attempted to correct Cahen's proof, but this also contained an inaccuracy which was pointed out by Landau in a letter to Hadamard. Finally, in his note *Rectification à la note 'Sur les séries de Dirichlet'* [I.147], Hadamard accepted Landau's criticism and gave correct necessary and sufficient conditions.

In the article [I.71], Hadamard found various relations between mean values of functions represented by Dirichlet series. In particular, he introduced the mean value

$$\lim_{T \to \infty} \frac{1}{2T} \int_{-T}^{T} |f(\sigma + it)|^2 dt$$

for absolutely convergent Dirichlet series $f(z)$. Such mean values play an important role in the theory of almost periodic functions, developed later by Harald Bohr.

**The real part theorem**. The real part theorem is an inequality which plays an important role in the factorization of entire functions. First obtained by Hadamard in 1892 [I.14], it gives an estimate of the maximum modulus of an analytic function in terms of its real part. We now give a proof of this result, close to that proposed by Hadamard, but with a better constant. Put

$$f(z) = \sum_{n \geq 1} a_n z^n$$

which means that we assume $f(0) = 0$. Define the maximum modulus

$$M(\rho) = \max_{|z|=\rho} |f(z)|$$

and the maximum term $\mu(\rho) = \max_n |a_n| \rho^n$ of $f(z)$. Clearly,

$$M(r) \leq \mu(R) \sum_{n \geq 1} \left(\frac{r}{R}\right)^n = \frac{r}{R - r} \, \mu(R). \tag{9.16}$$

Putting $u = \Re f$, one verifies that

$$a_n R^n = \frac{1}{\pi} \int_0^{2\pi} u(R, \theta) e^{in\theta} d\theta.$$

Since

$$\int_0^{2\pi} u(R, \theta) d\theta = 0, \quad \int_0^{2\pi} u(R, \theta) \sin n\theta \, d\theta = 0,$$

we obtain

$$|a_n| R^n = \max_{\phi \in [0, 2\pi]} \int_0^{2\pi} (1 + \cos(n\theta - \phi)) u(R, \theta) d\theta.$$

Using the identity

$$\frac{1}{\pi} \int_0^{2\pi} (1 + \cos(n\theta - \phi)) d\theta = 2,$$

we arrive at the inequality

$$\mu(R) \le 2 \max_{\theta \in [0, 2\pi]} u(R, \theta),$$

which, together with (9.16), yields the estimate for the maximum modulus

$$M(r) \le \frac{2r}{R - r} \max_{|z|=R} \Re f,$$

called the real part theorem.[8] It was used by Borel in his proof of Picard's theorem independent of modular functions [III.48].

**The three circles theorem.** We finish this chapter with another frequently cited inequality for analytic functions obtained by Hadamard in 1896. Let $f(z)$ be regular in the annulus $R_1 < |z| < R_2$ and let, as before, $M(r)$ be the maximum of $|f(z)|$ on the circle $|z| = r$, $R_1 < r < R_2$. In the short article [I.40] Hadamard claimed that $\log M(r)$ is a convex function of $\log r$, but he never published a proof. After the works of Landau, Hardy, and Harald Bohr it is referred to as Hadamard's three circles theorem, since it is often formulated in the form of the inequality

$$M(r_2)^{\log(r_3/r_1)} \le M(r_1)^{\log(r_3/r_2)} M(r_3)^{\log(r_2/r_1)},$$

where $R_1 < r_1 < r_2 < r_3 < R_2$.

---

[8]A different proof of this inequality can be found in [III.398, p. 175].

This result has important analogies. Hardy's theorem (1915) is of particular interest. He proved that the logarithm of the mean value

$$M_p(r) = \left( \frac{1}{2\pi} \int_0^{2\pi} |f(re^{i\phi})|^p d\phi \right)^{1/p}, \quad p > 0,$$

is an increasing function of $r$, which is convex in $\log r$. This property of $M_p$ is fundamental for the theory of Hardy spaces $H^p$ of analytic functions on the unit disc, subject to the condition

$$\sup_{r<1} M_p(r) < \infty.$$

Since

$$\lim_{p \to \infty} M_p(r) = M(r),$$

we see that Hadamard's result is a limiting case of Hardy's result.

Another generalization of Hadamard's three circles theorem is connected with the notion of subharmonicity, which can be considered as an extension of the notion of convexity to the $n$-dimensional space $\mathbf{R}^n$. Let $D$ be a domain in $\mathbf{R}^n$. A continuous function $f(P)$ given on $D$ is said to be subharmonic on $D$ if it satisfies the following condition: for any domain $G \subset D$ with boundary $\partial G$ the solution of the Dirichlet problem

$$\Delta u = 0, \quad u\big|_{\partial G} = f$$

satisfies the inequality $u \leq f$, i.e. there is a harmonic function $u$ whose graph is below that of the function $f$. In the one-dimensional case, the Laplace equation $\Delta u = 0$ takes the form $d^2u/dx^2 = 0$ and its solution is the linear function $u = kx + b$. The Dirichlet problem is reduced to drawing a straight line through two points lying on a curve. Thus, subharmonicity generalizes the notion of a convex curve.

The $n$-dimensional analogue of Hadamard's result is the following theorem proved by Montel in 1928. Let $f(P)$ be a subharmonic function in the layer $\{P \in \mathbf{R}^n : R_1 < |P| < R_2\}$. Then the function

$$\log(\max_{|P|=r} |f(P)|)$$

is convex on $(R_1, R_2)$ with respect to $\log r$ for $n = 2$ and convex with respect to $r^{2-n}$ for $n > 2$.

Extensions of the theorems of Hadamard, Hardy, and Montel for solutions of second order elliptic and parabolic partial differential equations (three balls, three cylinders, three curves theorems, and so on) can be found in the book by Protter and Weinberger [III.333] and in the paper [III.226, p. 201-204]. A variant of Hadamard's theorem, the so-called three lines theorem, forms the basis of the interpolation theorem of M. Riesz and G. Thorin for linear operators in $L^p$ [III.430]. The powerful method of complex interpolation, which is based on the Riesz-Thorin theorem, plays an important role in operator theory and analysis [III.400].

# Chapter 10

# Number Theory

## 10.1　The distribution of prime numbers

**Introduction.** One of the triumphs of the theory of analytic functions in the nineteenth century was the proof of the prime number theorem in 1896, due to Hadamard and de la Vallée-Poussin, who was a professor at the University of Louvain. This theorem has a special place in analytic number theory, a discipline dealing with number-theoretical problems using the techniques of mathematical analysis.

It was already known to Euclid that the number of primes is infinite. The ancient Greeks also knew that all the prime numbers

$$2, 3, 5, 7, 11, 13, 17, 19, 23, 29, \ldots \tag{10.1}$$

can be obtained using the following algorithm, known as the sieve of Eratosthenes. The first step consists in deleting 1 from the normal sequence of natural numbers, so that 2 becomes the first number in (10.1). Suppose now that the sequence of primes $p_1 = 2, p_2 = 3, \ldots, p_i$ has been found. Starting with $p_i$, we delete all subsequent natural numbers which are divisible by all $p_j$, $j \leq i$. Then the first number not divisible by the $p_j$, $j \leq i$, is the prime $p_{i+1}$.

Up to the eighteenth century, no regularity in the sequence (10.1) had been found. Euler wrote, in 1747: "Until now, mathematicians have tried in vain to discover any order in the sequence of prime numbers, and therefore they believed that it is a mystery which the human mind will never be able to penetrate. In order to convince oneself, one only needs to look at the table of primes, which many mathematicians made great efforts to extend beyond 100000. From this table, one can see that there is no law governing them"[III.125, p. 639]. The use of computers has made it possible to construct a table of prime numbers which extends to unbelievably large numbers, but this only confirms the impression of their chaotic distribution. Nevertheless, one can easily see that the density of prime numbers decreases. They are 40% of the first ten integers, 25% of the first hundred, about 17% of the first thousand, and almost 8% of the first million. On the one hand, there exist arbitrarily long intervals among the natural numbers where there are no primes. For instance, it is clear that none of the numbers $n! + 2, n! + 3, \ldots, n! + n$ are primes. On the other hand, one can meet so-called twins in the sequence (10.1), that is primes whose difference is equal to 2. For example, 3 and 5, 5 and 7, 11 and 13, 17 and 19 are twins. Moreover, such large primes as 8004119 and 8004121 are also twins. It is still not known whether the number of twins is infinite.

Let $\pi(x)$ denote the number of primes among the first $x$ natural numbers. Euclid's theorem only states that $\pi(x) \to \infty$ as $x \to \infty$. What can one say about the behaviour of $\pi(x)$ as $x \to \infty$? This problem, which occupied many famous mathematicians, became classic. After Hadamard and de la Vallée-Poussin simultaneously and independently justified the principal term in the asymptotic law of $\pi(x)$, other mathematicians contributed using various methods. Subsequent studies showed how wide is the range of theories connected with this problem: Tauberian theorems, harmonic analysis, Dirichlet series, to name only a few. In a rough form the final result is simple to formulate:

$$\pi(x) \sim \frac{x}{\log x}, \qquad (10.2)$$

where $u(x) \sim v(x)$ means that $\lim u(x)/v(x) = 1$ when $x \to \infty$.

**History.** We now trace the development of ideas of Hadamard's and de la Vallée-Poussin's predecessors. Let $n = 1, 2, \ldots$ and $p_1, p_2, \ldots$ be the sequences of natural and prime numbers in their usual order. In Euler's enormous legacy is the following identity, which he obtained in 1737:

$$\sum_{n=1}^{\infty} \frac{1}{n^s} = \prod_{i=1}^{\infty} (1 - \frac{1}{p_i^s})^{-1}, \qquad s > 1. \qquad (10.3)$$

This follows directly from the fact that the right-hand side in (10.3) is equal to

$$\prod_{i=1}^{\infty} (1 + \frac{1}{p_i^s} + \frac{1}{p_i^{2s}} + \ldots)$$

and from the representation of any natural number $n$ as a product $p_1^{k_1} \ldots p_m^{k_m}$ where $p_1, \ldots, p_m$ are different prime numbers, and $k_1, \ldots, k_m$ are positive integers.

The left-hand side of (10.3) is Riemann's zeta function which we have already met in Sections 1.10 and 9.2. From (10.3) it follows that

$$\frac{1}{\zeta(s)} = \prod_{i=1}^{\infty} (1 - \frac{1}{p_i^s}),$$

which, after expansion of the right-hand side, gives

$$\frac{1}{\zeta(s)} = \sum_{n=1}^{\infty} \frac{\mu(n)}{n^s}, \tag{10.4}$$

where $\mu$ is the Möbius function defined as follows: $\mu(1) = 1$, $\mu(n) = (-1)^k$ if $n$ is the product of $k$ different primes, and $\mu(n) = 0$ if $n$ contains any prime factor to a power higher than one.

From (10.3) Euler deduced the divergence of the series

$$\sum_{i=1}^{\infty} \frac{1}{p_i}. \tag{10.5}$$

Indeed, from (10.3) one finds

$$\log \sum_{n=1}^{\infty} \frac{1}{n^s} = -\sum_{i=1}^{\infty} \log(1 - \frac{1}{p_i^s}) = \sum_{i=1}^{\infty} (p_i^{-s} + \frac{p_i^{-2s}}{2} + \frac{p_i^{-3s}}{3} + \ldots).$$

As $s$ tends to 1, the left-hand side becomes infinite and the right-hand side is the sum of a convergent series and the series (10.5). The divergence of the sum of the reciprocals of the primes shows that the primes are distributed quite densely among the natural numbers, for instance more densely than the squares of natural numbers for which the sum of reciprocals gives a convergent series.

In 1798, while analysing the table of prime numbers, Legendre made the conjecture that for large $x$ the function $\pi(x)$ can be replaced by

$$\frac{x}{A \log x + B}$$

to a high degree of accuracy, where $A$ and $B$ are some constants. In 1808, in his monograph [III.232], Legendre formulates his conjecture in the form

$$\pi(x) \approx \frac{x}{\log x - 1.08366}. \tag{10.6}$$

*Carl Friedrich Gauss (1777-1855)*

Another empirical formula for $\pi(x)$ was obtained by Gauss, who was interested in the asymptotic law all throughout his life from when he was a youth. In his old age he said he liked to spend a quarter of an hour each day thinking about it. He never published any results on it, but in a letter to Encke, dated December 24, 1849 ([III.147, p.444-447]), he wrote that, while considering the table of primes during 1792-1793, he obtained convincing evidence for the approximate formula

$$\pi(x) \approx \mathrm{li}(x) = \int_2^x \frac{dt}{\log t}, \tag{10.7}$$

which implies (10.2).

Both conjectures (10.6) and (10.7) were analysed by Chebyshev in his memoir presented to the St. Petersburg Academy in 1848. His first result states that, for any natural number $n$, the sum

$$\sum_{x=2}^{\infty} \left(\pi(x+1) - \pi(x) - \frac{1}{\log x}\right) \frac{(\log x)^n}{x^{1+\rho}}$$

.

tends to a finite limit as $\rho$ tends to 0+.

From this result, Chebyshev deduces that if it exists, the limit of the ratio $\pi(x)/\mathrm{li}(x)$, as $x$ tends to infinity, should be equal to 1. He then concludes that the expression

$$\frac{x}{\pi(x)} - \log x$$

can only tend to $-1$, and thus he disproved Legendre's conjecture (10.6).

*P.L. Chebyshev (1821-1894)*

Later, in his paper *Sur les nombres premiers* [III.393], Chebyshev showed the validity of the inequalities (2.4) for all sufficiently large $x$. He introduced the function

$$\psi(x) = \sum_{p^m \le x} \log p,$$

where the sum is taken over all natural numbers $m$ and over all primes $p$ for which $p^m \leq x$ for each $m$, and based his proof on the identity

$$\log([x]!) = \sum_{n \leq x} \psi(\frac{x}{n}).$$

He further proved that, as $x$ tends to infinity,

$$\frac{\pi(x) \log x}{x}, \quad \text{and} \quad \frac{\psi(x)}{x}$$

have the same upper and lower limits. Thus, if one of the quotients has a limit, the same is true of the other, and so the two limits are equal. This result plays a significant role in subsequent proofs of the prime number theorem. Accordingly, the asymptotic law is often written in the form

$$\psi(x) \sim x. \tag{10.8}$$

*Joseph Bertrand (1822-1900)*

Another Chebyshev's achievement was a proof of the so-called Bertrand's postulate which claims that there is a prime number on any interval $(x, 2x)$, $x \geq 2$.

The inequalities (2.4) were sharpened by Sylvester in 1881 [III.382]. He showed that, for all sufficiently large $x$, the value $\pi(x)$ lies in a smaller interval, but the question of making this interval as small as possible remained open. Sylvester writes about this at the end of his article: "But to pronounce with certainty upon the existence of such possibility, we shall probably have to wait until someone is born into the world as far surpassing Tchebycheff in insight and penetration as Tchebycheff has proved himself superior in these qualities to the ordinary run of mankind". Commenting on this, Landau writes ([III.223, vol. 1, p. 44]): "Jacques Hadamard had already been born when Sylvester wrote these lines. He succeeded in achieving this goal, and by a way which had been used long ago for related, but different purposes, by Riemann".

*James Joseph Sylvester (1814-1897)*

**Riemann's work.** In 1859, Riemann published his paper *Über die Anzahl der Primzahlen unter einer gegebenen Grösse* [III.342] (see also [III.344]). This small article (it is only eight pages long) turned out to be prophetic for analytic number theory and function theory. The aim of this work was to derive a representation for $\pi(x)$ in the form of a series with principal term $\mathrm{li}(x)$. Unfortunately, one can not say that this goal was achieved: Riemann's formula was not completely proved; there was also no proof that $\mathrm{li}(x)$ is the principal term of the asymptotic formula for $\pi(x)$ as $x$ tends to infinity. However, Riemann's methods had a strong influence on further attempts to justify the prime number theorem. The principal innovation was to consider the function $\zeta(s)$ for complex values of $s = \sigma + it$. In doing this, he made the methods of analytic function theory, such as contour integration, analytic extension, residue theory and later, through the works of Hadamard and others, the theory of entire functions into a tool for number theory. Since the series $\sum n^{-s}$ converges for $\sigma > 1$ and each of its terms is a regular function in this half-plane, it follows by Weierstrass' theorem on the convergence of series of regular functions, that the sum $\zeta(s)$ of this series is regular for $\sigma > 1$. Riemann's idea leads to the extension of Euler's identity (10.3) for $s$ in the half-plane $\sigma > 1$.

Riemann proved that the function $\zeta(s)$ can be extended analytically to a function which is meromorphic in the whole of the $s$-plane, its only singularities being a simple pole $s = 1$ with residue equal to 1, that is

$$\zeta(s) = \frac{1}{s-1} + a_0 + a_1(s-1) + a_2(s-1)^2 + \ldots \qquad (10.9)$$

He also showed that $\zeta(s)$ satisfies the functional equation

$$\frac{\zeta(1-s)}{\zeta(s)} = 2\frac{\Gamma(s)}{(2\pi)^s}\cos\frac{s\pi}{2}. \qquad (10.10)$$

Hence the function $\xi$, introduced by (9.14), is entire and satisfies

$$\xi(\frac{1}{2} + it) = \xi(\frac{1}{2} - it).$$

In other words, the function $\xi(\frac{1}{2} + it)$ is an even entire function of $t$, which is real-valued on the real axis. This means that $-t_0$, $\overline{t_0}$, and $-\overline{t_0}$ are zeros of the function $\xi(\frac{1}{2} + it)$ whenever $t_0$ is a zero ; that is, its roots are symmetrical with respect to both coordinate axes.

Riemann understood that the complex roots of the zeta function play a special role in the problem of the distribution of the prime numbers. Since none of the factors in the right-hand side of (10.3) vanishes, and because the product converges for $\sigma > 1$, then $\zeta(s) \neq 0$ for $\sigma > 1$. From the equality (10.10) and the properties of the $\Gamma$-function, it follows that in the half-plane

$\sigma < 0$ the zeta function has only simple zeros at the real points $-2, -4, \ldots$, which are called trivial zeros. The identity

$$(1 - 2^{1-s})\zeta(s) = 1 - 2^{-s} + 3^{-s} - \cdots$$

immediately implies that $\zeta(s) \neq 0$ for $s \in (0, 1)$. Hence, non-trivial zeros of $\zeta(s)$ are complex, and all lie in the strip $0 \le \sigma \le 1$, which is called the critical strip. From what we have said above about the zeros of the function $\xi(\frac{1}{2} + it)$, one obtains that the complex roots of the zeta function are symmetrical about the real axis $t = 0$ and the vertical line $\sigma = \frac{1}{2}$.

Riemann made a precise conjecture. This is the famous *Riemann's hypothesis*: all the zeros of the zeta function contained in the critical strip are placed symmetrically on the line $\sigma = \frac{1}{2}$. An interesting discussion of heuristic arguments which could have led Riemann to this statement can be found in Edwards' book [III.118, p. 164].

Proving Riemann's hypothesis is an unsolved problem with which many famous mathematicians of the twentieth century have been involved. According to a well-known anecdote about Hilbert, in answer to the question "What would you ask if you were to rise from the dead in three hundred years' time?" he replied, "Has Riemann's hypothesis been proved?".

Riemann showed that the number $N(T)$ of zeros of the zeta function in the rectangle $0 < \sigma < 1$, $0 < t < T$ tends to infinity as $T$ tends to infinity. After his death, the following conjecture was found in his papers (see [III.344])

$$N(T) = \frac{T}{2\pi} \log \frac{T}{2\pi} - \frac{T}{2\pi} + O(\log T). \qquad (10.11)$$

For more details about Riemann's work and the development of his ideas, see [III.192], [III.397], [III.118].

**Hadamard's proof.** More than thirty years passed after Riemann's article until, in 1896, Hadamard and independently de la Vallée-Poussin found the first proofs of the prime number theorem. We describe here the main steps of Hadamard's proof.

**I.** Using the results on entire functions obtained in his memoir [I.15], Hadamard showed that the genus of the entire function $(s - 1)\zeta(s)$ is equal to 1 and therefore its Weierstrass product has the following form in the whole of the complex plane

$$\zeta(s) = \frac{1}{s - 1} e^{As + B} \prod_{\rho} (1 - \frac{s}{\rho}) e^{s/\rho}, \qquad (10.12)$$

where the product is taken over all the zeros (real and complex) of the zeta function.

ANALYSE MATHÉMATIQUE. — *Sur les zéros de la fonction $\zeta(s)$ de Riemann.*
Note de M. HADAMARD, présentée par M. Appell.

« On sait que la fonction $\zeta(s)$ ne s'annule pour aucune valeur de $s$ ayant sa partie réelle supérieure à $1$, ainsi que cela se voit par l'expression

(1)                    $\log\zeta(s) = -\sum_p \log(1 - p^{-s})$      (logarithmes naturels),

où $p$ représente successivement tous les nombres premiers. Stieltjes avait démontré que tous les zéros imaginaires de $\zeta(s)$ sont (conformément aux prévisions de Riemann) de la forme $\frac{1}{2} + ti$, $t$ étant réel; mais sa démonstration n'a jamais été publiée. Je me propose simplement de faire voir que $\zeta(s)$ ne saurait avoir de zéro dont la partie réelle soit *égale* à $1$.

» Pour cela, remarquons d'abord que $\zeta(s)$ admet pour pôle simple le point $s = 1$. L'expression (1) pouvant, à une quantité finie près, se remplacer par $S = \sum_p p^{-s}$, nous voyons que celle-ci ne diffère de $-\log(s - 1)$

que par une une quantité qui reste finie lorsque $s$ tend vers $1$.

» Remplaçons maintenant $s$ par $s + ti$. Si le point $s = 1 + ti$ était un zéro de $\zeta(s)$, la partie réelle de $\log\zeta(s + ti)$, c'est-à-dire à une quantité finie près, l'expression

(2)                        $P = \sum_p p^{-s}\cos(t \log p)$

devrait augmenter indéfiniment par valeurs négatives comme $\log(s - 1)$, c'est-à-dire comme $-S$, lorsque $s$ tendrait vers $1$ par valeurs supérieures, $t$ restant fixe. Soit alors $\alpha$ un angle choisi aussi petit qu'on le veut. Dans les sommes $S_n$, $P_n$, formées respectivement avec les $n$ premiers termes des séries S, P, distinguons deux parties : 1° les termes correspondant aux nombres $p$ qui vérifient l'une des doubles inégalités

(3)            $\frac{(2k+1)\pi - \alpha}{t} < \log p < \frac{(2k+1)\pi + \alpha}{t}$,      $(k = 1, 2, ..., \infty)$,

ces termes donneront, dans les sommes $S_n$, $P_n$, les parties $S'_n$, $P'_n$;

*The first page of Hadamard's note on the zeros of $\zeta(s)$. He writes: "Stiltjes has proved that all the zeros of $\zeta(s)$ are (in accordance with Riemann's prediction) of the form $\frac{1}{2} + ti$, $t$ being real; but his proof has never been published. I simply intend to show that $\zeta(s)$ should not have had zeros with real part equal to $1$".*

**II.** One may expect that the line $\sigma = 1$, which is the boundary of the half-plane in which $\zeta(s)$ can be expressed as the series $\sum n^{-s}$, plays a special role in this theory. Hadamard proved that the zeta function does not vanish on this line, that is

$$\zeta(1 + it) \neq 0. \qquad (10.13)$$

It later turned out that this assertion is equivalent to the prime number theorem (see [III.192, Ch. 2]).

We roughly outline Hadamard's proof of (10.13) following the book [III.397]. The starting point is the identity

$$\frac{\zeta'(s)}{\zeta(s)} = -\sum_{m=1}^{\infty} \sum_p \frac{\log p}{p^{ms}}, \quad \sigma > 1, \qquad (10.14)$$

obtained from (10.3). Equality (10.14) implies

$$\log \zeta(s) = \sum_m \sum_p \frac{1}{m\, p^{ms}},$$

where $p$ runs through all the primes, and $m$ through all the positive integers. Therefore the difference

$$\log \zeta(s) - \sum_p \frac{1}{p^s} \tag{10.15}$$

is regular for $\sigma > 1/2$. Since $\zeta(s)$ has a simple pole at $s = 1$ then

$$\sum_p \frac{1}{p^\sigma} \sim \log \frac{1}{\sigma - 1} \quad \text{as} \quad \sigma \to 1+. \tag{10.16}$$

By the identity (10.4) with $\Re s > 1$ it follows that

$$\frac{1}{|\zeta(s)|} \leq \sum_{n=1}^{\infty} \frac{1}{n^\sigma} = \zeta(\sigma) < \frac{\text{const}}{\sigma - 1}$$

for $\sigma$ near 1, that is

$$|\zeta(s)| > \text{const}(\sigma - 1).$$

Assume now that $\zeta(1 + it) = 0$, then for $s = \sigma + it$ we have

$$\log |\zeta(s)| \sim \log(\sigma - 1) \quad \text{as} \quad \sigma \to 1+.$$

Hence, by taking the real part of (10.15), we arrive at the asymptotic relation

$$\sum_p \frac{\cos(t \log p)}{p^\sigma} \sim \log(\sigma - 1). \tag{10.17}$$

Comparing (10.16) and (10.17) heuristically suggests that $\cos(t \log p)$ is close to $-1$ for most values of $p$. Then $\cos(2t \log p) \sim 1$ for the same $p$'s, and therefore

$$\log |\zeta(\sigma + 2it)| \sim \sum_p \frac{\cos(2t \log p)}{p^\sigma} \sim \log \frac{1}{\sigma - 1},$$

which implies that $\zeta$ has the pole $1 + 2it$. This is impossible since $\zeta$ has only one pole at $s = 1$, so we arrive at (10.13).

**III**. Next, Hadamard established the asymptotic relation

$$\sum_{p<x} \log p \log \frac{x}{p} \sim x \qquad \text{as} \quad x \to \infty. \qquad (10.18)$$

He first considered the integral

$$J(x) = -\frac{1}{2\pi i} \int_{a-i\infty}^{a+i\infty} s^{-2} \frac{\zeta'(s)}{\zeta(s)} x^s ds, \qquad (10.19)$$

where $a$ is an arbitrary number larger than 1. (Since the integrand is a regular function for $\sigma > 1$, there is no necessity to fix the value of $a$.) Using (10.14) and the easily verified formula

$$\frac{1}{2\pi i} \int_{a-i\infty}^{a+i\infty} s^{-2} x^s ds = \begin{cases} 0, & \text{if } x < 1, \\ \log x, & \text{if } x > 1. \end{cases}$$

Hadamard arrived at the identity

$$J(x) = \sum_k \sum_p \log p \log(\frac{x}{p^k}),$$

where $k$ ranges over all positive integers and $p$ ranges over all primes with $p^k < x$. One can easily check that the number of terms for which $k > 1$ is $o(x)$ as $x$ tends to infinity. Indeed, $2^k < p^k < x$ implies $k < \log x / \log 2$. Moreover

$$\sum_{p<x^{1/k}} \log p \log(\frac{x}{p^k}) \le \log x \log( \prod_{p<x^{1/k}} p)$$

$$\le \log x \log \Gamma(1 + x^{1/k}) \le \text{const } x^{1/k} (\log x)^2$$

and finally

$$J(x) = \sum_{p<x} \log p \log \frac{x}{p} + O(x^{1/2}(\log x)^3). \qquad (10.20)$$

The integral (10.19) is then calculated in a different manner, using the residue theorem and (10.14). Recall that if

$$f(z) = \sum_{-\infty}^{\infty} c_n (z - z_0)^n$$

then the residue $A$ of the function $f(z)$ with respect to the point $z_0$ is the coefficient of $(z - z_0)^{-1}$, i.e., $A = c_{-1}$. Let $f(z)$ be a single-valued function,

regular on the set $\overline{D}$, except perhaps at the points $z_k \in D, \quad k = 1, \ldots, m.$
Then

$$\int_{\partial D} f(z)dz = 2\pi i(A_1 + \ldots + A_m),$$

where $A_k$ is the residue of $f(z)$ at the point $z_k$.

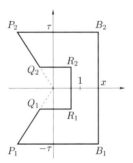

**Figure 2.**

Let $\tau$ be a sufficiently large positive number. Starting with the vertical line segment $B_1 B_2$ where $B_1 = a - i\tau$ and $B_2 = a + i\tau$ Hadamard formed the closed contour $\mathcal{L} = B_1 P_1 Q_1 R_1 R_2 Q_2 P_2 B_2$ (see Figure 2) so that the zeros of the zeta function are not contained in the domain D bounded by $\mathcal{L}$. By the residue theorem we have

$$\int_{a-i\tau}^{a+i\tau} = -\int_{\mathcal{L}} + 2\pi i\{\ldots\},$$

where $\{\ldots\}$ denotes the sum of the residues of the integrand in the domain D. Thus

$$\int_{a-i\infty}^{a+i\infty} = \lim_{\tau \to \infty} [-\int_{\mathcal{L}} + 2\pi i\{\ldots\}].$$

From (10.12) it follows that

$$\frac{\zeta'(s)}{\zeta(s)} = \sum_{\alpha}(\frac{1}{s-\alpha} + \frac{1}{\alpha}) + \sum_{\beta}(\frac{1}{s-\beta} + \frac{1}{\beta}) + C - \frac{1}{s-1},$$

where $C$ is a constant, the first sum is taken over the real zeros $\alpha$ of the zeta function and the second sum is over the complex zeros.

The residue of the integrand in (10.19) corresponding to the pole $s = 1$ of the function $\zeta'/\zeta$ is equal to $x$. The real zeros $\alpha$ lie outside of the domain D and so do not contribute to the sum of the residues. Hadamard further showed, using the methods of his paper on entire functions [I.15], that the integration over the segments $B_1 P_1$ and $B_2 P_2$ in (10.14) gives zero, in the limit $\tau \to \infty$. He then showed that the integral over the broken line $P_1 Q_1 R_1 R_2 Q_2 P_2$ and the contribution of the complex zeros $\beta$, are bounded by $\varepsilon x$, where $\varepsilon$ is an arbitrarily small number. Here it is important that the line $\Re s = 1$ contains no zeros. Therefore

$$J(x) = x + o(x), \tag{10.21}$$

which, together with (10.15), leads to (10.13).

**IV**. It is now easy to show that

$$\sum_{p<x} \log p \sim x. \tag{10.22}$$

In fact, by (10.18), we have, for any fixed $h$,

$$\sum_{p<x} \log p \log(1+h) + \sum_{x \le p < x(1+h)} \log p \log \frac{x(1+h)}{p} = x(h + o(x)).$$

Since $x(1+h)/p$ in the second sum lies between 1 and $1+h$, the following inequalities hold

$$\frac{1}{x} \sum_{p<x} \log p < \frac{h + o(x)}{\log(1+h)},$$

$$\frac{1}{x(1+h)} \sum_{p<x(1+h)} \log p > \frac{h + o(x)}{(1+h)\log(1+h)}.$$

Taking the upper and lower bounds and using the arbitrariness of $h$, we arrive at (10.22), and it remains only to use the equivalence of (10.2) and (10.8).

This is the general scheme of Hadamard's proof. At the end of his paper [I.45] he wrote that, while reading the galley proofs, he learned that de la Vallée-Poussin had also proved the prime number theorem (see [III.405]) and added "I believe that no one will deny that my method has the advantage of simplicity". Indeed, as has been noted in the literature (see for instance [III.192, p. 42]), Hadamard's proof is the simpler.

**Further results.** De la Vallée-Poussin went further in one important point: in 1899 he showed [III.406] that

$$\pi(x) = \text{li}(x) + O(xe^{-c\sqrt{\log x}}) \tag{10.23}$$

as $x$ tends to infinity, where $c$ is some positive constant. This result was the starting point for a succession of estimates for the behaviour of the difference $\pi(x) - \text{li}(x)$ obtained by other authors. A detailed analysis of the proofs of Hadamard and de la Vallée-Poussin theorem on the distribution of the primes is given in Chapter 4 of the book by Edwards [III.118].

An amusing remark concerning both these mathematicians can be found in the chapter on number theory in Dieudonné's book [III.105]. The authors of this chapter, William and Fern Ellison, commenting on the longevity of both mathematicians, wrote: "The fact that it took more than a hundred years to find a proof of the theorem on prime numbers had established the belief that those who discovered it would become immortal. This legend seemed to be true for a long time; unfortunately, it was undermined in 1962 when de la Vallée-Poussin died, aged 96, and completely destroyed in 1963, when Hadamard died at the age of 98!".

After 1896, the parallel work of Hadamard and de la Vallée-Poussin continued for some time in their studies of Dirichlet series. To de la Vallée-Poussin are also due many important results in potential theory, ordinary differential equations, approximation theory (including harmonic analysis), and descriptive and metric function theory.

All the above mentioned mathematicians, beginning with Euler, who worked on the problem of the distribution of the primes were also famous in other areas of mathematics. We mention another name, von Mangoldt, who was not an exception. He turned to analytic number theory when he had already been working for some time in analytic function theory. In 1895, in his long article [III.266], von Mangoldt gave a proof of formula (10.11) on the density of the distribution of the zeros of the zeta function.

In the celebrated talk by Hilbert at the Second International Congress of Mathematicians in Paris, in 1900, which influenced the development of twentieth-century mathematics so much, the problem of the distribution of the prime numbers was also considered. He said, in the eighth part of that talk: "The theory of the distribution of the prime numbers has quite recently been given a crucial stimulus by the works of Hadamard, de la Vallée-Poussin, von Mangoldt, and others. To solve completely the problem posed in Riemann's note *On the number of primes not exceeding a given value*, it is still necessary to prove Riemann's very important assertion that *the zeros of the function $\zeta(s)$ all have their real part equal to 1/2"* [III.176, p. 85].

A year later, in 1901, von Koch contributed to this problem [III.212]. Let $\tau$ be such that $\zeta(s) \neq 0$ to the right of the vertical line $\sigma = \tau$. Using the results of de la Vallée-Poussin and von Mangoldt, von Koch proved that

*Helge von Koch (1870-1924)*

$$\pi(x) = \mathrm{li}(x) + O(x^\tau \log x).$$

If the Riemann hypothesis is true, then $\tau = 1/2$ is admissible and von Koch's equality takes the form

$$\pi(x) = \mathrm{li}(x) + O(\sqrt{x} \log x).$$

This estimate for $\pi(x) - \mathrm{li}(x)$ is much better than the above estimate of de la Vallée-Poussin and all other estimates known up to this day which were proved without using Riemann's hypothesis.

In 1909 E. Landau published his fundamental treatise on prime numbers [III.223].[1] This work, containing many historical annotations, is an encyclopedia of results on the distribution of prime numbers, obtained up to 1909. He gave two new proofs of the prime number theorem, analysed the proofs known at the time, and showed that they can be simplified.

---

[1] His lectures [III.225] of 1927 contain the results of later research.

*Edmund Landau (1877-1938)*

Simultaneously with Landau, Hardy entered the mathematical scene, to be joined later by Littlewood. Landau's immense critical analysis seemed to exclude any outstanding new results concerning the asymptotic law or the zeta function. It turned out, however, that it was possible. In 1914, Hardy showed in [III.165] that $\zeta(s)$ has infinitely many zeros on the line $\sigma = 1/2$. Moreover, he proved that if $M(T)$ denotes the number of zeros of the zeta function of the form $s = 1/2 + it$ with $|t| \leq T$, then $M(T) > AT$ where $A$ is some positive constant independent of $T$. Selberg [III.360] improved the last estimate in 1942 by showing that

$$M(T) > AT \log T.$$

Considerable progress in the study of the zeta function was achieved by estimating the trigonometric sums

$$\sum_{n=k}^{l} e^{2\pi i f(n)}, \qquad (10.24)$$

where $f$ is a smooth function. In the case of the zeta function

$$f(n) = -\frac{s}{2\pi} \log n.$$

A. Weil and later Hardy and Littlewood were the first to apply estimates for (10.24) to number theory (see [III.397]).

Littlewood obtained the following refinement of the asymptotic formula (10.23) of de la Vallée-Poussin:

$$\pi(x) = \text{li}(x) + O\left(xe^{-c(\log x \log \log x)^{1/2}}\right) \tag{10.25}$$

(see [III.245]).

New proofs of the prime number theorem continued to appear. Wiener contributed to it by giving an argument based upon his Tauberian theorem [III.421]. "Elementary" proofs which did not use complex variable methods were proposed by Erdös and Selberg (see [III.124], [III.361]). However, their method does not give as good estimates for the remainder term in the prime number theorem as methods based upon the study of the zeta function.

Beurling considered the following generalization of the problem [III.34]. By calling the terms of an increasing sequence of real numbers $\{y_k\}_{k \geq 1}, y_1 > 1$, by "primes", he introduced the "integers" $y_{n_1} y_{n_2} \ldots y_{n_\nu}$ and denoted the corresponding counting functions by $\Pi(x)$ and $N(x)$. He extended the above mentioned Wiener's method to prove that the condition

$$N(x) = Ax + O\left(\frac{x}{(\log x)^\gamma}\right), \quad A = \text{const},$$

implies

$$\Pi(x) \sim \frac{x}{\log x}$$

if $\gamma > 3/2$ but does not imply this result if $\gamma = 3/2$.

Important improvements of formula (10.25) were obtained by using a method of estimating the sums (10.24) developed by Vinogradov and simplified by Hua Loo-Keng. The estimate

$$\pi(x) = \text{li}(x) + O\left(xe^{-c(\log x)^{3/5}(\log \log x)^{-1/5}}\right)$$

found in 1958 by Vinogradov and Korobov seems to be the best result of such a kind obtained so far (see [III.121]).

## 10.2   Prime numbers in arithmetical progressions

The methods developed by Hadamard in his paper [I.15] turned out to be useful in the problem of the distribution of prime numbers in arithmetical progressions. The problem goes back to Legendre who conjectured, as early as in 1795, that if $d$ and $l$ are mutually prime numbers, then the arithmetic progression $\{dm + l : m \geq 1\}$ contains infinitely many primes [III.232] (see also [III.417, p. 329]). In 1837, Dirichlet published a proof of this assertion in [III.106], in which an important role was played by the following generalization of the Riemann zeta function

$$L(s, \chi) = \sum_{n=1}^{\infty} \frac{\chi(n)}{n^s}. \tag{10.26}$$

Here, $s$ is a real number, $s > 1$, and the function $\chi(n)$, called the Dirichlet character modulo $l$, has the following properties:

1. $\chi(n) \neq 0$ if and only if $n$ is not divisible by $l$;

2. $\chi(n + l) = \chi(n)$ (periodicity);

3. $\chi(mn) = \chi(m)\chi(n)$ (multiplicativity).

The analytic extension of the series (10.26) defines the Dirichlet $L$-function, which has properties similar to the zeta function. In particular, the analogue of Euler's identity for the $L$-function is

$$\sum_{n=1}^{\infty} \frac{\chi(n)}{n^s} = \prod_{p} (1 - \frac{\chi(p)}{p^s})^{-1}.$$

In his article [I.45], Hadamard remarked that his proof of the absence of zeros of the zeta function on the line $\Re s = 1$ is based only upon the following two simple properties: first, that $\log \zeta(s)$ can be written as a Dirichlet series

$$\sum_{n} a_n e^{-\lambda_n s}$$

with positive coefficients $a_n$, and, second, that the zeta function has only one simple pole, which lies on the line $\Re s = 1$. Hadamard verified that the Dirichlet $L$-function has the same two properties, and thus arrived at the result $L(1 + it, \chi) \neq 0$. He further wrote that his arguments relating to the zeta function can be carried over to the Dirichlet $L$-function, which leads to the relation

$$\sum_{p \leq x, \, p \equiv l \,(\mathrm{mod}\, d)} \log p \sim \frac{x}{\varphi(d)} \quad \text{as} \quad x \to \infty, \tag{10.27}$$

where $\varphi$ is Euler's function, whose value $\varphi(d)$ is equal to the number of positive integers not exceeding $d$ and having no common divisors with $d$. The relation (10.27) leads to the following asymptotic formula for the function $\pi(x; d, l)$ which gives the number of primes in the progression $\{dm + l : m \geq 1\}$ not exceeding $x$:

$$\lim_{x \to \infty} \pi(x; d, l) \frac{\log x}{x} = \varphi(d). \tag{10.28}$$

In the paper [III.406], de la Vallée-Poussin obtained the formula

$$\pi(x; d, l) = \frac{\mathrm{li}(x)}{\varphi(d)} + O(xe^{-c\sqrt{\log x}}),$$

which was later improved upon (see [III.412, p. 182]).

It follows from (10.28) that

$$\lim_{x \to \infty} \frac{\pi(x; d, l_1)}{\pi(x; d, l_2)} = 1,$$

thus confirming Dirichlet's assumption that, in progressions with parameters $d, l_1$ and $d, l_2$, prime numbers are distributed asymptotically with equal density.

In later studies various authors (Landau, Hardy, Bohr) proved formulas, similar to the Riemann-Mangoldt formula (10.11), giving the number of zeros of the function $L(x, \chi)$ in the rectangle $0 < \sigma < 1$, $0 < t < T$ as $T$ tends to infinity. Other important results were obtained in 1940 by Linnik.

With an ingenious proof Bombieri arrived at a strong estimate for a certain average over $d$ of the remainder term in the prime number theorem for the arithmetical progression ([III.47],1965). This estimate may be used in some applications instead of Riemann's hypothesis for the Dirichlet $L$-function.

These are only a few results in a large area and the above sketch of the development of the prime number theorem after Hadamard's and de la Vallée-Poussin's break-through is by no means a systematic survey. For a detailed description of the domain, one can consult the monographs [III.397], [III.412], [III.121], and others.

# Chapter 11

# Analytical Mechanics and Geometry

## 11.1 Research in analytical mechanics

The great French mathematicians of the eighteenth and early nineteenth centuries, Lagrange, d'Alembert, Laplace, Cauchy, and Poisson were also the creators of analytical mechanics. As a result of the later works of Liouville, Poinsot, Bertrand, and others, the French school of theoretical mechanics kept its leading role in the middle of the nineteenth century. In the beginning of the 1880s there was also the brilliant work of Poincaré, but towards the end of the nineteenth century the French school was no longer dominant. The Italian school in the theory of elasticity, founded by Betti and Beltrami, and represented by Lauricella, Levi-Civita, Volterra, and others, had surpassed it. Great achievements in the mathematical methods of mechanics were made in researches of Bruns, Boltzmann, and Hertz. New directions in the theory of oscillations were discovered in the works of Rayleigh, and there were the significant ideas of Hill and Sundman, in celestial mechanics. The contribution of the Russian school was also of importance, especially with the theory of stability of motion by Liapunov.

Amongst those who, at the turn of the century, kept up the traditions of their great compatriots, one should mention, as well as Poincaré, the names of Darboux, Appell, Painlevé, and Hadamard. Darboux, following Lipschitz, developed the geometrical interpretation of mechanics. The works of Appell dealt with general principles, as well as many concrete problems, and the application of elliptic and other special functions in mechanics. Painlevé made important researches in the theory of orbits, friction, and the three-body problem.

347

Hadamard's works in analytical mechanics are characterized by extensive use of geometric methods and are very varied in their themes. Amongst these works is the memoir *Sur certaines propriétés des trajectoires en dynamique* [I.48], in which he considered a range of problems connected with the researches of Liapunov and Poincaré in the theory of stability of motion.

We only give a brief sketch of this area of Hadamard's work. Let us begin with the 1894 article *Sur les mouvements de roulement* [I.25] which contained one of the first investigations in non-holonomic mechanics. Suppose that the state of a dynamical system (a rigid body, a system of points) is defined by the functions $q_1, \ldots, q_n$ of time $t$, which are connected through the constraints

$$f_k(q_1, \ldots, q_n) = 0, \qquad k = 1, \ldots, m, \qquad n > m.$$

One can eliminate $m$ variables, say $q_{n-m+1}, \ldots, q_n$, from these (holonomic) constraints. Then the equations of motion in Lagrangian form are

$$\frac{d}{dt}\frac{\partial \mathcal{L}}{\partial \dot{q}_i} - \frac{\partial \mathcal{L}}{\partial q_i} = 0, \qquad i = 1, \ldots, n - m,$$

where $\mathcal{L} = T - U$ is the Lagrangian function, equal to the difference of the kinetic energy $T$ and potential energy $U$ of the system.

Consider now the case when the constraints depend also on the time derivatives $\dot{q}_i$, i.e.,

$$g_k(\mathbf{q}, \dot{\mathbf{q}}) = 0, \qquad \mathbf{q} = (q_1, \ldots, q_n), \quad k = 1, \ldots, m.$$

If these constraints are not equivalent to relations of the form

$$\frac{d}{dt}\Phi_k(\mathbf{q}) = 0,$$

then the constraints are called non-holonomic, thus distinguishing between holonomic and non-holonomic systems. This terminology, as well as the first serious work on non-holonomic problems, is due to Hertz (see his classic book [III.175]).

The simplest and most important case of non-holonomic constraints is when the functions $g_k$ depend linearly on $\dot{\mathbf{q}}$:

$$g_k(\mathbf{q}, \dot{\mathbf{q}}) = \sum_i c_{ik}(\mathbf{q})\dot{q}_i = 0. \tag{11.1}$$

The generic example of a system with non-holonomic constraints is that of rolling, which is characterized by the property that there is no relative motion at the point of contact. A simple example is the motion of a ball in a plane, without sliding. The relation (11.1) expresses the equality of velocities at the contact points. When studying rolling, one cannot introduce coordinates which correspond to the number of degrees of freedom of the body, as it can be done in the case of holonomic constraints.

In the presence of constraints of type (11.1), the equations of motion can be written in the form

$$\frac{d}{dt}\frac{\partial \mathcal{L}}{\partial \dot{q}_i} - \frac{\partial \mathcal{L}}{\partial q_i} = \mathcal{K}_i, \quad i = 1, \ldots, n, \tag{11.2}$$

where $\mathcal{K}_i$ are certain functions of $\mathbf{q}$ (correcting terms). If for some $i$ equation (11.2) is Lagrangian, i.e. $\mathcal{K}_i = 0$, then one says that $q_i$ is a holonomic coordinate.

Hadamard considers in detail the problem of rolling and studies general questions in the theory of non-holonomic systems in [I.21], [I.25]. The central question is: under which conditions do there exist $r$ holonomic coordinates? These conditions are expressed as a system of equations in the functions $c_{ik}$. Hadamard observed that the derivation of the equations of motion is much simplified if one notes that some combinations of the equations of constraints can be used as if they express holonomic constraints. This is the case for a curve rolling on a surface, as well as rolling without spinning.

Hadamard's paper on the theory of rolling [I.25] was republished five years later, as an appendix to the book by his teacher Appell. In particular, this book contains the derivation of the equations of motion for a system with acceleration energy in place of kinetic energy. Other important modifications of the equations of motion of non-holonomic systems were obtained at the end of the nineteenth century by Chaplygin and Volterra. Non-holonomic mechanics today is a wide discipline which has applications in electromechanics, robotics, and many other domains.

In [I.29] and [I.30] Hadamard proposed new proofs of some theorems on the motion of a rotating heavy body fixed at one of its points. For instance, in [I.30] he considered the problem of stability of such a motion, which is understood in the sense that an infinitesimal change in the initial conditions give an infinitesimal change in the parameters of motion. Sufficient conditions for stability are obtained by solving the problem of the conditional extremum of some function, which in its turn is obtained by considering the roots of an algebraic equation.

In the paper [I.165], Hadamard studied a class of dynamical systems which are integrable by quadratures. These are the so-called Liouville systems, for which the kinetic and potential energies can be expressed in the form

$$
\begin{aligned}
T &= \frac{1}{2}(u_1(q_1) + \ldots + u_n(q_n))(v_1(q_1)\dot{q}_1^2 + \ldots + v_n(q_n)\dot{q}_n^2), \\
U &= \frac{w_1(q_1) + \ldots + w_n(q_n)}{u_1(q_1) + \ldots + u_n(q_n)}.
\end{aligned}
$$

The procedure of integration of a Liouville system leads to relations of the form

$$
\begin{aligned}
H_{j1}(u_1) + \ldots + H_{jn}(u_n) &= b_j, \quad j = 1, \ldots, n-1, \\
H_{n1}(u_1) + \ldots + H_{nn}(u_n) &= t,
\end{aligned}
$$

where $b_1, \ldots, b_{n-1}$ are arbitrary constants, and $t$ is time. The $H_{jk}$ denote periodic functions. The analysis of the properties of solutions of this system consists in inverting the operator defined by the left-hand side. Applying the results of his earlier paper on point transformations [I.138], which we discuss in Section 13.2, Hadamard proved the global solvability of this system under the single condition that the corresponding functional determinant be different from zero.

*Joseph Liouville (1809-1882)*

Among the articles by Hadamard on mechanics is one concerning a problem of "marine kinematics" [I.164]. It can be interpreted as the problem of the protection of some marine object (a barrage) from enemy ships. The barrage is modelled as a line segment, and the defending and attacking vessels are represented by two points. All this takes place in the plane. The problem is stated as follows: how should the defending vessel with radius of vision $R$ and constant speed $V$ move, in a sea with vertical waves, in order to see the barrage before the enemy ship, moving with speed $v \leq V$, has reached it? If that happens, then one says that the barrage has been defended. Until the work of Hadamard, the assumption was always made, that the enemy vessel moves uniformly and rectilinearly. Hadamard dispensed with this requirement and found necessary and sufficient conditions for "the defence to be achieved" [I.164, p. 338].

Several papers of Hadamard were inspired by his interest in the methodological problems of teaching of mechanics ([I.54], [I.131], [I.267], [I.275]). In particular, in [I.54], following his lecture courses, he formulated general laws of mechanics by taking the principle of equality of action and reaction as a base for the definition of the mass.

## 11.2 The papers on geodesic lines

The memoir *Sur certaines propriétés des trajectoires en dynamique* [I.48] is one of Hadamard's most remarkable achievements in both analytical mechanics and geometry. His work was influenced by Poincaré's qualitative theory of solutions to systems of ordinary differential equations, which opened the way to study properties of the solutions without finding them explicitly. In connection with Poincaré's classification of all possible trajectories for first-order equations with two unknown functions Hadamard wrote: "This result remained unrivalled. Applied to other types of differential equations, Poincaré's methods gave a series of implications of great importance, but they did not lead him to a complete solution.[1] Since then, no geometer was more successful with this. Very few of them had even attempted this

---

[1]In 1889 Poincaré devoted a large memoir to the three-body problem of celestial mechanics, relating, for example, the motion of the Sun, the Moon, and the Earth under mutual attractions. He studied periodic solutions and showed the existence of solutions which are asymptotic to a periodic one. Especially striking was his discovery of the existence of trajectories with chaotic long-time behaviour. Poincaré came to this result under dramatic circumstances after getting a Prize of the Swedish king Oscar II for the memoir which contained an error. For the historical and mathematical details see the paper by K.G. Andersson [III.10], and the books by J. Barrow-Green [III.19] and F. Diacu and Ph. Holmes [III.103].

path which Poincaré had opened up. A problem proposed by the Academy for 1896 (the competition for the *Prix Bordin*) gave me the opportunity to present the first applications of a general and extremely simple method based on the consideration of the maximum and minimum of an arbitrary function $V$ of the unknowns and their derivatives (in other words, of the state of the system, if one uses the language of Dynamics)" [I.87, p. 11].

Hadamard's paper of 1896 begins with the study of first-order systems in the $n$-dimensional Euclidean space $\mathbf{R}^n$

$$\dot{\mathbf{x}} = \mathbf{f}(\mathbf{x}), \qquad (11.3)$$

where $\mathbf{f} = (f_1, \dots, f_n)$ is a vector-valued function of the variable $\mathbf{x} = (x_1, \dots, x_n)$. Let $V$ be a smooth scalar function of $\mathbf{x}$ . Then on any trajectory $\mathbf{x}(t), t \geq t_0$, two cases are possible:

(a) the point $t = \infty$ is a limit point of the sequence of maxima and minima of the function $V$;

(b) the function is monotone for $t$ greater than some particular value.

In the case (a), the following conditions hold at an extremal point:

$$\dot{V}(t) = 0, \quad \text{and either} \quad \ddot{V}(t) \leq 0 \quad \text{or} \quad \ddot{V}(t) \geq 0. \qquad (11.4)$$

Let the operator $D$ be defined by

$$D(\psi) = \sum_i f_i \frac{\partial \psi}{\partial x_i}$$

(differentiation along the trajectory). From (11.3) it follows that

$$D(V) = \dot{V} \quad \text{and} \quad D[D(V)] = \ddot{V}.$$

Hence (11.4) can be rewritten in the form

$$D(V) = 0, \qquad (11.5)$$

and either $\quad D[D(V)] \leq 0 \quad$ or $\quad D[D(V)] \geq 0.$ \qquad (11.6)

Equation (11.5) defines an $(n - 1)$-dimensional surface. On this surface, the points of transition of the trajectory from the region where $D(V) > 0$ to the region $D(V) < 0$ correspond to the inequality $D[D(V)] < 0$. The opposite inequality holds for the inverse transition. The common boundary of the sets in (11.6) consists of points where the trajectory is tangent to the surface (11.5). Thus, in the case (a), the trajectory intersects the surface (11.5) infinitely many times, passing successively through each of the sets defined in (11.6).

The case (b) needs more sophisticated consideration, and this is done under the additional assumptions of the boundedness of the functions $V$ and $f_i$, as well as their derivatives of order three for $V$ and order two for $f_i$. This condition ensures the boundedness of the derivatives of $V$ with respect to $t$, up to order three, on any trajectory. Then, according to Hadamard's result on differentiable functions (which we discuss in Section 13.5)

$$\lim_{t \to \infty} \dot{V} = 0; \quad \lim_{t \to \infty} \ddot{V} = 0.$$

Thus, at the limit points of the trajectory (as $t \to \infty$) one has

$$D(V) = D[D(V)] = 0. \tag{11.7}$$

Hence, in case (b) the trajectories asymptotically approach the $(n-2)$-dimensional manifold defined by (11.7). (We leave aside the degenerate cases when the sets (11.5) and (11.7) are of lower dimension.)

When considering the motion of a point mass on a surface, essential difficulties arise even in the two-dimensional case. The Lagrange equations for the motion of the point are second-order and can not be changed to first-order equations by introducing the velocities as new variables. Hadamard was able to obtain results for the case of two degrees of freedom, similar to the ones just described.

He considered a point of unit mass, moving without friction on a surface, under the action of a force with potential $U$. He defined a smooth function $V$ on the surface, in terms of which he formulated his results, just as in the case of (11.3). The arguments are simplified on taking $V = U$. Hadamard showed that, when $\dot{U} = 0$,

$$\ddot{U} = -(\nabla U)^2 + \frac{2T}{(\nabla U)^2}[(\nabla U)^2 \Delta U - \frac{1}{2}\nabla U \cdot \nabla(\nabla U)^2],$$

where $T$ is the kinetic energy of the point, $\nabla$ is the gradient and $\Delta$ is the Laplace operator on the surface. He then used some arguments similar to those used for system (11.3): he again considered the cases (a) and (b), and the role of $D(V)$ was played by the function

$$I_U = (\nabla U)^2 \Delta U - \frac{1}{2}\nabla U \cdot \nabla(\nabla U)^2.$$

The surface is divided into the two parts $I_U > 0$ and $I_U < 0$. The first contains all the points of the trajectories for which $U$ attains its minimum value, namely points where a stable equilibrium is possible. Thus in the case (a), this part contains infinitely many arcs of trajectories. Hadamard called this domain an attracting domain. The second domain, where the point cannot stay beyond a certain time, is called a repelling domain. All those points where $U$ attains its maximum belong to the repelling domain, and at these points the domain is convex in the direction of the force (this is due to the fact that the denominator in the formula for the geodesic curvature coincides with $I_U$). In particular, in the above example of a point moving on the sphere, the lower hemisphere is the attracting set.

In the case (b), when there are only a finite number of equilibrium positions on the surface, Hadamard established the following alternative, under the additional assumptions of smoothness on the surface and the potential: a trajectory either passes infinitely many times through the attracting domain, or asymptotically approaches the equilibrium position.

He obtained similar, but less general results for trajectories in $\mathbf{R}^n$ for $n > 2$. For instance, he showed that if the equipotential surfaces are convex in the direction of the force, then the trajectory either goes to infinity or asymptotically approaches a state of unstable equilibrium. He also included a proof of the converse of Dirichlet's theorem on the stability of equilibrium: an equilibrium position at which the potential does not attain its maximum, is unstable. Hadamard noted that this statement is contained in Liapunov's memoir "unfortunately written in Russian", and he added, "Since this proof is analogous, but not identical to Liapunov's, I believe it is useful to give it here" [I.406, p. 1783].

A large part of Hadamard's memoir is devoted to the problem of geodesics. These curves are such that sufficiently small arcs on them are the shortest distances between points on the surfaces. The geodesics in the plane are straight lines, and on the sphere they are great circles.[2] If $q_1, q_2$ are coordinates on the surface $S$ and the metric is given by the formula

$$dl^2 = \sum_{i,j} g_{ij}(\mathbf{q})dq_i dq_j,$$

then the length of the line $\mathbf{q}(t), \alpha < t < \beta$, on $S$ is expressed by the integral

$$l = \int_\alpha^\beta (\sum_{i,j} g_{ij}(\mathbf{q})\dot{q}_i \dot{q}_j)^{1/2} dt.$$

---

[2]The restriction to small arcs in the definition of geodesics is important. In fact, one of two geodesic paths joining two non-antipodal points on the sphere is not the shortest.

The geodesics on the surface are given by ordinary nonlinear differential equations which are necessary conditions for the extremum of this integral. The equations are of second order in $q_i$, which implies that through any point on $S$ one can draw a geodesic in an arbitrary direction, and that any points on $S$ sufficiently close to each other can be connected by a unique shortest geodesic.[3] The importance of the notion of geodesic for dynamics is due to the fact that the equation of a geodesic is essentially the equation of the motion of a point under the action of a force given by a potential function.

*Rudolf Lipschitz (1832-1903)*

Thus theorems about geodesics give answers to questions about the behaviour of trajectories of a dynamical system, and at the same time this relationship allows one to find new properties of geodesics. It was already anticipated by Jacobi, and later, Lipschitz and Darboux gave an explicit geometrical interpretation of the Hamilton-Jacobi formalism of mechanics.[4]

Hadamard's results supplemented the study of global properties of geodesics which at that time was a new field. One of the geometric theorems in his memoir states that on a closed surface with positive Gaussian curvature, any closed geodesic is intersected infinitely many times by any given geodesic (the latter has to be parametrized by the whole axis $-\infty < t < +\infty$). In particular, this implies that any two closed geodesics intersect. Hadamard also remarks that this assertion is a trivial corollary of the Gauss-Bonnet formula connecting the total Gaussian curvature of the domain and the integral of the geodesic curvature over the boundary. However, in order to apply this formula, it is necessary to know the topological type of the surface.

One should keep in mind that Hadamard, following the tradition of his time, thought of a surface as lying in a euclidean space. This imposes

---

[3]The first fact follows from the local solvability of the Cauchy problem for the second order ordinary differential equation when the solution and its first derivatives are prescribed at the initial time. The second assertion is a consequence of the solvability of the two point boundary value problem on a small interval.

[4]A historical analysis of interactions between mechanics and differential geometry during the nineteenth century is given by J. Lützen [III.252].

strong conditions on the topology of a closed surface of positive curvature: Hadamard proves that it is homeomorphic to a sphere and that the Gauss mapping is one-to-one (the Gauss mapping is the mapping of a surface $\Sigma$ into the unit sphere $S^2$ with centre at the origin, such that to each $x \in \Sigma$ there corresponds a point on $S^2$ with position vector $\nu(x)$ equal to the unit normal to $\Sigma$ at the point $x$).

In modern terms Hadamard's proof reduces to the following. He proves that the Gauss mapping is a local diffeomorphism and therefore a covering map of the sphere. Therefore, since the sphere is simply connected, it is a diffeomorphism. Unlike the essentially two-dimensional corollary on the intersection of two closed geodesics, Hadamard's theorem stimulated far-reaching multi-dimensional generalizations. It is sometimes included in textbooks on differential geometry (see for instance [III.40]).

In the same paper, Hadamard showed that on a surface of positive curvature no geodesic can approach a closed geodesic asymptotically while staying on one side of it. If a geodesic intersects a closed geodesic infinitely many times, then it passes through a set which is everywhere dense in some domain. It is possible that such geodesics have infinitely long arcs which are arbitrarily close to closed geodesics.

As an illustration, let us turn to geodesics on an ellipsoid with three axes. This was completely solved in a purely analytic way by Jacobi, who showed in his *Lectures on dynamics* [III.193], using complicated transformations, that the equations of geodesics on an ellipsoid can be integrated explicitly and that these integrals can be expressed in terms of elliptic functions.

*Carl Jacobi (1804-1851)*

Consider the ellipsoid

$$\frac{x^2}{a^2} + \frac{y^2}{b^2} + \frac{z^2}{c^2} = 1, \quad a > b > c.$$

Each one of the ellipses in the planes $x = 0$, $y = 0$, $z = 0$ is a closed geodesic. How does a geodesic behave which intersects the closed geodesic $z = 0$? It turns out that if the angle between the two curves at the point of intersection is small, then it is either closed or it forms an everywhere dense subset of the "ring" limited by the curves of intersection of the ellipsoid with some hyperboloid with one sheet (see Figure 3$a$). When the angle is large, then the geodesic forms an everywhere dense set in the "ring" limited by the curves of intersection of the ellipsoid with a hyperboloid with two sheets (see Figure 3$b$). More detailed information on the behaviour of the geodesics can be found in the book [III.12].

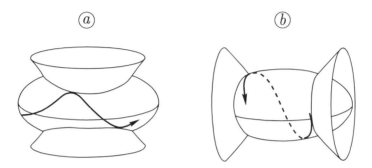

**Figure 3.** *Geodesics on the ellipsoid*

A year later, in 1897, Hadamard published a new work about geodesics. This time he considered surfaces with negative Gaussian curvature. He wrote "In this new case it is possible to arrive at more complete results than in the first case, and, without difficulty, to give a general discussion of geodesics"[I.406, vol. 2, p. 729]. However, the memoir is not limited to the study of geodesics; for instance, Hadamard shows that in $\mathbf{R}^3$ there exists a large variety of complete surfaces with negative curvature, and he gives examples of such surfaces of different topological types.

One of such surfaces is given by the equation $x_3 = u(X)$ with $X = (x_1, x_2)$ and

$$u(X) = c \log \frac{|X - P_1| \dots |X - P_m|}{|X - Q_1| \dots |X - Q_n|},$$

where $c = \text{const} > 0$ and $\{P_i\}$, $\{Q_j\}$ are collections of points on the plane $x_3 = 0$. Note that $u$ is harmonic and $u_{x_2} + iu_{x_1}$ is analytic outside the singular points $\{P_i\}$, $\{Q_j\}$. Hence the Gaussian curvature

$$K = \frac{u_{x_1x_1} u_{x_2x_2} - u_{x_1x_2}^2}{\left(1 + u_{x_1}^2 + u_{x_2}^2\right)^2} = -\frac{u_{x_1x_1}^2 + 2u_{x_1x_2}^2 + u_{x_2x_2}^2}{2\left(1 + u_{x_1}^2 + u_{x_2}^2\right)^2}$$

is negative outside a set of isolated points. The equation $x_3 = u(X)$ defines a surface with $m + n$ spikes reaching $\pm\infty$ at the points $\{P_i\}$, $\{Q_j\}$. The left surface on Figure 4 corresponds to the case $m = n = 1$.

Other examples are based on a geometrical construction which was later developed by other authors: one starts with the union of simple surfaces (for example, hyperboloids) and then smoothes out the edges (see [III.349]).

In passing, Hadamard claims that, in contrast with the surfaces with positive curvature, those with negative curvature necessarily go to infinity and gives a seemingly plausible argument in favour of this assertion. The argument proved to be insufficient for smooth non-analytic surfaces (see [III.349]), and moreover, the assertion itself, called Hadamard's conjecture, was recently disproved by Nadirashvili [III.290].

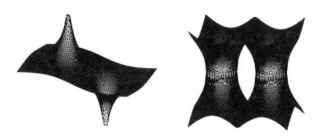

**Figure 4.** *Surfaces of negative curvature considered in Hadamard's memoir*

One can obtain an impression of the results on geodesics on surfaces with negative curvature from the following two theorems from the first part of the paper. In formulating Hadamard's theorems, we have to use some simple topological terms which are used nowadays. However, in the original paper, Hadamard introduced these notions and gave them other names.

**Theorem 1.** *In an arbitrary homotopy class of paths joining two points of a surface, there is only one geodesic.*

**Theorem 2.** *Each free homotopy class contains a unique closed geodesic.*

There are exceptions to these theorems, corresponding to the case of a surface with infinite spikes, but we do not discuss this.

Hadamard then divides all geodesic lines (geodesics extended to one of the semi-axes of $t$) into three classes. The first class consists of closed geodesic lines and geodesics which approach them asymptotically. The second class is formed by geodesics which go to infinity. They can also be characterized in another way: these, and only these, intersect some closed geodesic. Those geodesic lines which do not belong to the above two classes form the third class (these are geodesics which travel between neighbourhoods of different closed geodesics). The existence of geodesics of the third kind has to be proved, and Hadamard does this in this paper by using the following fact, which is of independent interest.

At an arbitrary point of a surface, the set of initial directions of geodesics which do not go to infinity is perfect and totally disconnected. In other words, it is closed and has neither isolated nor internal points, so it looks like the classic Cantor set which is constructed as follows.

Remove the middle third (1/3,2/3) from [0,1], then remove the open middle thirds from two intervals [0,1/3] and [2/3,1], remove again the open middle thirds from the remaining fours intervals, etc. The remainder of [0,1]at the end of this process is called the Cantor set (see Figure 5). One can also say that this is the set of all numbers

$$\sum_{j\geq 1} 3^{-j}\delta_j,$$

where $\delta_j$ equals 0 or 2.

**Figure 5.** *Construction of the Cantor set*

By assigning a certain sequence $\{\delta_j\}$ to each initial direction of the geodesic from the third class, Hadamard gave the first application of the so-called symbolic dynamics, the technique of characterizing the structure of trajectories of dynamical systems by infinite sequences of "symbols" (in Hadamard's case zeros and twos). Symbolic dynamics enables one to reduce the study of shifts along trajectories to that of homeomorphisms of sequences which is sometimes much easier. It was further developed by G.D. Birkhoff [III.38] in his studies of dynamical systems.

Hadamard showed that an infinitesimal change in the initial direction of a geodesic which does not go to infinity, can produce any variation in its final position; a perturbed geodesic can belong to any of the three above mentioned classes. Mandelbrojt ([II.42, p. 5]) has suggested that it was this phenomenon which could have led Hadamard to the concept of well-posedness. Indeed, in 1901, in his *Notice sur les travaux scientifiques* [I.87], commenting on his work on dynamical trajectories, Hadamard uses the term *questions mal posées* for the first time: "One of the fundamental problems of celestial mechanics, that of the stability of the solar system, falls probably into the category of ill-posed questions. In fact, if instead of investigating the stability of the solar system one treats the analogous question for the geodesics on a surface of negative curvature, one sees that each stable trajectory can be transformed, by an infinitesimal change in the initial data, into a completely unstable trajectory going to infinity. In fact, in astronomical problems, the initial data are only known to within a certain error. However small it is, this error could entail a total and absolute perturbation in the required result".

A picturesque account of Hadamard's memoir [I.62] can be found in Duhem's book *La théorie physique: son objet, sa structure* [III.116, Part 2, Ch. 3, Sec. 3] which appeared in 1906. Duhem takes the problem of geodesics on surfaces of negative curvature along with the problem of $n$ bodies as examples of deep mathematical problems "useless to the physicist". They are devoid of any direct physical meaning because of the oversimplification of the underlying physical models. Here is Duhem's description of surfaces of negative curvature considered by Hadamard:

> Imagine the forehead of a bull, with the protuberances from which the horns and ears start, and with the collars hollowed out between these protuberances; but elongate these horns and ears without limit so that they extend to infinity; then you will have one of the surfaces we wish to study.

Hadamard's study of these surfaces of negative curvature was the foundation of the theory of Riemannian manifolds of non-positive curvature. The next contribution to this theory, which is now one of the most developed and beautiful areas of Riemannian geometry was due to Élie Cartan. The foundation of the theory of manifolds of non-negative curvature is the following Hadamard-Cartan theorem. Its formulation contains the notion of the so called sectional curvature which coincides with Gaussian curvature for $n = 2$.

**Theorem.** *Each $n$-dimensional complete, simply connected Riemannian manifold with non-positive sectional curvature is diffeomorphic to $\mathbf{R}^n$.*

Hadamard's theorems on the asymptotic behaviour of geodesics on surfaces of negative curvature were extended in one of the areas of the theory of dynamical systems – the theory of Anosov flows [III.11]. The instability of geodesics on such surfaces is one of the principal examples of chaotic behaviour of trajectories of dynamical systems.

## 11.3  *Leçons de Géométrie élémentaire*

In 1898 the first volume of Hadamard's book on geometry, the *Leçons de Géométrie élémentaire (Géométrie plane)* appeared. One can judge the impression it made on the *lycéens* of that time from Denjoy's recollections:

> I applied myself to reading it with admiration, respect and devotion. Some friends, dogmatically sceptical, ironised: *"They call this geometry 'elementary' to make us believe that they know a lot more."* In fact, two notes came at the end of the book, one on the measurement of geometric figures, the other showing in an irrefutable way an important precursor to the theorem of Jordan on the division of the plane into two regions by every simple closed curve. These two expositions presented arguments which had never been considered before. At that time not much was known about them [II.56, p. 736].

Hadamard wrote the book at the instigation of Darboux, and it was the first of two volumes, the second dealing with geometry in three dimensions being published in 1901. These two volumes form the principal and best known part of Hadamard's pedagogical work. They came out in numerous editions in France and in many other countries.

In the preface, Hadamard stresses the special role which geometry plays in elementary mathematics:

> Placed at the beginning of mathematics, it is, in effect, the simplest and most accessible form of reasoning. The scope of the methods, their fecundity are more immediately tangible than those of the relatively abstract theories of arithmetic or of algebra. In this, it shows itself capable of exercising an undeniable influence on the activity of the mind. I have tried, above all, to develop this influence by wakening and assisting as much as possible the initiative of the student [I.67, p. V].

All this was facilitated by 1300 exercises throughout the whole book. Many were taken from original scientific publications, including Hadamard's own, and can be quite a serious challenge. He explained:

> ...it seemed to me necessary to increase the exercises in relation to the scope of the book. This necessity was, so to say, the only rule which guided me in this part of my work. I felt

I ought to give questions of different and gradually increasing difficulty: while the exercises at the end of each chapter, especially the first few of these, are very easy, those I included at the end of each book have a less immediate solution; finally I have relegated relatively difficult problems to the end of the volume [I.67, p. V].

The main body of the first volume is divided into four parts: the straight line, the circle, similarity, and area. In the "Note A"at the end of the first volume, Hadamard discusses the fundamental principles of the mathematical method, "principles with which beginners should be imbued from the first year of study and which are, however, too often badly known even by students of our *écoles supérieures*"[I.67, p. VI]. He then continues with this, later in the book:

> We wish to collect under this heading some advice which we believe useful for understanding mathematics in general and, in particular, for solving problems.
>
> The student should, in effect, convince himself that he will not be able to gain any profit from his mathematical studies, nor even pursue them without great efforts and obtain a true idea of what geometry is, if he is not able not only to understand the arguments which are taught to him but also to give other ones himself, to find, to a greater or lesser extent, proofs of theorems or solutions of problems.
>
> Contrary to a far too widespread prejudice, this can be achieved by everyone, or at least by all those who force themselves to think and to steer their thoughts methodically. The precepts which we shall indicate come from the commonest of common sense. There is not one of them that will not seem a sheer banality to the reader. However, experience shows that forgetting one or other of these obvious rules is the almost unique cause of the difficulties which arise in the solution of elementary problems, and it is also the case, more often than one would be led to believe, in research whose object is the more or less advanced parts of mathematical science [I.67, p. 261].

Further on, Hadamard formulates and then discusses three rules of proof of a theorem. First, following Pascal, it is necessary to give definitions for the objects used. The second rule is to use all the hypotheses, and the third is to give the result all its possible forms and to choose the most convenient one as the goal. Hadamard's deep interest in the creative process of mathematics is again manifested here, something he was fascinated by all his life.

Apparently, the treatment of the problem of the area of surfaces in Hadamard's book influenced Lebesgue's approach to measure theory to some extent. Lebesgue began his first paper devoted to the subject [III.229] by quoting the "Note D" *Sur la notion d'aire* (see [III.169, Ch.5]).

The second volume, *Géométrie dans l'espace*, is especially rich. In Parts 5 to 10 Hadamard discusses the plane and the straight line, polyhedra, displacement, symmetry, similitude, rotund bodies (the cylinder, the cone, and the sphere), conic sections, and the basic notions of topography. A lot of material going beyond the formal school programme is given in the form of supplements, whose length equals that of the main text. This supplementary material includes Euler's theorem on the connection between the number of vertices, edges, and faces of geometrical figures, the theory of regular polyhedra and their groups, and the famous theorem of Cauchy: two convex polyhedra with corresponding faces and the same orientation are either congruent or symmetric.

In subsequent editions of the book, Hadamard enlarged the number of Notes by including more non-traditional topics. For example, in the "Note M" he deals with the conformal geometry in $\mathbf{R}^3$, namely the theory of inversions and translations, which reflected Hadamard's own interests. He used the term *Géométrie anallagmatique*, which did not survive. Completing the material of this supplement, he wrote a number of papers on conformal geometry, the most significant being *Récents progrès de la géométrie anallagmatique* [I.259], published in Buenos Aires.

To this day, Hadamard's course remains one of the best texts for schoolchildren wanting to study geometry in depth.

# Chapter 12

# Calculus of Variations and Functionals

## 12.1 Some notions of the calculus of variations

The term "calculus of variations" is usually connected with problems concerning extremals, although this does not appear directly in the name. At present, the calculus of variations is a collection of theories which, although related to each other, lead in different directions. The one with which we shall mostly be concerned is the classical calculus of variations originating in the works of Euler, Lagrange, Legendre, and developed by Jacobi, Weierstrass, and their numerous successors. The history of the calculus of variations is outside of the scope of this book, and we refer the reader to Goldstine's monograph [III.152]. The theory is centred on the problem of finding the extremals for functions whose domain is a set of curves or surfaces.

Hadamard first encountered the calculus of variations at the end of the nineteenth century when working on wave theory, elasticity, and geometrical problems connected with geodesics. The second volume of his collected works begins with a cycle of articles which appeared, with some interruptions, during the period 1902-1913. He intended to write a treatise on this subject, but only one volume appeared, in 1910 [I.159].

Before we describe Hadamard's work, we give a few definitions. Let $C^l$ be the set of functions $y(x)$ defined on the segment $a \leq x \leq b$ and continuous there, together with the derivatives of all orders up to and including order $l$. A neighbourhood of order $l$ of the function $y$ in $C^\ell$ is the set of functions $y + \omega$ such that $\omega \in C^\ell$ and

$$\max_{a \le x \le b} |\omega(x)| < \varepsilon, \quad \max_{a \le x \le b} |\omega^{(k)}(x)| < \varepsilon, \qquad k = 1, \dots, l.$$

Let us now formulate the simplest problem of the calculus of variations.

*Gottfried Wilhelm Leibnitz (1646-1716)*

Let

$$U[y] = \int_a^b f(x, y(x), y'(x)) dx. \tag{12.1}$$

We seek functions $\tilde{y}$ in $C^1$ for which this integral attains relative maxima and minima. For instance, in the first case this means that the inequality $U[\tilde{y}] \ge U[y]$ holds for all $y$ in some neighbourhood of $\tilde{y}$ of order 1. Two further conditions are imposed: we require that

$$y(a) = A, \quad y(b) = B, \tag{12.2}$$

where $A, B$ are given numbers. The integral $U[y]$ is a function of the curve given by the graph of $y$, and we seek a curve with endpoints $(a, A)$ and $(b, B)$ which is an extremum of the functional $U[y]$ (in the following, we consider the minimum for the sake of definiteness).

Particular cases of this problem had already been considered by Newton, Leibnitz, Jacob I Bernoulli, and Johann I Bernoulli towards the end of the seventeenth century, and it was studied for the first time in its general formulation by Euler (1741). In 1755, Euler received a letter from Lagrange, then a nineteen-year-old youth in Turin unknown to him, in which the following approach to the problem was suggested.

Let $\omega = \alpha\eta$, where $\alpha$ is a numerical parameter and $\eta$ is a function in the class $C^1$, for which $\eta(a) = \eta(b) = 0$. One then has

*Isaac Newton (1643-1727)*

$$U[y + \alpha\eta] = U(\alpha) = \int_a^b f(x, y + \alpha\eta, y' + \alpha\eta')dx.$$

From the expansion

$$U(\alpha) = U(0) + \alpha U'(0) + \frac{\alpha^2}{2}U''(0) + \dots$$

it follows that

$$U'(0) = \int_a^b (f_\eta \eta + f_{\eta'} \eta') dx, \qquad (12.3)$$

and

$$U''(0) = \int_a^b [f_{\eta\eta} \eta^2 + 2 f_{\eta\eta'} \eta\eta' + f_{\eta'\eta'} (\eta')^2] dx$$

(where we assume the existence of all the derivatives which appear in the expressions). The expressions $U'(0)$ and $U''(0)$ are respectively called the first and second variations of $U[y]$. If $U(0)$ is an extremal value of $U(\alpha)$, then necessarily $U'(0) = 0$. Integrating by parts the second term in (12.3), and taking into account $\eta(a) = \eta(b) = 0$, we obtain

Leonhard Euler (1707-1783)        Joseph Louis Lagrange (1736-1813)

$$U'(0) = \int_a^b (f_y - (f_{y'})') \eta \, dx = 0.$$

Due to the arbitrariness of the function $\eta$, it follows that

$$f_y - (f_{y'})' = 0 \qquad (12.4)$$

which had earlier been obtained by Euler, and is called the Euler equation.

This elegant argument made a profound impression on Euler. He replied with an enthusiastic letter, and said that he would not publish anything on this problem until Lagrange published this result. The term "calculus of variations" is due to Euler, and the function $f(x, y, y')$ in (12.1) is called a Lagrangian function. The solution of equation (12.4) under condition (12.2) will be denoted $\tilde{y}$, and by $\tilde{\varphi}$ we mean $\varphi(x, \tilde{y}, \tilde{y}')$.

A minimization problem for the integral depending on a vector-valued function

$$U[\mathbf{y}] = \int_a^b f(x, \mathbf{y}, \mathbf{y}')dx, \quad \mathbf{y} = (y_1, \ldots, y_m), \quad \mathbf{y}' = (y_1', \ldots, y_m'), \quad (12.5)$$

is formulated in a similar way. A necessary condition for the minimum is that the components of the vector $\mathbf{y}$ satisfy the system of differential equations

$$f_{y_i} - (f_{y_i'})' = 0, \quad i = 1, \ldots, m,$$

with the boundary conditions $y_i(a) = A_i, y_i(b) = B_i$.

Let us return to the scalar case $m = 1$, i.e. to the integral (12.1). After some manipulation with the second variation $U''(0)$, Legendre arrived (in 1786) at the conclusion that the inequality $\tilde{f}_{y'y'} > 0$ is a sufficient condition for $y = \tilde{y}$ to minimize (12.1). In a few years Lagrange showed that this is only true if the interval $(a, b)$ is sufficiently small. In turn, Lagrange was wrong in claiming that, just as in the usual calculus, the conditions $U'(0) = 0, U''(0) \neq 0$ are sufficient for an extremum. There is hardly any other part of analysis whose history is as full of mistakes as the calculus of variations. Mentioning Lagrange's error, L.C. Young wrote: "A full discussion of the fallacy of Lagrange's argument, together with a counterexample that shows that the answer is no, is given in Hadamard (Sections 38-43).[1] We invite the reader to study this, at first hand, as explained with remarkable clarity by one of the great thinkers of a past era" [III.426, p. 73].

In fact, Legendre's condition is only necessary for the minimum of the integral (12.1). Its analogue for the integral (12.5) takes the form

$$\sum_{i,j=1}^m f_{y_i'y_j'}(x, \mathbf{y}, \mathbf{y}')\eta_i\eta_j > 0$$

for all $x \in [a, b]$ and $\eta = (\eta_1, \ldots, \eta_m) \in \mathbf{R}^n \backslash \{0\}$.

---

[1] In the book [I.159].

For functions of two or more independent variables, the problem is formulated in the following way. Let $\mathcal{D}$ be a domain in $\mathbf{R}^m$ bounded by a surface $\Gamma$. We seek a function $z$, differentiable in $\mathcal{D}$, continuous on $\overline{\mathcal{D}} = \mathcal{D} \cup \Gamma$, satisfying the boundary condition $z|_\Gamma = \varphi$, such that the integral

$$U[z] = \int_{\mathcal{D}} F(x, z(x), \operatorname{grad} z(x)) dx$$

takes an extremal value. The extremum is to be local: if $\tilde{z}$ is a solution, then we vary $z$ in the neighbourhood of order one given by

$$|z - \tilde{z}| < \varepsilon, \quad \text{and} \quad |\operatorname{grad} z - \operatorname{grad} \tilde{z}| < \varepsilon.$$

Similarly to the one-dimensional case, we put $z = \tilde{z} + \alpha\eta$. Using the necessary condition for an extremum $U'(0) = 0$, and the same arguments as before, we find that the function $\tilde{z}$ satisfies the Euler equation

$$\sum_{j=1}^{m} \frac{\partial}{\partial x_j} \left( \frac{\partial F}{\partial z_{x_j}} \right) - \frac{\partial F}{\partial z} = 0. \tag{12.6}$$

In his first article on the calculus of variations [I.95], published in 1902, Hadamard stated an analogue of Legendre's condition for variational problems with $m$ unknown functions of $n$ independent variables.

It is appropriate to discuss one principal question. In order that the left-hand side of (12.6) have a meaning in the classical sense, we need the existence of the second derivatives of the solution. In 1879, du Bois-Reymond showed that this condition is superfluous for $m = 1$: any function $y(x)$ for which $U'(0) = 0$ has, of necessity, a second derivative in the interval $(a, b)$ if $f_{y'y'} \neq 0$. This is not the case for multiple integrals, as was shown by Hadamard ([I.144]) with the following counterexample.

Let $m = 2$ and

$$U[z] = \iint_{\mathcal{D}} (z_x^2 - z_y^2) dx dy. \tag{12.7}$$

The first variation of this functional is

$$U'(0) = 2 \iint_{\mathcal{D}} (z_x \eta_x - z_y \eta_y) dx dy.$$

Let $z = \psi(x + y)$ where $\psi$ is a continuously differentiable function. Then $z_x = z_y$ and

$$U'(0) = \iint_{\mathcal{D}} [(z\eta_x)_y - (z\eta_y)_x]dxdy = \int_\Gamma z(\eta_x n_x - \eta_y n_y)ds,$$

where $n$ is the normal vector to $\Gamma$. The last integral is equal to zero, since $\eta = 0$ on $\Gamma$. Thus $U'(0) = 0$ does not guarantee the existence of $\psi''(x + y)$.

Let us compare (12.7) with the Dirichlet integral

$$\iint_{\mathcal{D}} (z_x^2 + z_y^2)dxdy,$$

for which the Euler equation is $z_{xx} + z_{yy} = 0$. In the case of the Dirichlet integral, the solution of the variational problem not only possesses second derivatives, it is also real analytic. Thus the integrands $F(p, q) = p^2 + q^2$ and $F(p, q) = p^2 - q^2$ are essentially different from the point of view of the smoothness of the solutions of the variational problem, and formally they can be distinguished from each other by the sign of the function $K = F_{pp}F_{qq} - F_{pq}^2$. In the first case, the variational problem is said to be regular, i.e., it satisfies $K > 0$. For integrals with analytic $F$ the proof of the analyticity of solutions is Hilbert's twentieth problem. The solution of this was found in 1904 by the young Sergeï Bernstein, in his thesis, presented at the *Sorbonne*.

Another counterexample by Hadamard concerning a variational problem for the Dirichlet integral will be discussed in Section 12.3.

## 12.2 A method in the calculus of variations

In the article *Sur une méthode de calcul des variations* [I.135], Hadamard formulated a new constructive approach to the problem of the minimum of the integral (12.1), under the conditions (12.2). The detailed presentation of this approach forms the fourth part of his memoir on elastic plates [I.145], for which he obtained the *Prix Vaillant* in 1907 (see Section 14.3).

Hadamard starts with a prescribed function $y_0(x)$ and looks for a solution $y(x)$ of the variational problem. In order to do this he proposes joining $y_0(x)$ and $y(x)$ by a family of functions $y_t(x)$ depending on a positive parameter $t$. (One may think about a family of graphs varying with time.) The idea is to construct $y_t$ in such a way that the integral $U[y_t]$ decreases for every $t$. Then there is a hope that as $t \to \infty$, $y_t$ approaches the function $y$ minimizing (12.1) or at least providing a local minimum.

Hadamard notes that the first derivative of $U[y_t]$ with respect to $t$ is given by

$$\frac{d}{dt}U[y_t] = \int_a^b \frac{\partial f}{\partial y'}(x, y_t, y'_t)\frac{\partial y'_t}{\partial t}dx + \int_a^b \frac{\partial f}{\partial y}(x, y_t, y'_t)\frac{\partial y_t}{\partial t}dx,$$

where $y'_t = \partial y_t/\partial x$. Hence

$$\frac{d}{dt}U[y_t] = \int_a^b Q_t\frac{\partial y'_t}{\partial t}dx, \tag{12.8}$$

where

$$Q_t(x) = \frac{\partial f}{\partial y'}(x, y_t(x), y'_t(x)) - \int_a^x \frac{\partial f}{\partial y}(s, y_t(s), y'_t(s))ds - h_t$$

with an arbitrary $h_t$ independent of $x$. Hadamard determines $h_t$ from the orthogonality condition

$$\int_a^b Q_t dx = 0$$

and suggests finding $y_t(x)$ as a solution of the equation

$$\frac{\partial y'_t}{\partial t} = -\rho_t Q_t, \tag{12.9}$$

where $\rho_t$ is an arbitrary positive function of $x$ and $t$. By (12.8) and (12.9)

$$\frac{d}{dt}U[y_t] = -\int_a^b \rho_t\left(\frac{\partial y'_t}{\partial t}\right)^2 dx \leq 0,$$

i.e., $U[y_t]$ is a decreasing function of $t$.

Hadamard chooses $\rho = 1$ and shows that (12.9) can be solved by successive approximations in the same manner as an ordinary differential equation for a function of the variable $t$. The procedure gives $y_t$ on a certain interval $0 < t < t_0$ and this is the only solution of (12.9) starting with $y_t|_{t=0} = y_0$.

The trajectory obtained can be extended to the semi-axis $t > 0$ if the boundedness of $y_t$ and $y'_t$ is established beforehand. The boundedness holds under certain assumptions on the function $f$, of which the simplest is the condition

$$|f(x, z, z')| = O(|z'|^q), \qquad q > 1.$$

In obtaining a majorant for the function $|y'_t|$, Hadamard proved an interesting inequality for differentiable functions, which we shall describe in more detail at the beginning of Section 13.5. Finally, he showed that as $t \to \infty$, $y_t(x)$ tends to the solution of the variational problem.

## 12.3  The Dirichlet principle

*Friedrich Prym (1841-1915)*

Hadamard's four-page article *Sur le principe de Dirichlet* [I.139] appeared in 1906 in the *Bulletin de la Société Mathématique de France*. This work contains a perceptive remark concerning the variational method for the solution of the classical Dirichlet problem for harmonic functions (see Section 3.1). Hadamard did not know that the same observation on this subject had already been made in 1871 by the German mathematician Prym in the article *Zur Integration der Differentialgleichung $\frac{\partial^2 u}{\partial x^2} + \frac{\partial^2 u}{\partial y^2} = 0$* [III.335]. Prym's work was not noticed by his contemporaries, and his priority was established quite recently [III.105]. On the other hand, Hadamard's article had a happier fate: the counterexample constructed there became standard, and under the name 'Hadamard's counterexample', it is included in many books on the calculus of variations.

In order to understand the meaning of the examples of Prym and Hadamard, it is useful first to recall some dramatic events in mathematics which happened in the middle of the last century, and then some notions from the foundations of the general theory of boundary value problems created in the middle of this century. Efforts to establish the solvability of problems of mathematical physics for domains of arbitrary shape were made by the great names of mathematical analysis during the first half of the nineteenth century. However, at that time rigorous proofs were absent and the arguments were based either on physical considerations (Green's function method) or on the variational principle which was always taken to be valid. This principle had already been proposed in the 1840s by Gauss and Kelvin, who noticed that the harmonic function in a plane domain $\Omega$ with prescribed values, say $g$, on the boundary $\partial\Omega$ minimizes the integral

$$D(u) = \iint_\Omega \left[ \left(\frac{\partial u}{\partial x}\right)^2 + \left(\frac{\partial u}{\partial y}\right)^2 \right] dxdy \tag{12.10}$$

for differentiable functions $u$ with the same boundary values. Since the existence of the minimizer was said to be obvious, the variational principle served as the justification for the solvability of the Dirichlet boundary value problem for the Laplace equation

$$\Delta u = 0 \quad \text{in} \quad \Omega, \quad u = g \quad \text{on} \quad \partial\Omega. \tag{12.11}$$

The same arguments were used by Riemann as a basis for his geometric theory of functions of a complex variable. It was Riemann who, having learned these ideas from Dirichlet's lectures, named the variational principle and the boundary value problem in honour of his teacher. The terms "Dirichlet principle" and "Dirichlet problem" are commonly used today.[2]

The prestige of the Dirichlet principle was called into question by Weierstrass in 1869, when he showed by an example that the infimum of the non-negative integral (12.10) may not be attained for functions of one variable. Weierstrass' criticism made a great impact on mathematicians: belief in the variational principle was severely undermined for almost thirty years. The fundamental results obtained by this principle needed justification, so that mathematicians began developing new approaches to the solvability of boundary value problems.

One of the first rigorous methods for the solution of the Dirichlet problem was the method of arithmetic means of C. Neumann (in 1870), which made it possible to solve the problem for convex domains. At the same time the alternating method of Schwarz was introduced, enabling one to obtain the solution of the problem for the union of two domains, if it is solved in each of them. In 1887, Poincaré proposed the method of *balayage*, with the help of which one can solve the Dirichlet problem for a fairly wide class of domains. In the works of C. Neumann, Robin, Poincaré, Hölder, Liapounov, Steklov, Fredholm, Radon, Zaremba, Plemelj, Carleman, and others, the method of boundary integral equations was developed and applied to boundary value problems for the Laplace operator. This method, as it has been made clearer with the advent of computers, turned out to be quite effective, and sometimes the only possible method for numerical solution of problems of mathematical physics for bodies having a complicated form.

---

[2]For the history of the "Dirichlet principle" see Appendix in the book of Bottazzini [III.53].

Thus, at the turn of the century, the theory of boundary value problems had made significant achievements. It could already deal with linear elliptic equations with variable analytic coefficients, and even with nonlinear elliptic equations for two independent variables. One can see this, for instance, in the paper of Bernstein [III.32] published in 1904, in which he created an elaborate technique of a priori estimates, whose importance has not diminished even today.

As for the Dirichlet principle, this attractive idea had to wait for many years until its time came. It was Arzelà in 1896, and Hilbert in 1897, in his lecture to the First International Congress of Mathematicians, who independently gave a justification of the variational principle for the solution of the Dirichlet problem.[3]

Hadamard's note *Sur le principe de Dirichlet* [I.139] was a response to Hilbert's work, whose starting point was the possibility of extending the boundary function to the whole of the domain in such a way that the integral (12.10) is finite for the extension. It is clear that this property is necessary for the applicability of the variational principle. The question was whether it was possible to use it for arbitrary, continuous boundary data. Prym, and later Hadamard, gave a negative answer to this question. Their counter-examples are quite different, and each is instructive in itself.

Prym investigated the function

$$u_0(r,\theta) = [(\log r)^2 + \theta^2]^{1/4} \sin\left[\frac{1}{2}\arctan\frac{\theta}{\log r}\right]$$

in the disc with center $(1, 0)$ and the radius 1, i.e., $|z - 1| < 1$ with $z = r\exp i\theta$. This function is harmonic and equal to $-\Im((-\log z)^{1/2})$. The only "bad" point for $u_0$ is the origin. On the circle $r = 2\cos\theta$, $|\theta| < \pi/2$, which bounds the domain, $u_0$ has the behaviour $u_0 = O(|\log r|^{-1/2})$, and consequently the boundary values of the function are continuous. The Dirichlet integral can be estimated as follows

$$D(u_0) = \int_{|r\exp i\theta - 1| < 1}\left[\left(\frac{\partial u_0}{\partial r}\right)^2 + \frac{1}{r^2}\left(\frac{\partial u_0}{\partial \theta}\right)^2\right]r\,dr\,d\theta$$

$$> \frac{1}{4}\int_0^1\frac{dr}{r}\int_{-\pi/4}^{\pi/4}\frac{d\theta}{\sqrt{(\log r)^2 + \theta^2}}.$$

---

[3]A historical commentary on this event appears in the article by Lebesgue *En marge du calcul des variations*, which was found after his death and published in 1963. Concluding the article, Lebesgue writes that Hadamard's paper on the Dirichlet principle justifies him dedicating "this small critical study to his memory".

For small values of $r$ the inner integral is bounded below by $c|\log r|^{-1}$, where $c$ is a positive constant. Hence, $D(u_0) = \infty$ and so the Dirichlet principle is not applicable for the boundary value $g = u_0|_{\partial\Omega}$.

Hadamard considered a function $g$ defined on the boundary $|z| = 1$ of the unit disc $\Omega = \{z : |z| < 1\}$, which corresponds to a Fourier series

$$g(\theta) = \frac{a_0}{2} + \sum_{n \geq 1}(a_n \cos n\theta + b_n \sin n\theta).$$

A harmonic function in $\Omega$ whose values on the circumference are $g(\theta)$, is given by the Fourier series

$$u(r, \theta) = \frac{a_0}{2} + \sum_{n \geq 1} r^n(a_n \cos n\theta + b_n \sin n\theta),$$

which is convergent for each $r < 1$ and is differentiable term by term. Writing the integral (12.10) in polar coordinates

$$D(u) = \int_0^1 \int_0^{2\pi} \left[\left(\frac{\partial u}{\partial r}\right)^2 + \frac{1}{r^2}\left(\frac{\partial u}{\partial \theta}\right)^2\right] r\,dr\,d\theta$$

and applying Parseval's formula, Hadamard obtained the identity

$$D(u) = \pi \sum_{n \geq 1} n(a_n^2 + b_n^2) \tag{12.12}$$

and concluded that the Dirichlet principle is not applicable to the function $g(\theta)$ when the series on the right-hand side of (12.12) diverges. An example of an "inadmissible" function $g(\theta)$ is given by the uniformly convergent series

$$\sum_{n \geq 1} 2^{-n} \cos(2^{2n}\theta).$$

Thus the examples of Prym and Hadamard show that the variational principle is not true although the boundary value problem (12.11) is solvable.

The finiteness of the right-hand side of formula (12.12) determines the boundary values (in terms of their Fourier coefficients) for which the Dirichlet principle is applicable. In an article on Plateau's problem and minimal surfaces, written in 1931, Douglas proved that when $u$ is a solution of the Dirichlet problem (12.11) where $\Omega$ is a unit disc, then

$$D(u) = \frac{1}{8\pi} \int_0^{2\pi} \int_0^{2\pi} \frac{|g(\theta) - g(\varphi)|^2}{(\sin[(\theta - \varphi)/2])^2}\,d\theta\,d\varphi. \tag{12.13}$$

The identity (12.13) was used in the paper by Beurling [III.35] in his investigation of so-called exceptional subsets of the circle.

Double integrals of the type (12.13) give a characterization of boundary values of functions with finite integral $D(u)$ for an arbitrary domain whose boundary is not too bad. This was realized only in the 1950s, when the representation (12.13) was rediscovered, and together with the identity (12.12) was a starting point for the extensive theory of spaces of functions with "fractional smoothness" (Aronszajn, Babich, Slobodetskiĭ, Gagliardo, Peetre, Lions and Magenes, and many others).[4] The modern theory of boundary value problems of partial differential equations is unthinkable without these spaces.

## 12.4 Lectures on the calculus of variations

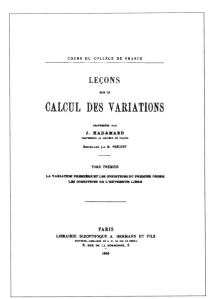

Hadamard's main contribution to the calculus of variations was his treatise [I.159] which contained material from lectures given at the *Collège de France*. From the middle of the nineteenth century, the calculus of variations developed mainly under the influence of the German mathematicians Jacobi, Weierstrass, Clebsch, Mayer, and Hilbert. Kneser's monograph on the subject [III.211] published in 1900, and Bolza's textbook of 1909 [III.46], both were written in German. Although some French treatises in analysis contained chapters on the calculus on variations, the only book in French devoted to this subject was published in 1861 by Lindelöf. Hadamard's book therefore filled a gap in the French literature. Carathéodory in his detailed review [III.67] called it "a book which marked a date in the history of this chapter of Analysis".

The book begins with the simplest problem of the calculus of variations, i.e., with the study of extremum of the integral (12.1). Then isoperimetric problems are studied, in which the extremum of $U[y]$ is sought under

---

[4]In modern terms, formula (12.13) means that $H^{1/2}(\partial\Omega)$ (the space of functions defined on $\partial\Omega$ with derivatives of order $1/2$ which are square summable on $\partial\Omega$) is the trace space of $H^1(\Omega)$ (the space of functions defined on $\Omega$ which have finite Dirichlet integral). The examples of Prym and Hadamard state that the space $C(\partial\Omega)$ is not imbedded in $H^{1/2}(\partial\Omega)$.

the additional conditions $V_j[y] = A_j, 1 \leq j \leq N$, where the $A_j$ are known numbers and $V_j[y]$ are functionals similar to $U[y]$. Hadamard next considers the Mayer problem, which can be stated as follows. Suppose that the vector function $\mathbf{y} = (y_1, \ldots, y_m)$, defined on the interval $[0, a]$, satisfies the equations

$$g_j(x, \mathbf{y}, \mathbf{y}') = 0, \qquad j = 1, \ldots, N.$$

The vector $\mathbf{y}$ is given at the endpoint 0 and the components $y_2, \ldots, y_m$ are prescribed at the endpoint $a$. One seeks the extremum of $y_1(a)$. The first two parts of Hadamard's book contain "first order" necessary conditions which have to be satisfied by any minimizer (or maximizer) of this and other variational problems. The second variation appears at the end of the book. Hadamard also gives necessary and sufficient conditions for extrema, and in particular, he describes the Weierstrass theory. Furthermore, he considers variational problems with discontinuous solutions, higher order derivatives, and one-sided variations. This exposition is illustrated by examples from geometry and mechanics.

One could say that here Hadamard remained within the traditional calculus of variations, but one should not trust this superficial impression. Although he did not indicate his own results or proofs, Hadamard added something new to the exposition of almost all difficult questions, analysed special cases, and filled in the gaps in the arguments of his predecessors. Paul Lévy writes:

> [...] to resolve a difficult problem, he did not hesitate to change its statement. The problem was unsolvable because it was ill-posed; it became solvable when it was well-posed, and Hadamard was well able to devise the necessary methods. In his calculus of variations, there is no theorem which one can point out as being his, in contrast with other parts of his work. Rather, one can say that the calculus of variations was an edifice whose framework had been erected by Euler, Lagrange, Jacobi and Weierstrass. However, it lacked cohesion. Despite the efforts of Weierstrass, the real conditions for the extremum were not well-known. Even for a simple problem such as that of a plane curve joining two points, rotating about a straight line in the same plane to generate the least possible surface area, the cases where the minimum is realized by a regular curve, which was a solution of the Euler equation, had not been precisely defined. Hadamard came, examined the difficulties with a rare clarity, and left a finished work where previously there had only been an outline [II.37, p. 16].

Furthermore, Hadamard included in the book an absolutely non-traditional chapter on the calculus of functionals, the topic to be considered in the next section.

## 12.5 Functional analysis

In 1887 Vito Volterra started publishing a series of articles, in which he laid the foundations of what he called "the theory of functions of lines". By this he understood mappings defined on certain sets of function with real values. Volterra extended many notions of algebra and analysis, including that of derivative, to these functions of lines. He considered various types of equations with such "functional"(or "variational") derivatives, and showed that the differential, integral and integro-differential equations arising in the calculus of variations, can be interpreted as different types of equations with functional derivatives.

From the very beginning, Hadamard was an ardent exponent of Volterra's theory of functions of lines. He enriched it with many results which are important in mathematical physics, and his *Leçons sur le calcul des variations* was the first book in which the ideas of modern functional analysis appear.[5]

In his talk *Le développement et le rôle scientifique du calcul fonctionnel* [I.278] at the International Congress of Mathematicians in 1928, in Bologna, Hadamard described the achievements in this area during the first few decades of its existence. Nowadays, the functional analysis includes such areas as the theory of Banach and Hilbert spaces, normed algebras, analytic semigroups, and distributions. All of these, despite their individual features, are unified by the fact that they are infinite-dimensional spaces endowed with a certain algebraic structure and topology.

Let $\mathcal{A}$ be a set of functions. If, for any $y \in \mathcal{A}$ there corresponds a number $U[y]$, then one says that $U$ is a functional defined on $\mathcal{A}$. This term was introduced by Hadamard to replace the earlier terminology "function of lines"of Volterra. In his paper *Sur les opérations fonctionnelles* [I.106], published in 1903, Hadamard used the term *opération fonctionnelle*. However, he soon replaced even it by the term *fonctionnelle* (functional), as he noted in a letter to Fréchet written at some time in 1904 or 1905: "I decided

---

[5]G. Fichera's paper [III.132, p. 171], contains interesting counter-arguments to the following Dieudonné's remark: "We must finally mention the first attempt at 'Functional Analysis' of the young Volterra in 1887, to which, under the influence of Hadamard, has been attributed an exaggerated historical importance".

to call *functionals* the 'functions of functions' or 'functions of lines'. $U$ is a functional of $\phi(x)$. In any case, I propose to you this term" [III.391, p. 251].

The subject of Volterra's theory was non-linear functionals. However, even the simpler theory of linear functionals was far from being thoroughly investigated. The problem is that, as well as linearity $U[c_1 x + c_2 y] = c_1 U[x] + c_2 U[y]$, the definition of a linear functional should include a continuity condition, which can have a different meaning, depending on the sense of convergence in the space of functions upon which the functional is given. This topological aspect makes the description of linear functionals on various function spaces rather non-trivial.

The first answer to such a question was given by Hadamard. In his paper *Sur les opérations fonctionnelles* [I.106], Hadamard showed that an arbitrary linear functional $U[y]$ on the space $C[a,b]$ of functions $y$, which are continuous on the interval $[a,b]$ can be represented in the form:

$$U[y] = \lim_{\lambda \to \infty} \int_a^b F(t,\lambda)y(t)dt,$$

where $F$ is independent of $y$ and defined by the functional $U$ (not uniquely), and is continuous on the half strip $\{(t,\lambda) : a \le t \le b, \lambda > 0\}$. This result preceded the well-known Riesz representation [III.345] obtained in 1909, for the same functional which states that:

$$U[y] = \int_a^b y(t)d\psi(t).$$

Here $\psi$ is a function of bounded variation on $[a,b]$, which is uniquely defined by the functional.

The problem of describing a linear functional $U[w]$ on the set of analytic functions $w(z)$ of the complex variable $z$, is of great interest. It is generally accepted that it was solved in the 1920s by the Italian mathematician and theologian Fantappiè, whose work dealt with analytic functions, differential equations and their applications to mathematical physics. Indeed, his main contribution is a vast theory of analytic functions of one and several variables. In fact, the first representation of $U[w]$ as a line integral was obtained by Hadamard, not later than 1910 [I.159, p. 293-294]. Here, the notion of the indicator of a functional plays an important role. This indicator is the function

$$\varphi(\zeta) = U\left[\frac{1}{\zeta - z}\right]. \tag{12.14}$$

Hadamard's argument is the following. Let $w(z)$ be a regular function on a domain and its boundary $\gamma$. Multiplying both sides in (12.14) by $w(\zeta)$ and integrating over the contour $\gamma$, we obtain, from the linearity of $U$ that

$$U\Big[\frac{1}{2\pi i}\int_\gamma \frac{w(\zeta)}{\zeta - z}d\zeta\Big] = \frac{1}{2\pi i}\int_\gamma w(\zeta)\varphi(\zeta)d\zeta,$$

from which, using the Cauchy integral formula

$$\frac{1}{2\pi i}\int_\gamma \frac{w(\zeta)}{\zeta - z}d\zeta = w(z),$$

we find the required representation of the functional

$$U[w] = \frac{1}{2\pi i}\int_\gamma w(\zeta)\varphi(\zeta)d\zeta. \tag{12.15}$$

Here we can take $\gamma$ to be any closed curve separating the domains where the singularities of $w$ and $\varphi$ lie. Let us note that later, Fantappiè used another indicator for the same purpose, namely

$$\psi(\zeta) = U\Big[\frac{1}{1 - z\zeta}\Big],$$

which he called the symmetric indicator.

Hadamard's representation (12.15) is a special case and precursor of general theorems on the structure of the dual of the space of holomorphic functions on a plane compact set. A somewhat naive form of such a theorem was proved in 1928 by Fantappiè [III.128], who studied mainly non-linear functionals. He was interested in the so-called analytic functionals, which transform an arbitrary analytic curve in a certain space into an analytic function of one complex variable. By an analytic curve one means a mapping $\gamma : \Omega \to X$ which depends analytically on its argument, where $\Omega$ is a domain in the complex plane $\mathbf{C}$ and $X$ is the space under consideration. As a by-product, Fantappiè described the structure of linear analytic functionals where $X$ is the space of functions which are analytic in a neighbourhood of a compact set $K \subset \mathbf{C}$. The naivity of Fantappiè's approach lies in the fact that he, as Hadamard before him, avoided the topology and assumed analyticity in the sense of his general theory, instead of using continuity. A survey of Fantappiè's theory of analytic functionals was written by Pellegrino and is included in the second edition of P. Lévy's book *Problèmes concrets d'analyse fonctionnelle* [III.240] (see also [III.249]).

The natural topology of $X(K)$ is rather complicated and the modern ideas on this object appeared in the 1950s, when the general theory of linear topological spaces was created. The first result to be stated in modern terms, a description of the dual of $X(K)$, was given by Köthe in 1949, and Grothendieck obtained the result for vector-valued holomorphic functions in 1953. The answer is quite similar to that obtained by Hadamard: an arbitrary continuous functional $U$ is given by (12.15), where $\gamma$ is a system of contours depending on $w$, inside which $w$ is analytic, and $\varphi$ is an analytic function in $\mathbf{C} \backslash K$, that vanishes at infinity. Neither Köthe nor Grothendieck cite Hadamard as the first to prove this result.

A linear functional can be considered as the principal part of a small change (variation) of a nonlinear functional. This idea played an important role in Hadamard's studies of the variation of Green's function under variation of the domain (cf. Section 14.3). Under Hadamard's influence, a whole new area in non-linear analysis was created. In particular, Fréchet and Gâteaux gave definitions of the differential of a nonlinear functional which have become classical. A theory of integration on function spaces was developed by Gâteaux and P. Lévy, an account of which is given in Lévy's *Leçons d'analyse fonctionnelle* [III.239] published in 1922. The above mentioned Lévy's book *Problèmes concrets d'analyse fonctionnelle* [III.240] contains, along with other material, a survey of definitions of the differential due to Hadamard, Gâteaux, Fréchet, and himself.

Both representations for linear functionals and variational formulas for Green's function are discussed in the chapter on the calculus of functionals in Hadamard's book [I.159] described briefly in Section 12.4. The inclusion of this material reflected Hadamard's general concept, which he stated in the preface to the book:

> The calculus of variations is nothing else than the first chapter of the theory which is nowadays called the calculus of functionals, and whose development will undoubtedly be one of the first tasks of analysis in the future. It is this idea which inspired me above all, in the course of lectures I gave on this topic at the *Collège de France* as well as in the preparation of this work.

The role of Hadamard as a "critic and catalyst" in the development of functional analysis and even in a broader context is discussed in the papers of J.D. Gray [III.156] and R. Siegmund-Schultze [III.364].

# Chapter 13

# Miscellaneous Topics

In this chapter we write about some of Hadamard's works in analysis, algebra, probability, topology, and set theory. These studies were not central for him, and his influence on their future development was either small or indirect, mainly through the works of others as, for example, in the case of functional analysis. Among these dispersed results and ideas are many quoted as well as forgotten, and they are also of different depth and value. The variety reflected Hadamard's insatiable curiosity and the breadth of his interests which were spread through almost all mathematics. He even left a three-page-long paper [I.301] on Lie's contact transformations despite his own confession that he felt "the insuperable difficulty of mastering more than a rather elementary and superficial knowledge" of the group theory and, in particular, of Lie's theory [I.372, p. 115]. Although there is no mention of great theorems by Hadamard in this chapter, it adds interesting features to his mathematical portrait.

## 13.1   A determinantal inequality

**The inequality**. In Hadamard's works on the theory of analytic functions, an important role is played by determinants which are defined by the coefficients of the series representing the functions. It is quite possible that, while writing his thesis, he considered the problem of estimating an arbitrary $n$th order determinant. In the article *Sur le module maximum que puisse atteindre un déterminant* [I.16] which appeared in 1893, soon after his thesis, he proved that for any $n$th order determinant $\Delta = \det(a_{ij})$ with complex entries

$$|\Delta| \le \prod_j (\sum_i |a_{ij}|^2)^{1/2}, \qquad (13.1)$$

which became a classical inequality. Beckenbach and Bellman wrote: "This result of Hadamard has attracted, and continues to attract, considerable attention. There are perhaps a hundred proofs available in published and unpublished form" ([III.22, p. 89]). It follows from (13.1) that

$$|\Delta| \le M^n n^{n/2} \qquad (13.2)$$

when $|a_{ij}| \le M$. This estimate is considerably sharper than $|\Delta| \le M^n n!$, which is obtained directly if we expand $\Delta$ in the usual way.

*Lord Kelvin (1824-1907)*

In fact, the inequality (13.1) was first conjectured by W. Thomson (Lord Kelvin) and proved by Muir, for real $a_{ij}$'s (1885) (see [III.287], [III.288]). For such $a_{ij}$ this inequality has a simple geometrical meaning. If one denotes $\mathbf{r}_i = (a_{i1}, \ldots, a_{in})$, then $\Delta$ is, up to the sign, the volume $V$ of the parallelepiped defined by the vectors $\mathbf{r}_i$. The inequality (13.1) means that, for arbitrary rotations of the vectors $\mathbf{r}_i$, the volume $V$ attains its largest value when they are orthogonal. For $n = 3$, this follows from the formula $V = Sh$ where $S = |\mathbf{r}_1||\mathbf{r}_2|\sin\varphi, h = |\mathbf{r}_3|\cos\theta$, where $\varphi$ is the angle between $\mathbf{r}_1$ and $\mathbf{r}_2$ and $\theta$ is the angle between $\mathbf{r}_3$ and the plane containing $\mathbf{r}_1, \mathbf{r}_2$.

In the case of an arbitrary $n$ a simple proof of inequality (13.1) can be obtained with the help of the orthogonalisation process of linear independent vectors $\mathbf{r}_1, \ldots, \mathbf{r}_n$, which consists in the construction of vectors $\mathbf{s}_1, \ldots, \mathbf{s}_n$ by the formula

$$\mathbf{s}_1 = \mathbf{r}_1, \ \mathbf{s}_j = \mathbf{r}_j - \frac{(\mathbf{r}_j, \mathbf{s}_{j-1})}{|\mathbf{s}_{j-1}|^2} \mathbf{s}_{j-1} - \ldots - \frac{(\mathbf{r}_j, \mathbf{s}_1)}{|\mathbf{s}_1|^2} \mathbf{s}_1, \ j = 2, \ldots, n.$$

Since $(\mathbf{s}_i, \mathbf{s}_j) = 0$ for $i \neq j$, then $|\mathbf{s}_i| \leq |\mathbf{r}_i|$. Clearly,

$$\begin{aligned} \Delta &= \det(\mathbf{r}_1, \ldots, \mathbf{r}_n) = \det(\mathbf{s}_1, \mathbf{r}_2, \ldots, \mathbf{r}_n) = \\ &= \det(\mathbf{s}_1, \mathbf{s}_2, \ldots, \mathbf{r}_n) = \det(\mathbf{s}_1, \ldots, \mathbf{s}_n). \end{aligned}$$

So $\Delta = |\mathbf{s}_1| \ldots |\mathbf{s}_n| \det U$, where $U$ is a unitary matrix. Since $\det U = 1$, then

$$|\Delta| \leq |\mathbf{r}_1| \ldots |\mathbf{r}_n|,$$

which is equivalent to (13.1).

It is well known that any positive definite matrix $B = (b_{ij})$ can be represented in the form $AA^*$, where $A$ is an $(n \times n)$-matrix and $A^*$ is its adjoint. From this we obtain

$$\det B \leq \prod_{i=1}^{n} b_{ii}. \tag{13.3}$$

This inequality is also often called Hadamard's inequality [III.288]. The following elegant argument leading to (13.3) yields at the same time one more proof of (13.1).

Let $D$ be the diagonal matrix with elements $d_i = b_{ii}^{-1/2}$ at the main diagonal. Clearly,

$$\det B = d_1 \ldots d_n \det(DBD). \tag{13.4}$$

Let $\lambda_1, \ldots, \lambda_n$ denote the eigenvalues of the matrix $DBD = (c_{ij})$. The sum of all $\lambda_i$ equals the trace of the matrix $DBD$. Hence by the inequality for the arithmetic and geometric means one has

$$\lambda_1 \ldots \lambda_n \leq (\frac{1}{n} \sum_{i=1}^{n} \lambda_n)^n = (\frac{1}{n} \sum_{i=1}^{n} c_{ii})^n = 1$$

which together with (13.4) gives (13.3). From this proof it follows that equality in (13.3) is valid if and only if $B$ is a diagonal matrix, i.e., if and only if $A$ is unitary.

In 1917 Szász [III.383] proved the following extension of (13.3) for a positive definite matrix $B$. Let $P_k$ denote the product of all principal $k$-th order minors of $B$. Then

$$P_1 \geq P_2^{1/\binom{n-1}{1}} \geq P_2^{1/\binom{n-1}{2}} \geq \cdots \geq P_{n-1}^{1/\binom{n-1}{n-2}} \geq P_n.$$

The inequality $P_1 \geq P_n$ between the extreme terms is clearly equivalent to (13.3).

The generalized Hadamard inequality for determinants is the inequality whose geometric meaning is that the volume of a parallelepiped does not exceed the product of volumes of any two of its complementary faces. In particular,

$$|\det(\mathbf{r}_1, \ldots, \mathbf{r}_n)| \leq |\mathbf{r}_n||\det(\mathbf{r}_1, \ldots, \mathbf{r}_{n-1})|,$$

which immediately leads to the Hadamard inequality (cf. [III.142]). References to works on other generalizations of Hadamard's inequality, as well as different proofs of it, can be found in the books [III.22], [III.142].

**Application to integral equations**. Our account would be incomplete if we do not mention the role of inequality (13.1) in the theory of integral equations, developed at the beginning of this century. In 1900, Fredholm studied equations of the form

$$\psi(x) - \lambda \int_a^b K(x,t)\psi(t)dt = f(x), \tag{13.5}$$

which later were named after him. Here $\psi$ is an unknown function, the kernel $K$ and $f$ are given, and $\lambda$ is a complex parameter. The kernel is assumed to be continuous on the square $a \leq x, t \leq b$ and the function $f$ is piecewise continuous for $a \leq x \leq b$. Approximating the integral by a finite sum, Fredholm reduced equation (13.5) to a system of $n$ linear algebraic equations, derived an approximate expression for $\psi(x)$ and then, passing to the limit as $n$ tends to infinity, he obtained the exact solution in the form

$$\psi(x) = f(x) + \lambda \int_a^b \Gamma(x,t;\lambda)f(t)dt,$$

where

$$\Gamma(x, t; \lambda) = \frac{\mathcal{D}(x, t; \lambda)}{\mathcal{D}(\lambda)},$$

is called the resolvent of equation (13.5). Here, the numerator and the denominator are given by the series

$$\mathcal{D}(x, t; \lambda) = \sum_{x=0}^{\infty} \beta_n(x, t) \lambda^n, \qquad \mathcal{D}(\lambda) = \sum_{n=0}^{\infty} \alpha_n \lambda^n$$

with coefficients $\alpha_n, \beta_n$ expressed as certain determinants of order $n$. Starting from inequality (13.1), Fredholm proved that both these series are entire functions of $\lambda$, and that, consequently, the quotient $\Gamma$ is a meromorphic function of $\lambda$.

Fredholm obtained inequality (13.1) independently, unaware of Hadamard's paper of 1893. Indeed, in a letter to Mittag-Leffler on August 8, 1899, he wrote [III.137]:

> The convergence may be proved with the aid of a theorem on determinants which I have not seen cited anywhere, and which is as follows:
>
> $$\begin{vmatrix} a_{11} & \cdots & a_{1n} \\ \cdots & \cdots & \cdots \\ a_{n1} & \cdots & a_{nn} \end{vmatrix} < \prod_{i=1}^{n} \sqrt{a_{i1}^2 + a_{i2}^2 + \cdots + a_{in}^2}.$$

However in a short article *Sur une nouvelle méthode pour la résolution du problème de Dirichlet* [III.138], presented on January 10, 1900, he stated: "I now assert that $\mathcal{D}$ is an entire function of $\lambda$. This is an almost immediate consequence of a theorem by Hadamard".

The following letter of Hadamard to Fredholm is undated, but judging from its content it appears to have been written about 1900:

> Dear Monsieur Fredholm,
> Painlevé tells me that you have recently obtained a new proof of the existence of the minimum for an integral of the type
>
> $$\iiint \left[ \left( \frac{\partial V}{\partial x} \right)^2 + \left( \frac{\partial V}{\partial y} \right)^2 + \left( \frac{\partial V}{\partial z} \right)^2 \right] dx\, dy\, dz$$
>
> and, consequently, a new proof of the Dirichlet principle.

This year, I am again giving my course on the calculus of variations and I intended presenting Hilbert's method, or at least what we know of it, because I do not know of a complete article on this subject. It would be very interesting for me to have what you have done on this important subject. From what Painlevé tells me, I did not know if your work already appeared. Could you be so kind as to send me, for example, the galley proofs when you receive them, or inform me in some other way? I would be very grateful to you for this.

I have received your latest Note and thank you for it. As you know, I always follow your work on this point with great interest.

Yours sincerely,

J. Hadamard. [IV.48]

Describing his missed opportunities in the book [I.372, p. 52], Hadamard wrote:

> To continue about my failures, I shall mention one which I particularly regret. It concerns the celebrated Dirichlet problem which, for years, I tried to solve in the same initial direction as Fredholm did, i.e., by re-writing it as a system of an infinite number of equations of the first degree in an infinite number of unknowns. But physical interpretation, which is in general a very sure guide and had been most often such for me, misled me in this case.

However, he could console himself with the fact that his inequality was very important for Fredholm's theory.

**Hadamard's matrices**. After proving (13.1) in [I.16], Hadamard turned to the question of the sharpness of inequality (13.2). He noticed that equality holds in (13.2) if and only if the absolute values of elements $a_{ij}$ are equal and the vectors $\mathbf{r}_i$ and $\mathbf{r}_j$ are orthogonal for $i \neq j$.

This class of determinants had already been considered by Sylvester in 1867 [III.381] who, in particular, remarked that it contains the Vandermonde determinant

$$\Delta = \begin{vmatrix} 1 & \varepsilon_1 & \cdots & \varepsilon_{n-1} \\ 1 & \varepsilon_1^2 & \cdots & \varepsilon_{n-1}^2 \\ . & . & \cdots & . \\ 1 & \varepsilon_1^{n-1} & \cdots & \varepsilon_{n-1}^{n-1} \end{vmatrix},$$

where $1, \varepsilon_1, \ldots, \varepsilon_{n-1}$ are the $n$-th roots of unity.

For $n = 3$ this determinant exhausts all the extremal determinants (up to permutation of rows or columns and multiplication of each row or column by the same number). However, for $n \geq 4$, Hadamard showed that the situation becomes more complicated. For example, for $n = 4$ the extremal determinants are

$$\begin{vmatrix} 1 & 1 & 1 & 1 \\ 1 & -1 & e^{i\theta} & -e^{i\theta} \\ 1 & 1 & -1 & -1 \\ 1 & -1 & -e^{i\theta} & e^{i\theta} \end{vmatrix}$$

for any $\theta \in [0, 2\pi)$.

Especially important for applications is the case of extremal determinants of matrices with real-valued elements $\pm 1$ whose columns are orthogonal. Contrary to the case of complex matrices, they may not exist for all $n$. Such matrices are nowadays called Hadamard matrices or H-matrices.

In his paper *Résolution d'une question relative aux déterminants* [I.17] Hadamard proved that if $n$ is the order of a H-matrix, then $n = 1$ or $n = 2$ or $n$ is divisible by four. His argument is very short. Note first that the class of H-matrices is invariant under the change of sign of columns or rows and under their permutations. Hence, one may assume that the first row and the first column contain only $+1$. It is easily seen that for $n = 1$ and $n = 2$ Hadamard matrices are

$$1 \qquad \begin{pmatrix} 1 & 1 \\ 1 & -1 \end{pmatrix}$$

We turn to the case $n \geq 3$. Let the matrix have the following three columns

$$\begin{array}{llrrr} p & \text{rows of the form} & 1 & 1 & 1 \\ q & \text{rows of the form} & 1 & 1 & -1 \\ r & \text{rows of the form} & 1 & -1 & 1 \\ s & \text{rows of the form} & 1 & -1 & -1. \end{array}$$

Then

$$p + q + r + s = n,$$

and multiplying the columns pairwise we obtain

$$p + q - r - s = 0,$$

$$p - q + r - s = 0,$$

$$p - q - r + s = 0.$$

By summing up these four equalities we find $4p = n$, which completes the proof.

In the same article [I.17] Hadamard noticed that H-matrices exist for $n = 2^m, m = 2, 3, \ldots$, and gave a procedure for the construction of H-matrices of orders $n = 12$ and $n = 20$. The well-known conjecture that there exist Hadamard matrices for every order divisible by 4 has not been proved or disproved. The problem is discussed in [III.171].

Hadamard had no idea that these matrices would find applications in coding theory (cf., for instance, [III.261]), in design theory [III.33], communication engineering [III.328], statistics [III.336], and optics [III.168]. This has resulted in the appearance of many special terms bearing Hadamard's name which would greatly surprise him if he were still alive. For example, there is the Hadamard transform in optics, dealing with the construction of masks blocking or transmitting light, there are Hadamard computers, Hadamard imaging spectrometers, Hadamard chemical balance weighing designs, Hadamard interferometers, etc.

Hadamard's name is sometimes associated with quite another class of matrices (cf. [III.22], [III.25]) namely matrices satisfying the condition of dominance of its main diagonal

$$|a_{ii}| > \sum_{j \neq i} |a_{ij}|. \tag{13.6}$$

In Chapter I, p.13–14 of his book [I.109] Hadamard shows that the determinant of such a matrix does not vanish. Earlier statements of such a kind are due to L. Lévy, 1881 [III.237], Desplanques, 1887 [III.101], and Minkowski, 1900 [III.277].

Here is Hadamard's proof. Suppose that $\Delta = 0$. Then the system

$$\sum_{j=1}^{n} a_{ij} x_j = 0, \qquad i = 1, \ldots, n,$$

has a non-trivial solution. Let $x_k$ be a component of such a solution with the largest absolute value. Then

$$|a_{kk}||x_k| \leq |\sum_{j \neq k} |a_{kj} x_j| \leq \sum_{j \neq k} |a_{kj}||x_j|$$

and so,

$$|a_{kk}| \leq \sum_{j \neq k} |a_{kj}|,$$

which contradicts (13.6).

A. Ostrowski [III.298], 1937, using similar arguments, obtained the inequality

$$|\Delta| \geq \prod_{i=1}^{n} (|a_{ii}| - \sum_{j \neq i} |a_{ij}|),$$

and now quite a number of estimates of such a type are known.

Matrices with a "dominant" main diagonal, that is matrices satisfying condition (13.5), arise in various applications, for instance, in statistics, in the theory of stability of systems, in the theory of networks (for bibliography see O. Taussky [III.277], R.A. Horn, and C.R. Johnson [III.181]).

## 13.2 Some works in analysis and algebra

**Convergent and asymptotic series.** From time to time, Hadamard turned to different problems of the theory of series, those areas of classical mathematical analysis which one learns during a basic course in mathematics. Even in this domain, which was seemingly complete towards the end of the nineteenth century, he was able to add something new.

In 1894, Hadamard's paper [I.19] on the convergence with positive numerical series appeared in *Acta Mathematica*. This was his first publication in a foreign journal. The starting point of the article was Abel's result, which says that for any divergent series $\sum u_n$ with positive terms, one can construct a sequence $\varphi(n)$ converging to zero, such that the series $\sum u_n \varphi(n)$ diverges.

*Niels Henrik Abel (1802-1829)*

Du Bois-Reymond complemented this result by showing that however slowly a positive series $\sum u_n$ converges, one can always multiply its terms by the elements of a sequence $\varphi(n)$ going to infinity and such that the series $\sum \varphi(n)u_n$ converges.

Roughly speaking, as a corollary of these two theorems one can state the following result: given any convergent positive series, one can construct a second series converging more slowly than the first.

Hadamard generalized this fact to sequences of series, proving that one can always find a convergent series which converges more slowly than any one of the series in a given sequence of series.

One more result that Hadamard proved in [I.19] is the following: let $\{\varphi_p(n)\}_{p\geq 1}$ be a sequence of functions (with integer argument $n$), which are increasing and tend to infinity, with rates of growth which decrease as $p$ increases. Then one can find positive numbers $u_n$ such that the series $\sum u_n$ converges, whereas every series $\sum u_n\varphi_p(n)$ diverges.

Several years later, Hadamard returned to the topic in the article [I.91], where he answered the following question: how can one choose a sequence $\{\xi_n\}_{n\geq 1}$ such that given any convergent series $\sum u_n$, the series $\sum u_n\xi_n$ is also convergent? It turned out that the necessary and sufficient condition for this is that the series $\sum(\xi_{n+1} - \xi_n)$ converges.

Hadamard addressed the so-called asymptotic series in the articles [I.149], [I.184]. A formal series $\sum_{n=1}^{\infty} f_n(x)$ is called asymptotic for a function $f(x)$ as $x$ tends to $x_0$, if for $m = 1, 2, \ldots$

$$f(x) - \sum_{n=1}^{m} f_n(x) = o(f_m(x)) \qquad \text{as } x \to x_0.$$

Asymptotic series are not necessarily convergent. For example, if $f$ is infinitely differentiable but not analytic, its Taylor series is asymptotic without being convergent.

Poincaré, who gave the above definition in 1886 [III.317], started his study of asymptotic series with the following one due to Stirling (1730)

$$\frac{1}{2x} - \frac{1}{360x^3} + \cdots + (-1)^{n-1}\frac{B_n}{2n(2n-1)}x^{1-2n} + \ldots, \qquad (13.7)$$

where $B_n$ are the so-called Bernoulli numbers. Although the series is always divergent the sum of its first few terms represents the function

$$\log\left(\frac{e^x\Gamma(x+1)}{(2\pi)^{1/2}x^{x+1/2}}\right)$$

very accurately for large values of $x$.

In [I.184] Hadamard dealt with the problem of the transformation of divergent asymptotic series into convergent ones, a subject which had been treated earlier by Borel. Hadamard showed that if one adds $\Phi_n(x^{-1}, e^{-2\pi x})$ to the $n$-th term of series (13.7), where $\Phi_n$ is a certain polynomial of degree $2n-1$ in the first variable and of degree $n$ in the second, then the new series is asymptotic for the same function and is convergent.

A similar construction was proposed by Hadamard in [I.149] for the asymptotic series representing the Bessel function $J_0$ with real or purely imaginary arguments.

**Memoir on elimination**. In 1896 a large memoir by Hadamard [I.35] appeared in *Acta Mathematica*. It dealt with the theory of elimination of unknowns from a system of algebraic equations. Let us consider the simplest case of the system of two equations

$$f(x, y) = 0, \ g(x, y) = 0, \qquad (13.8)$$

where

$$\begin{aligned} f(x, y) &= a_0(y)x^n + a_1(y)x^{n-1} + \cdots + a_n(y), \\ g(x, y) &= b_0(y)x^m + b_1(y)x^{m-1} + \cdots + b_m(y), \end{aligned}$$

and the coefficients $a_i$ and $b_j$ are polynomials in $y$. If $a_0b_0 \neq 0$, then $f$ and $g$ have at least one common root when the resultant, $R(f, g)$ vanishes, the resultant being the determinant of a certain square matrix of order $n + m$ whose non-zero elements are $a_i(y)$ and $b_j(y)$. It is said then that the polynomial $R(f, g)$ is obtained as a result of the elimination of $x$ from the system (13.8). When the roots of the resultant are found, it remains only to substitute them into (13.8) and calculate the roots of two polynomials in $x$ obtained through the elimination of $y$.

Similar arguments can be used for the solution of the system

$$f_i(x_1, \ldots, x_n) = 0, \qquad i = 1, \ldots, n+1,$$

considered by Hadamard, but the analysis becomes substantially more complicated. In particular, the process of elimination depends on the way of numeration of the equations. Hadamard, however, showed that a change in the order of elimination only results in a change of sign in the resultants arising from the elimination. The final part of the paper deals with applications to the calculations of expressions

$$\prod Q(x_i, y_i) \text{ and } \sum Q(x_i, y_i),$$

where $Q$ is a ratio of polynomials and $(x_i, y_i)$ are intersection points of two given algebraic curves.

**Invertibility of point transformations.** In his paper [I.138], Hadamard formulated and solved the problem of the global inversion of point mappings. He considered the system of equations

$$X_k = f_k(x_1, \ldots, x_m), \quad k = 1, \ldots, m, \tag{13.9}$$

where the collection of functions $\{f_1, \ldots, f_m\}$ defines a mapping of $\mathbf{R}^m$ into itself. What are the conditions for the unique solvability of (13.9)? At first glance, the answer seems trivial and is known by an undergraduate student: the Jacobian

$$\frac{D(f_1, \ldots, f_m)}{D(x_1, \ldots, x_m)}$$

must be of constant sign. However, this conclusion neglects the restriction of locality in the inverse function theorem; moreover, it is erroneous because the system (13.9) can be unsolvable while the determinant is positive, even for $m = 1$. Indeed, one can not solve equation $X = \arctan x$ for an arbitrary $X$ although arctan is strictly increasing. In fact, the equation $X = f(x)$ with $f'(x) > 0$ on $-\infty < x < \infty$ is solvable if and only if

$$\int_{-\infty}^{0} f'(x)dx = \int_{0}^{+\infty} f'(x)dx = \infty.$$

In the multidimensional case, the positivity of the Jacobian also guarantees neither the global solvability of (13.9), nor the uniqueness of the solution, and an analogue of the above mentioned one-dimensional condition of global invertibility is required. In order to obtain this analogue Hadamard considered the length of the smallest axis of the ellipsoid of deformation:

$$\lambda(x) = \min\left(\frac{\sum_{1 \le i \le m}(dX_i)^2}{\sum_{1 \le i \le m}(dx_i)^2}\right)^{1/2}$$

(in the language of the time), i.e., the square root of the smallest eigenvalue of the matrix with $(j, k)$ element

$$\sum_{1 \leq i \leq m} \frac{\partial f_i}{\partial x_j} \frac{\partial f_i}{\partial x_k}.$$

He found the following sufficient conditions for unique solvability of (13.9):

$$\mu(\rho) > 0, \qquad \int_0^\infty \mu(\rho)d\rho = \infty, \tag{13.10}$$

where $\mu(\rho)$ is the minimum of $\lambda(x)$ on the sphere $x_1^2 + \ldots + x_m^2 = \rho^2$.

Hadamard obtained this result as a corollary of the following two sufficient conditions for solvability of system (13.9), where the functions $f_j$ are continuous and such that the mapping $x \to X$ transforms any continuously differentiable curve into a rectifiable one:

($i$) the mapping $x \to X$ is a local homeomorphism, i.e., is a one-to-one mapping of a neighbourhood of any point $x$ onto a neighbourhood of the corresponding point $X = f(x)$ which is continuous, together with its inverse;

($ii$) the image of any continuously differentiable curve going to infinity has infinite length.

Hadamard later showed in a short paper [I.198], that the conditions ($i$), ($ii$) are also necessary. This criterion is called Hadamard's global inverse function theorem. It was proved by Palais [III.302] in the following equivalent formulation: a $C^\infty$-map is a diffeomorphism on $\mathbf{R}^n$ if and only if its Jacobian matrix is nowhere singular and preimages of compact sets are compact.

New attention has recently been given to the development and application of these results of Hadamard (see [III.199], [III.314], [III.61], [III.329]). In particular, far-reaching generalizations to non-smooth mappings between pairs of arbitrary Banach spaces have been obtained. Applications of Hadamard's global inverse function theory to systems of ordinary differential equations are given in the paper [III.184], where references to earlier work in the same area can be found.

## 13.3   Hadamard and set theory

The birth of set theory goes back to the early 1870s, when Georg Cantor became interested in properties of infinite sets in connection with his study of trigonometric series. He noticed at the very beginning that the set of all rational numbers is countable, i.e. there exists a one-to-one correspondence between this set and the sequence of natural numbers $1, 2, 3, \ldots$. The countability of the set of algebraic numbers (1874) proved to be even more surprising. At the same time Cantor was able to prove the non-countability of all real numbers. Taken together, the two facts revealed that transcendentals are incomparably more numerous than algebraic numbers. (Note that about the same time Hermite proved the transcen-

*Georg Cantor (1845-1918)*

dence of $e$ and that already in 1844 Liouville had constructed a certain infinite class of transcendentals.) This is only one example of how the notion of one-to-one correspondence became in Cantor's hands the tool for comparison of infinite sets. Cantor said that two sets are of the same power (have the same transfinite number) if such a correspondence between their elements does exist. Hence, the rational and algebraic numbers are of the same power as are the integer numbers, whereas the real numbers are of a greater one, called the power of continuum. After vain attempts to prove the converse Cantor also showed that the plane and the space have the power of continuum. On the other hand he constructed infinitely many different transfinite numbers, thus introducing an "hierarchy of infinities". However, even a short account of Cantor's theory is out of the question here.

In the 1890s, when Cantor summarized his life work, his unorthodox ideas were not easily accepted. Hermite was the first in France to appreciate them, and Poincaré applied them to classical analysis in his studies of automorphic functions and differential equations in the 1880s.

As for Hadamard, he became an ardent advocate of set theory very early: his thesis (1892) already contained references to it. In Section 11.2 we mentioned that he found a striking application of the set theory in his study of geodesics on surfaces of negative curvature (1896), where Cantor's set theoretical construction appeared naturally in a classical geometrical problem.

In 1897, at the First International Congress of Mathematicians in Zurich, Hadamard gave a short talk entitled *Sur certaines applications possibles de la théorie des ensembles* [I.68], where he emphasized the utility of the study of sets of functions with properties different from those of numbers or $n$-dimensional vectors. Underlining the importance of the theory of such sets for mathematical physics and the calculus of variations, he wrote:

> But it is specially in the theory of partial differential equations of mathematical physics that studies of this kind will, without doubt, play a fundamental role. To cite only one example, it is thanks to these researches that one can give a solid foundation to the well-known reasoning which reduces the definition of integrals of these equations to questions of minima [I.406, v. 1, p 311].

Sur certaines applications possibles de la théorie des ensembles.

Par

J. HADAMARD à Paris.

Quoique la théorie des ensembles fasse abstraction de la nature des éléments, on a surtout considéré, jusqu'à présent, les ensembles composés de nombres, ou, tout au plus, de points dans l'espace à $n$ dimensions.

Il ne me semble pas inutile de signaler l'intérêt qu'il y aurait à étudier des ensembles composés de fonctions. De tels ensembles peuvent d'ailleurs présenter des propriétés tout autres que les précédents.

C'est ainsi que la question de la convergence des séries conduirait à rechercher un ensemble de fonctions d'une puissance supérieure à la première et bien ordonné, c'est-à-dire tel que l'on puisse non seulement indiquer l'ordre de deux éléments quelconques, mais assigner l'élément qui précède et celui qui suit immédiatement un élément donné.

Mais c'est principalement dans la théorie des équations aux dérivées partielles de la physique mathématique que des études de cette espèce joueraient, sans nul doute, un rôle fondamental. Pour n'en citer qu'un exemple c'est grâce à ces recherches qu'on arriverait à donner un fondement solide aux raisonnements bien connus qui ramènent la définition des intégrales de ces équations à des questions de minimum.

Il est clair, en effet, que de telles questions sont intimement liées à la nature du domaine dans lequel le minimum est recherché. Par exemple, le minimum d'une quantité $f(x, y, s)$ qui dépend continument des coordonnées d'un point existe toujours sur une surface fermée (ou dont les bords sont considérés comme faisant partie de la surface); il n'existe pas nécessairement si une ligne ou un point donné sont exclus.

Il faut toutefois remarquer que le problème présente des difficultés spéciales dans le cas du calcul des variations, la solution dépendant à

202    II. Teil: Wissenschaftliche Vorträge.

la fois de la nature du domaine et de celle de l'expression dont on étudie la variation. C'est ainsi qu'une intégrale étendue à un arc de courbe assujetti à avoir ses extrémités en deux point donnés, avec des tangentes données en ces points, admet en général un minimum si la fonction sous le signe $\int$ contient des dérivées secondes et n'en admet pas si cette fonction ne contient que des dérivées premières.

Il n'en est pas moins évident qu'il y aurait lieu d'étudier l'ensemble $E$ formé par les fonctions continues d'une variable comprise dans l'intervalle $(0, 1)$ et prenant aux extrémités des valeurs données, ainsi que les ensembles analogues.

Un des premiers problèmes qui se poseraient dans cette étude me paraît être le suivant:

Divisons l'ensemble considéré en ensembles partiels $E'$ tels que deux fonctions inférieures à l'un quelconque d'entre eux aient une distance (au sens de Weierstrass) moindre qu'un nombre déterminé $\delta$. Considérant tous ces $E'$ comme autant d'individus, on peut dire que l'ensemble qui a ces $E'$ pour éléments numère l'ensemble $E$. C'est cet ensemble numérant dont il conviendrait d'étudier les propriétés et, en premier lieu, la puissance.

*Hadamard's talk at the First International Congress of Mathematicians in Zurich*

As one of the first problems arising in the study of such classes, Hadamard proposed the calculation of the cardinality of the set of subsets of continuous functions having small prescribed diameter. There are neither metric spaces nor compact sets in Hadamard's report, but Fréchet, influenced by it, introduced both notions a few years later [III.391, p. 259]. Hadamard's intuitive idea found its exact formulation much later. In 1956 Kolmogorov, proceeding from certain concepts in information theory, defined the $\varepsilon$-entropy of a compact metric space as the logarithm of the smallest number of sets of diameter $2\varepsilon$ covering the space [III.215]. The study of this and other quantitative characteristics of compact sets which appeared later now constitutes a large part of approximation theory.

At the turn of the century, inspired by the discovery of paradoxes (antinomies) of set theory, a discussion unfolded on foundations of mathematics, and in particular, on mathematical proofs of existence. Poincaré, Zermelo, Hilbert, Russell, Borel, Baire, Lebesgue, Richard, and other mathematicians took part in it and there was no unanimity on these questions. It would have been surprising if Hadamard, with his remarkable energy, had kept away from the debates.

Logical paradoxes were known from ancient times. An example is the story of the Cretan Epimenides who said that all Cretans were liars. So, if he said the truth, he was a liar. About 1895 Cantor discovered the first paradox in set theory (too technical to be quoted here) and described it in a letter to Hilbert. Soon after that, the same paradox was rediscovered and published in 1897 by Burali-Forti. In the following years other paradoxes appeared, among them the following question, due to Rusell (1902): is the set of all sets, which are not their own elements, an element of itself? Both positive and negative answers are wrong.

Poincaré vividly describes his discussions with Hadamard on the paradox of Burali-Forti:

> One day Mr. Hadamard came to see me and the talk fell upon this antinomy. "Burali-Forti's reasoning," I said, "does it not seem to you irreproachable?" "No, and on the contrary I find nothing to object to in that of Cantor. Besides, Burali-Forti had no right to speak of the aggregate of *all* the ordinal numbers".

The rest of the conversation requires special knowledge, so we omit it, but here is its conclusion in Poincaré's words: "It was in vain, I could not convince him (which besides would have been sad, since he was right)" [III.321, p. 459].

Ernst Zermelo (1871-1953)          René Baire (1874-1935)

In 1904 Zermelo published his paper [III.427] in which he proved the possibility of well-ordering of an arbitrary set. His proof was based on the so-called axiom of choice (Zermelo's axiom), according to which one can construct from an arbitrary system of sets, a new set $M$, by choosing one element from each set of the system. Although this axiom is used in the proofs of various theorems, it is open to criticism, since one cannot always make explicit the rule according to which the set $M$ is constructed.

Borel reacted to this paper by an article [III.51] in which he expressed his objections to the legitimacy of applying Zermelo's axiom to a system of subsets of the continuum. The possibility of well-ordering was also doubtful for him, since in his opinion, Zermelo showed the equivalence of such a possibility with the axiom of choice. Borel wrote letters to Hadamard, Baire, and Lebesgue, asking them to give their opinions concerning his criticism of Zermelo's paper. Their answers, together with one more letter from Borel, formed the article *Cinq lettres sur la théorie des ensembles* [I.123] (1905).

In particular, Lebesgue wrote to Borel:

> The question reduces to the following not novel one: is it possible to prove the existence of a mathematical object without defining it? This is obviously a matter of convention; but I believe that one can only build solidly admitting that existence is proved only for an object which is defined.

Hadamard argued, insisting that the assertion of the existence of objects did not require describing them: "The existence ... is a fact like any other". He was Zermelo's only staunch defender. He urged that the axiom of choice be accepted on the same ground that he used to defend Cantor's work: its usefulness for the progress of mathematics was decisive. In a letter to Borel he wrote: "Thus, there are being present two conceptions of mathematics, two mentalities. I do not see, in all that has been said until now, any reason to change mine. I do not intend to impose it" [I.123].

This polemic about Zermelo's axiom was an interesting part in a larger picture of confrontation of opinions on cardinal questions of the foundations of mathematics.

M. Henri POINCARÉ,
Membre de l'Académie Française,
Membre de l'Académie des Sciences.
Cliché Henri Manuel, phot., Paris.

## 13.4  Hadamard and topology

Topology is the part of mathematics dealing with geometric properties which are invariant under continuous transformations. An example of such properties is provided by Jordan's theorem (1893), which is natural but not easy to prove, and which states that any plane curve which does not intersect itself divides the plane into two components. The subject was in the embryonic state until the end of the nineteenth century, with separate facts being found by Euler, Riemann, Möbius, Jordan, Kronecker, Betti, and others.[1] In the 1880s Klein and Poincaré used topological concepts in their work on automorphic functions. Motivated by these applications as well as his own studies of dynamical systems, Poincaré realized the importance of topology as a special discipline, and between 1895 and 1904 he devoted to it a series of memoirs, in which the foundation of modern topology was laid down.

Camille Jordan (1838-1922)          Enrico Betti (1823-1892)

Hadamard was one of the few mathematicians attracted to topology immediately after Poincaré. In one of the lectures he gave at Columbia University [I.194], he explained that the number of solutions of system (13.9), where the functions $f_j$ map a surface onto itself, depends on the topological properties of the surface. His interest in topology manifested itself even earlier in connection with his studies of geodesics (see Sec. 11.2).

Hadamard's contacts with Brouwer date from 1909. Brouwer turned to Poincaré's *analysis situs* after his fundamental work in mathematical

---

[1]For an account of the prehistory and history of topology see the books by J.C. Pont [III.326] and J. Dieudonné [III.105].

logic and the foundations of mathematics, and in 1909-1913 he obtained important topological results. Among them one finds his celebrated fixed point theorem: any continuous mapping of the $n$-dimensional ball into itself leaves fixed at least one point,[2] as well as his theorem on the invariance of dimension under one-to-one continuous mappings. Note that Cantor tried to obtain this invariance for arbitrary one-to-one mappings but discovered that there is none.

LA

## GÉOMÉTRIE DE SITUATION

ET SON

### RÔLE EN MATHÉMATIQUES

LEÇON D'OUVERTURE

*du cours de Mécanique Analytique et de Mécanique céleste, faite au Collège de France, le 18 mai 1909.*

En prenant possession de cette chaire, je veux que mes premières paroles expriment ma profonde gratitude à tous ceux dont la confiance m'appelle à l'occuper: aux savants éminents du Collège de France et de l'Académie des sciences dont les suffrages m'ont désigné ; aux pouvoirs publics qui ont bien voulu ratifier leur choix.

Comment pourrais-je ne pas nommer ici tout de suite pour lui offrir l'hommage particulier de ma reconnaissance, M. Maurice Lévy ? Cette confiance qui m'honore si profondément, c'est lui qui me l'a témoignée tout d'abord. Je ne saurais oublier les années déjà nombreuses pendant lesquelles il m'a jugé digne de poursuivre une tâche qu'il avait remplie avec tant d'éclat, de continuer un enseignement dont tous ses auditeurs gardent le durable souvenir.

Ajouterai-je que cette charge, où je ne pouvais voir qu'un très grand honneur, a été pour moi, par surcroît, au point de vue scientifique, un très grand profit. L'enseignement du Collège de France, par la place qu'il réserve aux dernières découvertes, comme par les recherches originales qu'il suscite, obtient

*In 1909 Hadamard's opening lecture at the Collège de France was La géométrie de situation et son rôle en mathématiques, a talk on the history of topology.*

However, Hadamard's role in the development of topology proved to be rather modest. Still, there is something about his contribution to be recounted. Besides popular articles he wrote only one work with his own topological results. That was the *Note sur quelques applications de l'indice de Kronecker* published in 1910 [I.163]. The index of a system of $n$ functions given on a $(n-1)$-dimensional closed surface was introduced by Kronecker in 1869 in the form of a certain integral and is a multi-dimensional generalization of rotation of a vector field on a plane.[3] Applying the notion

---

[2]Infinite dimensional versions of the last statement due to Schauder and Leray-Schauder are instrumental in the modern theory of non-linear differential and integral equations, where they provide existence theorems for solutions considered as fixed points in function spaces.

[3]Imagine that a non-zero continuous vector-valued function is prescribed on a closed oriented plane curve. The increment of the angle between the vector and a fixed direction when moving along the curve, divided by $2\pi$, is called rotation of the vector field.

of index, Hadamard found new, simpler proofs of some known topological facts, in particular, for Jordan's curve theorem.

ЗАМѢТКА О НѢКОТОРЫХЪ ПРИМѢНЕНIЯХЪ
УКАЗАТЕЛЯ КРОНЕККЕРА.

Жака Гадамара.

Доказательство, по Эису (Ames), теоремы Жордана о сомкнутыхъ кривыхъ безъ двойной точки (nᵒnᵒ 306, 307) покоится на разсмотрѣнiи порядка точки, или, если угодно, на разсмотрѣнiи измѣненiя аргумента.

Обобщенie, для случая, когда число измѣренiй превосходитъ два, даетъ указатель Кронеккера. Понятiе это стало теперь классическимъ (¹).

Во многихъ современныхъ работахъ оно получило новыя примѣненiя. Я ставлю себѣ задачей изложить нѣкоторыя изъ нихъ.

Всѣ послѣдующiя разсужденiя легко выражаются въ чисто арифметической формѣ, даже и тогда, когда, для краткости, они не редактированы въ такой формѣ непосредственно; впрочемъ, они и должны удовлетворять этому условiю, для пригодности въ допускаемыхъ нами общихъ предположенiяхъ, которыя вводятъ только непрерывность употребляемыхъ функцiй.

I. Теорема Жордана въ плоскости.

1.—Сначала, вернусь на минуту къ доказательству теоремы Жордана для случая плоскости и постараюсь, для одной изъ частей этой теоремы, пойти немного дальше того, что потребовалось при введенiи понятiя о порядкѣ.

Дана плоская линiя (*) (C), опредѣляемая двумя уравненiями

(1)                    $x = x(t), \quad y = y(t)$,

(¹) Особенно посля Traité d'Analyse, Пикара (т. I, стр. 123 и т. II, стр. 193).
(*) Въ текстѣ она была названа простой сомкнутой линiей (nᵒ 290).

*At the beginning of the twentieth century, the Russian translation of J. Tannery's Introduction à la théorie des fonctions d'une variable was a popular university textbook. Its second volume contains Hadamard's paper on Kronecker's index.*

In [III.200, p. 139] D. M. Johnson writes:

> Hadamard's *Note* is markedly similar to Brouwer's classic paper defining the degree of a mapping, *Über Abbildung von Mannigfaltigkeiten* (1911), which appeared a short time after the *Note*. This resemblance is to be expected. The two men met in Paris around New Year 1910 and discussed the ideas that are basic to both works. Yet there is hardly any doubt that Brouwer's is the superior work. Whereas Hadamard's *Note* stands at the end of a great line of mathematical development, in a sense putting the finishing touches on the presentation of an important mathematical idea, Brouwer's great paper looks forward to new avenues of topological thinking. Brouwer's work is much more revolutionary than the one by Hadamard.

An interesting observation on Hadamard's influence on Brouwer's topological work has been made by H. Freudenthal [III.139] who analysed a part of Brouwer's unpublished correspondence. He writes:

> In 1909 he [Brouwer] published investigation on continuous mappings of surfaces and on vector fields upon surfaces. He

proved the existence of a fixed point in any mapping of the 2-sphere belonging to the homotopy class of the identity (to say it in modern terminology), and the existence of a singularity in vector fields on the 2-sphere (without noticing that the last property, albeit under much rougher assumptions, had already been proved by H. Poincaré). At that moment Brouwer did not yet see the connection between these two theorems. A remark in a letter to J. Hadamard shows that this idea had come into his mind in the course of their correspondence. Probably it was also Hadamard who indicated to Brouwer Poincaré's priority with respect to the theorem on spherical vector fields [III.139, p. 495].

*L.E.J. Brouwer (1881-1966)*

Freudental also remarks: "Compared with Brouwer's revolutionary methods, Hadamard's were quite traditional, but Hadamard's assistance at the birth of Brouwer's ideas certainly resembled more that of midwife than a spectator" [III.139, p. 501].

## 13.5  Gagliardo-Nirenberg inequalities

The well-known saying that everything new is actually something old and forgotten can also be applied to one of Hadamard's results. In the fourth part of his memoir on elastic plates of 1908 [I.145], he proved the inequality

$$\max_{0\le x\le a} |u(x)|^{2+q} \le 2^{2+q} \int_0^a |u'(t)|^2 dt \int_0^a |u(t)|^q dt, \qquad (13.11)$$

for every continuously differentiable function $u$ on $[0, a]$, vanishing at least at one point of this interval,[4] and for every $q > 0$. It is this inequality we had in mind when in Section 12.2 we described Hadamard's new method for the solution of variational problems. The inequality was obtained as a particular case of the following lemma.

Let $f$ be a non-negative increasing function on the positive semi-axis, and let

$$A = \int_0^a f(|u(t)|)dt, \qquad B = \int_0^a |u'(t)|^2 dt.$$

If $u$ is continuously differentiable on $[0, a]$ and if $u$ vanishes at least at one point of $[0, a]$, then

$$|u(x)| \le 2C \qquad \text{for} \quad \text{all} \quad x \in [0, a],$$

where $C$ is a non-negative solution of the equation

$$C^2 f(C) = AB.$$

Hadamard's proof of this takes only a few lines. Let $x \in [0, a]$ be such that $|u(x)| > C$, and let

$$M(C) = \{t \in [0, a] : |u(t)| > C\}.$$

From the definition of the integral $A$, it follows that

$$f(C) \operatorname{mes} M(C) \le A.$$

Since the function $|u|$ equals zero at some point of $[0, a]$, it takes the value $C$. Applying the Fundamental Theorem of Calculus to the positive part $(|u| - C)_+$ of the function $|u| - C$, one obtains

---

[4]The last condition is omitted in Hadamard's statement, but it is used in the proof.

$$|u(x)| - C \le \int_0^a |(|u(t)| - C)'_+| dt = \int_{M(C)} |u'(t)| dt.$$

By the Cauchy inequality, the last integral does not exceed $(B \operatorname{mes} M(C))^{1/2}$, and so we obtain the estimate

$$|u(x)| \le C + \left( \frac{AB}{f(C)} \right)^{1/2}. \tag{13.12}$$

To complete the proof, one uses the definition of $C$.

Inequality (13.11) and its generalization (13.12) remained unnoticed. In 1941, Szőkefalvy-Nagy obtained inequality (13.11) with the best constant $(1 + q/2)^2$, together with other inequalities of a similar nature [III.387], unaware of Hadamard's work.

In their most general form, the multi-dimensional versions of (13.11), or Gagliardo-Nirenberg inequalities, as estimates of the form (13.11) are often called, were proved independently by Gagliardo and Nirenberg [III.141], [III.294] in 1959. They described the admissible values for the exponents $m, l, p, q, r, \theta$ in the inequality

$$\left( \int_\Omega \sum_{|\alpha| \le m} |D^\alpha u(x)|^q dx \right)^{1/q}$$
$$\le c \left( \int_\Omega \sum_{|\alpha| \le l} |D^\alpha u(x)|^p dx \right)^{\theta/p} \left( \int_\Omega |u(x)|^r dx \right)^{(1-\theta)/r},$$

where $\Omega$ is a domain in $\mathbf{R}^n$, $D^\alpha = \partial^{|\alpha|}/\partial x_1^{\alpha_1} \dots \partial x_n^{\alpha_n}$ and $\alpha$ is a multi-index $(\alpha_1, \dots, \alpha_n)$ with $|\alpha| = \alpha_1 + \dots + \alpha_n$.

The usefulness of such estimates, which give information about intermediate derivatives when the properties of the function and its derivatives up to a certain order are known, appear frequently in the theory of boundary value problems for linear and non-linear partial differential equations.

Hadamard's much earlier paper *Sur certaines propriétés des trajectoires en dynamique* [I.48], for which he got the *Prix Bordin*, contains the following lemma which belongs to the same circle of ideas.

**Lemma.** *If the function $V(t)$ tends to a limit as $t$ tends to infinity and if all its derivatives up to and including order $n + 1$ exist and remain bounded, then the first $n$ of them tend to zero.*

Here is the proof of this assertion, given by Hadamard [I.48]:

Let us restrict ourselves to the first derivative, for simplicity. We have to show that the absolute value of this derivative is less than an arbitrarily given $\varepsilon$ for sufficiently large $t$.

To this effect, let $l$ be an arbitrarily chosen number. In the set of values of $t$, there can not exist an infinite number of intervals of length greater than $l$ on which $|dV/dt|$ is greater than $\varepsilon/2$: for, on such an interval, $V$ would change by more than $l\varepsilon/2$, which cannot occur indefinitely because $V$ tends to a limit. From the time these intervals cease to occur, the modulus of $dV/dt$ will manifestly be less than $\varepsilon$ if we take $l$ to be a number which, when multiplied by the limit superior of $|d^2V/dt^2|$, will give a value less than $\varepsilon/2$.

The same property of differentiable functions was discovered and used independently by Kneser, in 1897, in a paper concerning motion in the neighbourhood of an unstable point of equilibrium [III.210]. The Hadamard-Kneser lemma was also proved by Littlewood, who used it to establish the following Tauberian theorem on power series. If

$$\sum_{n\geq 0} a_n x^n \to s \qquad \text{as} \quad x \to 1-$$

and if $a_n = O(n^{-1})$, then

$$\sum_{n\geq 0} a_n = s.$$

This was proved by Littlewood in 1911; a weaker theorem of this type, in which $a_n = o(n^{-1})$, was first found by Tauber in 1897. The above lemma appears twice in Littlewood's book *A mathematician's miscellany* with the comment "Not, alas, by J. E. Littlewood, though my rediscovery of it was an important moment in my career" ([III.246, p. 36]). Littlewood's proof was diagrammatic, but essentially the same as Hadamard's. He remarks further: " ...without it I should never have got the main theorem. (The derivative theorem was actually known, but buried in a paper by Hadamard on waves [5])" ([III.246, p. 83]).

Two years later, Hardy and Littlewood, in a long paper on power series and Dirichlet series [III.166], used the following result: if $|f|$ and $|f''|$ have positive, increasing majorants $\varphi$ and $\psi$, then $|f'| = O(\sqrt{\varphi\psi})$.

---

[5]This is an error. It is in Hadamard's paper on dynamical trajectories.

A further development of the Hadamard-Kneser lemma is due to Landau [III.224], who proved that the inequalities $|f(x)| \leq M$, $|f''(x)| \leq N$ valid on an interval $I$ of length $|I| \geq 2\sqrt{M/N}$ imply the inequality

$$\sup |f'| \leq 2\sqrt{MN} \tag{13.13}$$

with 2 being the best constant.

In 1914, Hadamard replied to this paper by Landau, in his article [I.190], in which, having in mind possible applications to dynamics, he treated the case $|I| < 2\sqrt{M/N}$. He also proved that, for functions given on an arbitrary interval $I$, one has either

$$\sup_I |f'| < c_1 \Big( \int_I f(x)^2 dx \Big)^{(l-1)/(2l+1)} \sup_I |f^{(l)}|^{1/(2l+1)}$$

or

$$\sup_I |f'| \leq c_2 \Big( \frac{1}{|I|} \int_I f(x)^2 dx \Big)^{1/2},$$

where $c_1, c_2$ are explicit absolute constants.

In 1938, Kolmogorov ([III.213], [III.214]) obtained the following estimate with best possible constant:

$$\sup |f^{(m)}| \leq \frac{t_{l-m}}{t_l^{1-m/l}} \sup |f|^{1-m/l} \sup |f^{(l)}|^{m/l},$$

where $f$ is a function on $\mathbf{R}$, $m < l$ and

$$t_l = \frac{4}{\pi} \sum_{p=0}^{\infty} \frac{(-1)^{p(l+1)}}{(2p+1)^{l+1}}.$$

Inequalities of the type (13.13) and their generalizations are discussed in the books by Hardy, Littlewood and Pólya [III.167], and Beckenbach and Bellman [III.22], both with the same title *Inequalities*. If one is not interested in the best possible constants, then one can obtain these inequalities as particular cases of the above-mentioned theorems of Gagliardo and Nirenberg.

## 13.6 Articles on Markov chains

Hadamard's papers [I.265], [I.292] as well as his joint paper with Fréchet [I.313] deal with probability theory. In [I.265], [I.292] he studied the problem of shuffling cards, which can be described as follows.

During a card game (without cheating) the changes in the deck of cards are random. Let $p_{ij}(m)$ denote the conditional probability of the outcome: in exactly $m$ steps the system passes from the state $i$ into the state $j$. One has to prove the existence of the limit of the sequence $\{p_{ij}(m)\}_{m=1}^{\infty}$ and to calculate it.

This problem was solved by Poincaré in [III.320, p. 301-313], by means of an ingenious but complicated method. He showed that the above limit equals $1/N!$, where $N$ is the number of cards in the deck. Poincaré's result is valid only if $p_{ij}(m) > 0$ for any $i, j$ and for sufficiently large $m$.

In the article [I.265] of 1927, Hadamard proposed a simpler method for solving the problem of shuffling cards and took into consideration the opposite case, which he called singular, namely, that there exist states $i \neq j$ with $p_{ij}(m) = 0$ starting from some $m$.

*A.A. Markov (1856-1922) worked at Saint-Petersburg University. His principal results are in number theory, probability, and mathematical analysis. Markov chain, Markov process, Markov inequality for polynomials are called after him.*

Neither Hadamard, nor Poincaré before him, were aware of the fundamental work of A.A. Markov of 1907. Judging by a reference in the article

[I.292], it was Pólya who attracted Hadamard's attention to Markov's paper. Markov considered sequences of $r$ random variables which represent the result of $r$ trials (in the case of shuffling cards, $r = N!$). He assumed that the outcome of any trial depends only on the outcome of the directly preceeding trial. Such sequences are called Markov chains. Since shuffling cards is an example of such a chain, Poincaré's result followed from the general theorems of Markov, who, moreover, considered not only the regular case.

However, Hadamard's method turned to be new and useful. In his book [III.136] Fréchet gave an exposition of the method and described its further development by Kolmogorov. He wrote:

> Mr. Hadamard is the first person to have used, for the problem of probabilities in a chain, a direct method allowing him to use exclusively the language of probability and avoiding any borrowing from algebraic theories. His method, which he applied to the problem of shuffling cards, extends without modification to the case of Markoff's problem where one has the condition
>
> $$\sum_i p_{ik} = 1. \qquad (13.14)$$
>
> It allowed him to establish results which had until then not been obtained by the algebraic method, and which show the distinction between sure properties (those with probability equal to 1) of the variations of the system under consideration and those which are strictly due to chance.
>
> Mr. Kolmogoroff, inspired by Hadamard's method, was able to modify this method in such a way as to extend it to the case where condition (13.14) does not hold, which necessitates important and not evident modifications of Mr. Hadamard's reasoning.

In [I.265] and [I.292], Hadamard considered iterations of a random mapping of an interval $(a, b)$ into itself, complementing the work of Hostinsky which had just appeared. He noticed the analogy between the study of its limit distribution and ergodic problems of statistical mechanics.

These works of Hadamard of 1927–1928 appeared before the remarkable advances in the theory of Markov processes which were achieved by Kolmogorov, réchet, P. Lévy, Feller, et al. Paul Lévy wrote:

Hadamard had the intimation of this prodigious development. It is certain, in any case, that he understood the importance of the ergodic principle in probability theory as well as in statistical mechanics; memories of conversations going back to that time permit me to confirm this [II.37, p. 18].

# Chapter 14

# Elasticity and Hydrodynamics

## 14.1 Aspects of the history of mathematical physics

Laurent Schwartz wrote:

> Already through his association with his physicist fellow-students at the *École Normale*, he [Hadamard] had been fascinated by the close connections between mathematics and physics. He always had excellent relations with Duhem, Langevin and other physicists, and he always said how much these contacts had inspired him [II.5, p. 15].

However, Hadamard only began working in mathematical physics, which at that time was more or less synonymous with the theory of partial differential equations, towards the end of the last century. This was to dominate his later research, which he continued to the end of his life. "The centre of modern mathematics is in the theory of partial differential equations," he wrote in his article on the mathematical work of Poincaré [I.208, p. 388].

By the time Hadamard became interested in the theory of partial differential equations, it had already made substantial progress, and had a rich 150-year-old history, which we dwell on briefly.

Partial differential equations appeared in the 1740s in the works of Euler and d'Alembert (see [III.97], [III.122]). For instance, d'Alembert, in 1747, introduced and integrated the equation for a vibrating string which can be obtained by the following argument.

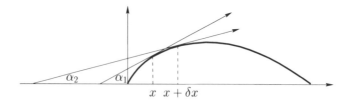

**Figure 6.**

Let a string, i.e. a flexible elastic thread, perform small transverse oscillations in the plane of the page. Let $(x, u(x,t))$ be Cartesian coordinates of its arbitrary point at the moment $t$ and let $T$ denote the tension of the string which is supposed to be constant. The force acting on the interval $[x, x + \delta x]$ of the string in the direction orthogonal to the $x$-axis is

$$T \sin \alpha_2 - T \sin \alpha_1 \simeq T(\tan \alpha_2 - \tan \alpha_1) = T(\frac{\partial u}{\partial x}(x + \delta x, t) - \frac{\partial u}{\partial x}(x, t)),$$

*Jean Le Rond d'Alembert (1717-1783)*

where the angles $\alpha_1$ and $\alpha_2$ are shown in fig. 6. By Newton's law (force equals mass times acceleration)

$$\rho \delta x \frac{\partial^2 u}{\partial t^2} \simeq T(\frac{\partial u}{\partial x}(x + \delta x, t) - \frac{\partial u}{\partial x}(x, t)),$$

where $\rho$ is the linear density of the string. Dividing both parts of the equality by $\delta x$ and passing to the limit as $\delta x \to 0$ we arrive at the equation

$$\frac{\partial^2 u}{\partial t^2} - \omega^2 \frac{\partial^2 u}{\partial x^2} = 0, \qquad \omega = (T/\rho)^{1/2}. \tag{14.1}$$

Another classical equation of mathematical physics

$$\frac{\partial^2 u}{\partial x^2} + \frac{\partial^2 u}{\partial y^2} = 0 \tag{14.2}$$

appeared in 1755 in Euler's work on the flow of a homogeneous incompressible liquid, and special solutions of the equivalent system

$$\frac{\partial v}{\partial x} = \frac{\partial w}{\partial y}, \qquad \frac{\partial v}{\partial y} = -\frac{\partial w}{\partial x}$$

(now called the Cauchy-Riemann system) had already been found by d'Alembert in 1747. Equation (14.2) and its three-dimensional generalization

$$\frac{\partial^2 u}{\partial x^2} + \frac{\partial^2 u}{\partial y^2} + \frac{\partial^2 u}{\partial z^2} = 0 \tag{14.3}$$

were later called the Laplace equation.

To justify the name, let us remark that Laplace, while developing the theory of gravity, obtained many properties of the solutions of equations (14.2) and (14.3). Solutions of Laplace's equations are called harmonic functions. The importance of this class of functions in the case of two independent variables is also due to its connection with the notion of analyticity: indeed, the real and imaginary parts of an analytic function are harmonic.

Both equations (14.2) and (14.3) are frequently written as $\Delta u = 0$, where $\Delta$ is the Laplace operator in $\mathbf{R}^n$ defined by

$$\Delta u = \sum_{i=1}^{n} \frac{\partial^2 u}{\partial x_i^2}.$$

*Pierre Simon Laplace (1749-1827)*

The wave equation and the heat equation already mentioned in Section 3.1 are written with the notation $\Delta$ as

$$\frac{\partial^2 u}{\partial t^2} - \omega^2 \Delta u = 0, \qquad (14.4)$$

$$\frac{\partial u}{\partial t} - \omega^2 \Delta u = 0. \qquad (14.5)$$

On one hand, it turned out that the same equation can describe different processes. For example, the equation for vibrations of a string is also satisfied by the electric force and potential in wires as well as by the longitudinal oscillations of a rod and the wave processes in liquids and gases. On the other hand, there are many equations coming from mathematical models of physical phenomena. For instance, the problem of bending of thin plates leads to the biharmonic equation

$$\Delta^2 u = 0,$$

or, which is the same,

$$\frac{\partial^4 u}{\partial x^4} + 2\frac{\partial^4 u}{\partial x^2 \partial y^2} + \frac{\partial^4 u}{\partial y^4} = 0.$$

The motion of a viscous incompressible liquid is described by the non-linear system of Navier-Stokes equations, whereas the Lamé system describes the stress-strain state of elastic bodies, and Maxwell's equations characterize the propagation of electromagnetic waves.

Each of these equations was studied in the framework of some discipline in physics. However, some general principles were established at the very beginning, for instance that solutions of partial differential equations depend on an arbitrary function of a smaller number of variables. Thus, to obtain a unique solution, one has to supply initial or boundary conditions. Such conditions were dictated by the physical meaning of the process described by the equation under consideration. For example, it is natural to consider the equation of an infinite, vibrating string together with the initial conditions

$$u(x,0) = f(x), \qquad \frac{\partial u}{\partial t}(x,0) = g(x), \qquad (14.6)$$

that is, to assume that the position and velocity of the string at $t = 0$ are known. The problems of finding the solution of a partial differential equation satisfying a certain initial condition are called Cauchy problems.

If the string is semi-infinite and its coordinate $x$ varies on the half-axis $0 \leq x < \infty$, then together with the initial conditions (14.6) for any $x > 0$ it is reasonable to give some information about the movement of its endpoint. Thus, the condition $u(0,t) = 0$ means that the end of the string is fixed, whereas $\partial u/\partial x(0,t) = 0$ means that it is free. In the same way, one can consider a finite string. Such problems were called mixed problems by Hadamard, but nowadays it is more common to call them initial boundary value problems.

The first formulations for the boundary value problems for the Laplace equation appeared in the nineteenth century. They were considered in a two- or three-dimensional domain $\Omega$ bounded by a curve or surface $\partial\Omega$.

It was self-taught mathematician Green who, while studying a problem of electrostatics, as early as 1828 considered the so-called first boundary value problem, which was later called the Dirichlet problem:

$$\Delta u = 0 \quad \text{in} \quad \Omega, \qquad u = \varphi \quad \text{on} \quad \partial\Omega, \qquad (14.7)$$

where $\varphi$ is given. The so-called second boundary value problem for the Laplace equation arises when the derivative of the unknown function with respect to the normal is given on the boundary, i.e.,

$$\Delta u = 0 \quad \text{in} \quad \Omega, \qquad \frac{\partial u}{\partial \nu} = \varphi \quad \text{on} \quad \partial\Omega,$$

where $\nu = (\nu_1, \ldots, \nu_n)$ is the unit normal to $\partial\Omega$ and

$$\frac{\partial u}{\partial \nu} = \sum_{i=1}^{n} \frac{\partial u}{\partial x_i} \nu_i.$$

This problem had essentially been formulated by Kirchhoff in 1845 when he was studying problems of electrodynamics. The term "Neumann boundary condition", used for the equality $\partial u / \partial \nu = \varphi$ on $\partial\Omega$, is named after Carl Neumann, who made a systematic study of boundary value problems for the Laplace equation in his book [III.293] in 1877.

In the eighteenth century and in the beginning of the nineteenth century, mathematicians were mostly interested in finding explicit solutions of problems of mathematical physics. The general solution of the equation (14.1) for the vibrating string had been given by d'Alembert in 1747 in the form

$$u(x, t) = \varphi(x + \omega t) + \psi(x - \omega t),$$

where $\varphi$ and $\psi$ are arbitrary functions of one variable. The arbitrariness in the choice of $\varphi$ and $\psi$ can be eliminated if initial conditions (14.6) are prescribed. Then the solution is given by the d'Alembert formula

$$u(x, t) = \frac{1}{2}[f(x + \omega t) - f(x - \omega t)] + \frac{1}{2} \int_{x-\omega t}^{x+\omega t} g(\tau) d\tau.$$

About the same time (not later than 1754) Daniel Bernoulli gave a solution of equation (14.1) in the form of a trigonometric series. In their correspondence, Euler and d'Alembert began a famous discussion on the vibrating string, in which D. Bernoulli, Lagrange, and others soon became involved.[1] D'Alembert took the view that the initial form of the string can only be given by a unique analytic expression, but Euler thought that it could be an arbitrary curve "traced with a free movement of the hand". In 1753 D. Bernoulli claimed that trigonometric series gives the general solution. This discussion played an important role in the subsequent clarification of the notion of function in the ninenteenth century and anticipated the problem of finding an appropriate function space in which one should seek the solution (this was finally formulated in the present century).

---

[1]See Section 1.3 in Bottazzini's book [III.53] and Part 1 in Lützen's book [III.249].

*Siméon-Denis Poisson (1781-1840)*

Representations for solutions of the Cauchy problem for the wave equation (14.4) and the heat equation (14.5) were found by Poisson. He also gave explicit solutions of the Dirichlet problem (14.7) in a disc and in a ball. Poisson's formulas became prototypes for solutions of other problems of mathematical physics.

Green introduced a special function $g(P,Q)$ of two points (which was named after him) which enabled him to give the solution of the Dirichlet problem

$$\Delta u = f \quad \text{in} \quad \Omega, \qquad u = \varphi \quad \text{on} \quad \partial\Omega \qquad (14.8)$$

as the sum of two integrals. Green's function for the Laplace operator in the domain $\Omega \subset \mathbf{R}^n$, $n = 2, 3$, is defined by

$$g(P,Q) = \frac{1}{2\pi} \log |P - Q| + h(P,Q) \quad \text{for} \quad n = 2,$$

$$g(P,Q) = -\frac{1}{4\pi |P - Q|} + h(P,Q) \quad \text{for} \quad n = 3,$$

where $h$ is a function which is harmonic in each of the variables $P$ and $Q$, and $g = 0$ whenever $P \in \partial\Omega$ or $Q \in \partial\Omega$. With the help of Green's formula

$$\int_{\Omega} (u\Delta v - v\Delta u)dQ = \int_{\partial\Omega} (u\frac{\partial v}{\partial \nu_Q} - v\frac{\partial u}{\partial \nu_Q})ds_Q, \qquad (14.9)$$

where $\nu_Q$ is the unit outward normal to $\partial\Omega$ at the point $Q$ and $ds_Q$ is the element of length or surface area on $\partial\Omega$ at the point $Q$, one can prove that the solution of the problem (14.8) can be represented in the form

$$u(P) = \int_{\Omega} g(P, Q)f(Q)dQ + \int_{\partial\Omega} \frac{\partial g(P, Q)}{\partial \nu_Q}\varphi(Q)ds_Q.$$

For a disc, ball and some other special domains, Green's function can be written explicitly.[2]

*Jean Baptiste Joseph Fourier (1768-1830)*

In the nineteenth century, Fourier's method which enabled one to express solutions as series in eigenfunctions of certain differential operators became well developed. The application of this method gave rise to the theory of special functions. These latter functions occupied a central position in mathematical analysis, and mathematicians attained great virtuosity in applying them.

---

[2]For information on Green's life and his function, see Cannell's book [III.66] and a popular essay by J.J. Gray [III.158].

In spite of these achievements, it became clear quite early that there were limits in the possibility of solving the problems of mathematical physics explicitly. For instance, the attempts to find solutions of boundary value problems in a closed form, in more or less arbitrary domains, were in vain, even for the Laplace equation. By the end of the nineteenth century methods to prove the solvability of boundary value problems for the Laplace equation were developed (see Section 12.3).

The solution of boundary value problems for equations with variable coefficients, which arise naturally in the description of physical processes in inhomogeneous media, also posed great difficulties. It was sometimes possible to obtain some success using a change of variables and to reduce the initial equation to one solvable by quadratures, i.e., explicitly. This led to the classification of second-order partial differential equations, which was done by du Bois-Reymond in his paper [III.111]. He distinguished three types of equations: elliptic, parabolic and hyperbolic, of which the simplest examples are the Laplace equation, the heat equation, and the wave equation.

*Paul du Bois-Reymond (1831-1889)*

The equation

$$a\frac{\partial^2 u}{\partial x^2} + 2b\frac{\partial^2 u}{\partial x \partial y} + c\frac{\partial^2 u}{\partial y^2} + F(x, y, u, \frac{\partial u}{\partial x}, \frac{\partial u}{\partial y}) = 0, \qquad (14.10)$$

where $a, b, c$ are functions of the variables $x, y$, is of elliptic, hyperbolic, or parabolic type at a point $(x, y)$ depending on which of the following conditions are satisfied

$$b^2 - ac < 0, \qquad b^2 - ac > 0, \qquad b^2 - ac = 0.$$

Using a non-degenerate change of variables $x = x(\xi, \eta), \quad y = y(\xi, \eta)$, one can transform the equation (14.10) of one of the above three types in a domain, to one of the following canonical forms

$$\frac{\partial^2 u}{\partial \xi^2} + \frac{\partial^2 u}{\partial \eta^2} + f(\xi, \eta, u, \frac{\partial u}{\partial \xi}, \frac{\partial u}{\partial \eta}) = 0 \quad \text{(elliptic)}$$

$$\frac{\partial^2 u}{\partial \xi^2} - \frac{\partial^2 u}{\partial \eta^2} + f(\xi, \eta, u, \frac{\partial u}{\partial \xi}, \frac{\partial u}{\partial \eta}) = 0 \quad \text{(hyperbolic)}$$

$$\frac{\partial^2 u}{\partial \xi^2} + f(\xi, \eta, u, \frac{\partial u}{\partial \xi}, \frac{\partial u}{\partial \eta}) = 0 \quad \text{(parabolic)}$$

The generalization of these three types of equations to the case of $m$ independent variables is performed in the following way. The equation

$$\sum_{i,j=1}^{m} A_{ij}(x)\frac{\partial^2 u}{\partial x_i \partial x_j} + F(x, u, \text{grad} u) = 0,$$

where grad stands for the gradient, is of elliptic type at the point $x$ if all the eigenvalues of the quadratic form

$$\sum_{i,j=1}^{m} A_{ij}(x)\xi_i \xi_j$$

are non-zero, and have the same sign. If the sign of one of the eigenvalues is opposite to that of the others, then the equation is of hyperbolic type. If all eigenvalues are of the same sign, except one which is zero, then the equation is of parabolic type.

The general properties of second-order partial differential equations depend largely on the concept of characteristic surface when $m \geq 3$ and characteristic curve when $m = 2$, also called characteristics. This is an $m - 1$-dimensional surface $S$ defined by a function $\mathcal{F}(x) = \text{const.}$, such that $\text{grad}\mathcal{F} \neq 0$ on $S$, and the function $\mathcal{F}$ satisfies the differential equation

$$\sum_{i,j=1}^{m} A_{ij}(x)\frac{\partial \mathcal{F}}{\partial x_i}\frac{\partial \mathcal{F}}{\partial x_j} = 0.$$

*Viktor Bäcklund (1845-1922)*

The characteristics are distinguished by the fact that one cannot assign Cauchy data to the solution arbitrarily on them. An elliptic operator does not have any (real) characteristic, whereas a second-order hyperbolic operator in two variables has two families of characteristics, and a parabolic operator has one. The concept of characteristic was used in a crucial way by Riemann [III.343] in 1860, when he obtained the representation of the solution of the Cauchy problem for hyperbolic equations of second order in two independent variables. The term "characteristic" first appeared in the works of Monge (1807) who developed a geometric theory of the equation

$$F\left(x, y, u, \frac{\partial u}{\partial x}, \frac{\partial u}{\partial y}\right) = 0.$$

Previously Lagrange showed that this equation can be reduced to a system of ordinary differential equaitons; before him this was known only for equations which are linear with respect to derivatives. The theory of characteristics was pushed forward by du Bois-Reymond, Bäcklund, Lie, Goursat, and others. Lie's group-theoretic point of view enabled him to unify various known methods for integration of first order partial differential equations.[3]

*Sophus Lie (1842-1899)*

In concluding this brief survey, let us mention one further discovery of the nineteenth century, which is a fundamental property of linear and nonlinear partial differential equations of arbitrary order with analytic coefficients. This is the Cauchy-Kovalevskaya theorem on the existence of analytic solutions of the Cauchy problem with given data on a non-characteristic surface. We shall deal with this result later on and describe how Hadamard corrected some misconceptions about this universal result.

## 14.2    Lectures on the propagation of waves

The book *Leçons sur la propagation des ondes et les équations de l'hydrodynamique* [I.109] appeared in Paris in 1903. It contained an expanded exposition of the course Hadamard gave in 1898-99 and 1899-1900 at the *Collège de France* (a resumé of his results appeared in his article on the propagation of waves [I.81] in 1901). "My main intention was to study how

---

[3]For the history of partial differential equaitons of first order in the eighteenth and nineteenth centuries, see [III.98].

boundary conditions influence the motion of a fluid", Hadamard wrote in the preface.

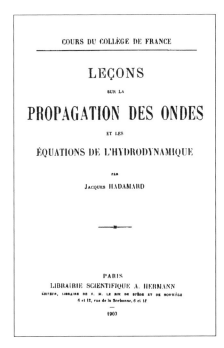

COURS DU COLLÈGE DE FRANCE
_____

LEÇONS
SUR LA

PROPAGATION DES ONDES
ET LES

ÉQUATIONS DE L'HYDRODYNAMIQUE

PAR

JACQUES HADAMARD

Chapter 1 of this rather large book (375 pages) is devoted to the proof of the solvability of the main boundary value problem of the hydrodynamics of an ideal, incompressible fluid, i.e., the Neumann problem for harmonic functions, when the normal derivative of the solution is given on the boundary. The required harmonic function is the velocity potential of the irrotational flow of an incompressible fluid,[4] and the boundary condition means that the normal component of the velocity of the flow is given. This material was probably included in the book because the important papers of Poincaré, Liapunov, and Steklov, justifying the methods of potential theory, had just appeared. Hadamard's exposition essentially follows Steklov's memoir *Les méthodes générales pour résoudre les problèmes fondamentaux de la physique mathématique* [III.375]. Although the discussion of the Neumann problem for the Laplace equation in the book on hydrodynamics is quite natural, Chapter 1 stands apart from the main topic of the book, which concerns non-stationary processes.

PARIS
LIBRAIRIE SCIENTIFIQUE A. HERMANN
ÉDITEUR, LIBRAIRE DE S. M. LE ROI DE SUÈDE ET DE NORWÈGE
6 et 12, rue de la Sorbonne, 6 et 12
—
1903

In Chapter 2, Hadamard considers some general problems of the propagation of waves in deformable media from the kinematic point of view. The history of this problem in the nineteenth century is somewhat confused, since many results had been rediscovered several times. The dynamics of ruptures of physical characteristics of gases, propagating rectilinearly in a cylinder with moving bases, was studied for the first time by Riemann in his memoir [III.343] of 1860. The works of Riemann as well as the papers of Rankine [III.337] and Christoffel [III.80], concerning phenomena of the same nature, did not attract any attention for some time. These studies were also unknown to Hugoniot, who in 1885 constructed the general theory

_____

[4]The motion is said to be irrotational if the fluid moves in such a way that, denoting by $V$ the velocity vector, the vorticity vector curl $V$ equals zero. When the motion is irrotational there exists a scalar function (called the velocity potential) $\Phi$ such that $V = \mathrm{grad}\,\Phi$. For the incompressible fluid, one further has div $V = 0$ which implies the harmonicity of the potential.

of one-dimensional motions of deformable media, and in particular rediscovered most of Riemann's results. The two parts of Hugoniot's memoir [III.183] appeared in 1887 and 1889, after his death.

*Duhem's lectures in Lille*

Hugoniot's papers were brought to Hadamard's attention through lectures on hydrodynamics, elasticity, and acoustics, given by Duhem, in Lille, and published in 1891 [III.114]. Conversations with Duhem in Bordeaux also played an important role.

When speaking of waves, Hadamard, just as Hugoniot before him, meant a perturbation propagating in a medium, with a sharp front. The main emphasis of Chapter 2 of Hadamard's book is on the notion of compatibility of two motions, which had already appeared in Riemann's papers, and played an important role in the work of Hugoniot. This concerns the motion of the disturbed medium behind the wave front, and of the undisturbed medium before the front. Hadamard wrote in [I.109, p. viii]:

> Thus I had to identify those properties of a purely kinematic nature by distinguishing them from those depending on the dynamical properties of the motion. With this distinction, many questions are clarified, as one might expect. In particular, a geometrical representation [of the process] appears immediately as a result. This representation in its turn allows us to make closer the analogy between waves in the sense of Hugoniot and those considered in the mechanics of vibrations.

In order to give an account of the results which are mentioned in this quotation, we shall recall some notions from hydrodynamics. The motion of a deformable medium (for instance, a liquid or a gas) is considered to be determined if the position of each of its particles is defined at any time $t$. Let the coordinates $(x_1, x_2, x_3)$ of the particle be considered as functions of the so-called Lagrangian variables $(a_1, a_2, a_3)$ and of the time $t$, where $(a_1, a_2, a_3)$ are the coordinates of the same particle at the initial time.

Suppose that a wave propagates in the medium, so that its front $F(t)$ is a smooth surface defined by the equation $f(a_1, a_2, a_3, t) = 0$. Here, $f$ is a smooth function of the Lagrangian coordinates such that

$$\sum_{i=1}^{3}\left(\frac{\partial f}{\partial a_i}\right)^2 > 0 \quad \text{on} \quad F(t).$$

Then $F(t)$ divides the medium into two parts, $\Omega_+(t)$ and $\Omega_-(t)$ (the disturbed and undisturbed parts), defined respectively by $f(a_1, a_2, a_3, t) > 0$ and $f(a_1, a_2, a_3, t) < 0$.

We denote the restrictions of a function

$$\varphi(a_1, a_2, a_3, t) \quad \text{to} \quad \Omega_{\pm}(t) \quad \text{by} \quad \varphi_{\pm}(a_1, a_2, a_3, t),$$

the same notation being retained for the limiting values of $\varphi_+$ and $\varphi_-$ on $F(t)$, and we use $[\varphi]$ to denote the difference $\varphi_+ - \varphi_-$ of these limiting values, that is, the jump of the function $\varphi$ at the wave front.

In gas dynamics, $\varphi$ can be the components of the velocity $\mathbf{v}$, the pressure $p$, the density $\rho$, the temperature $T$ and their derivatives. The quantities $\mathbf{v}, p, \rho, T$ satisfy certain relations which follow from the basic physical laws for the motion of a gas. If these physical parameters are continuously differentiable, then these relations can be rewritten as differential equations. In the case of discontinuous $\mathbf{v}, p, \rho, T$, the relations give rise to constraints on the jumps at the points of the surface $F(t)$. These constraints are called the conditions of dynamic compatibility. In connection with the previous comments on the general notion of compatibility of two motions, here we mean the motions characterized by the quantities $\mathbf{v}_+, p_+, \rho_+, T_+$ and $\mathbf{v}_-, p_-, \rho_-, T_-$.

A surface $F(t)$ on which at least one of the dynamic quantities has a jump, is called a surface of strong discontinuity. When all of these quantities are continuous, and at least one of their first order derivatives has a jump, the surface is called a surface of weak discontinuity. If the wave front is a surface of weak discontinuity, then there will be other conditions of compatibility on it, of a geometric and kinematic nature. They are independent of the equations of the motion of the medium, and are derived from purely geometric or kinematic considerations. Thus, these compatibility conditions are applicable to waves in various media, and follow only from the existence of the wave.

It is not difficult to understand how kinematic and dynamic conditions arise. Suppose that the function $\varphi(a_1, a_2, a_3, t)$ is continuous, and that its first derivatives $\partial\varphi/\partial a_i$, $i = 1, 2, 3$, have jumps at the surface $F(t)$. We shall consider the function $\varphi_+$ extended continuously to $\Omega_-(t)$, so that its derivatives are also continuous. Similarly, we extend $\varphi_-$ to $\Omega_+(t)$. Then by definition of the jump we have on the surface $F(t)$:

$$\left[\frac{\partial\varphi}{\partial a_i}\right] = \frac{\partial\varphi_+}{\partial a_i} - \frac{\partial\varphi_-}{\partial a_i}, \qquad i = 1, 2, 3.$$

Since

$$(\varphi_+ - \varphi_-)(a_1, a_2, a_3, t) = 0 \qquad \text{on} \quad F(t),$$

it follows that the gradients of the functions $f$ and $\varphi_+ - \varphi_-$ (with respect to the Lagrangian variables) are parallel, i.e.,

$$\left[\frac{\partial\varphi}{\partial a_i}\right] = \lambda\frac{\partial f}{\partial a_i}, \quad i = 1, 2, 3, \qquad \text{on} \quad F(t),$$

where $\lambda$ is some function. These equalities are the above mentioned geometric ("identical", in Hadamard's terminology) compatibility conditions on a surface of weak discontinuity.

Hadamard was the first to carry out a precise classification of the conditions of compatibility according to their geometric, kinematic, and dynamic origins. Hugoniot considered first order compatibility conditions, that is, conditions for jumps of the first derivatives, but Hadamard, in Chapter 2 of his book, outlined the general theory of geometric and kinematic conditions of arbitrary order, and in particular he obtained expressions for jumps of the second and third derivatives.

The theory developed in Chapter 2 became the basis for more concrete studies concerning the motion of a gas in Chapters 4 and 5, and concerning waves in elastic media in Chapter 6. Moreover, these compatibility conditions enable one to obtain some parameters of the motion of the medium at the wave front without solving differential equations, for example, the velocity of wave propagation.

The short Chapter 3 has an auxiliary character: it contains information on the equations of gas dynamics and the corresponding boundary conditions. Chapter 4, "The rectilinear motion of a gas", deals mainly with an exposition of the Riemann-Hugoniot theory in the one-dimensional case when the equations of gas dynamics can be integrated.

The three-dimensional motion of a gas is treated in Chapter 5, in which Hadamard studies the direction and velocity of propagation of second order discontinuities (acceleration waves). He shows that such waves do not change the conservation of the circulation of the velocity, that is the integral

$$\int_C v_1 dx_1 + v_2 dx_2 + v_3 dx_3,$$

where $C$ is a closed contour. Thus, the acceleration waves do not influence the vortices. For shock waves (velocity waves), i.e., for first order discontinuities, the situation is quite different. In one of the appendices to the book, Hadamard proves that shock waves can give rise to vortices where there were none prior to the passing of the front.

In Chapter 6, Hadamard studies wave propagation in elastic bodies under finite deformations. The differential equations for the dynamics of such media, with the displacements as the unknown functions, form a non-linear hyperbolic system of second order. Inside the medium, the wave propagates so that the displacements and their first derivatives are continuous on the front, whereas the second derivatives have jumps. Substituting into this system the second order kinematic compatibility conditions obtained in Chapter 2, Hadamard arrives at the eigenvalue problem

$$A(\nu)\lambda = c^2\lambda.$$

The square of the velocity of propagation $c$, is an eigenvalue, and the normal to the wave front, $\lambda$, is an eigenvector. The matrix $A(\nu)$ is called the acoustic tensor of the body in the direction of propagation, $\nu$, and is defined at every point of the body, and for each fixed vector $\nu$, by the elastic characteristics of the material and the state of stress. The components of the tensor $A(\nu)$ have the form

$$a_{ij}^{\alpha\beta}(x)\nu_\alpha\nu_\beta,$$

where $x$ is an arbitrary point of the body $\Omega$, $\nu_1, \nu_2, \nu_3$ are the components of the vector $\nu$, and $a_{ij}^{\alpha\beta}$ are continuous functions in the closure $\overline{\Omega}$. In accordance with classical elasticity theory, it is assumed that the condition $a_{ij}^{\alpha\beta}(x) = a_{\alpha\beta}^{ij}(x)$ holds, which ensures the symmetry of the matrix $A(\nu)$. The non-negativity of this matrix is equivalent to the existence of a wave.

The theorem proved in Chapter 6 of the book shows the possibility of wave propagation in $\Omega$, in any direction $\nu$, if the so-called condition of infinitesimal stability is valid:

$$\int_\Omega \sum_{\alpha,\beta,i,j=1}^3 a_{ij}^{\alpha\beta}(x)\frac{\partial u_i}{\partial x_\alpha}\frac{\partial u_j}{\partial x_\beta}dx \geq 0,$$

for every smooth real vector-valued function $(u_1, u_2, u_3)$ defined on $\Omega$, which vanishes near the boundary.

Two years later, in 1905, Duhem [III.115] remarked on the inadequacy of the arguments Hadamard used in the proof of these facts. It appears that the first rigorous proof was given in 1946 by Cattaneo [III.72]. Twenty years later, Fichera found a simple proof by a technique used in the general theory of elliptic systems [III.131]. He also showed by an example, that the non-negativity of the matrix $A(\nu)$ is not sufficient for the infinitesimal stability, although this is the case for the homogeneous body, when $a_{ij}^{\alpha\beta} = \text{const}$ [III.130].

We do not want to finish the discussion of Chapter 6 of *Leçons sur la propagation des ondes* without mentioning the activity in the same field and circle of ideas during the second half of this century. In a paper dedicated to Hadamard on the sixtieth anniversary of the appearance of this book Truesdell wrote:

> After the classical researches of Christoffel, Hugoniot, Hadamard, and Duhem on waves in elastic materials, it might seem that little remains to be learned. Such is not the case. As for most parts of mechanics, it has been necessary in the last decade to go over the matter again, not only so as to free the conceptual structure from lingering linearizing and to fix it more solidly in the common foundation of modern mechanics, but also so as to derive from it specific predictions satisfying modern needs for contact between theory and rationally conceived experiment [III.401, p. 263].

Truesdell listed Hadamard's contributions to general continuum mechanics as:

1. The basic lemma (half anticipated by Maxwell) by which compatibility in general is distinguished from compatibility of particular, kinematically defined waves.

2. Recognition of levels of compatibility: geometric, kinematic, dynamic, energetic, material (though only the first two are clearly developed in his book).

3. Classification of kinematical singular surfaces, and construction of a general theory, including an outline of higher-order conditions.

4. Calculation of the exact wavespeeds in finite elastic strain, and proof that they are all real and non-vanishing if and only if the equations of equilibrium, for the particular strain, are strongly elliptic.

5. Proof that weak singular surfaces in gas dynamics do not destroy the circulation-preserving property, and in particular such waves do not invalidate the Lagrange-Cauchy velocity-potential theorem.

6. Proof that an oblique curved shock wave in a gas generates vorticity.

7. The first rigorous definition and analysis of stability in finite elastic strain, and proof that in stable equilibrium the inequality defining strong ellipticity must hold, provided '$\geq$' be replaced by '$>$'.

Then Truesdell continued:

> All this in his *Propagation des ondes*. No. 4 is made clear by my article in the *Archive*. No. 7 was not noticed, apparently, until 1952; the modern American literature on continuum mechanics begins to take note of it, and a new proof, filling a lacuna noted by Duhem in Hadamard's argument, is given in the second article for the *Handbuch*, now in press. When I asked Hadamard about No. 7 in 1955, on my way to Berlin to deliver a lecture in which his work on elasticity was summarized in part, he said he had forgotten it (see [II.4, p. 90]).

Let us note further that the condition of strict positivity of the tensor $A(\nu)$, often called the Legendre-Hadamard condition, also turned out to be useful in mathematics, in the modern theory of quasi-linear systems of partial differential equations. Being less restrictive than the usual condition of ellipticity, it is more natural in a range of problems (see for instance, [III.148] [III.285]).

The final Chapter 7 of the book *Leçons sur la propagation des ondes* concerns a general theory of characteristics for the second order hyperbolic equation

$$\sum_{i,k} a_{ik} \frac{\partial^2 z}{\partial x_i \partial x_k} + a = 0,$$

where $a_{ik}$ and $a$ are functions of $x, z(x)$ and grad $z(x)$. Later Hadamard made Chapter 7 a starting point for his famous investigation of the Cauchy problem which will be discussed in Section 15.4. In the preface of his book *Lectures on the Cauchy problem in linear partial differential equations* [I.223] of 1923, he writes: "...the present work can be considered as a continuation of my *Leçons sur la propagation des ondes et les équations de l'hydrodynamique*, and even as replacing several pages of the last chapter. The latter, indeed, was a first attempt, in which I only succeeded in showing the difficulties of the problem, the solution of which I am now able to present."

In the first of four Notes placed at the end of the book on propagation of waves Hadamard made an important remark on the uniqueness of a solution to the Cauchy problem in a class of differentiable functions. It was inspired by Holmgren's theorem of 1901 [III.178] for linear equations and systems with holomorphic coefficients. Hadamard noticed that the proof of the uniqueness for the general nonlinear equation

$$F\left(x, z, \{\partial z/\partial x_i\}_{i=1}^n, \{\partial^2 z/\partial x_i \partial x_k\}_{i,k=1}^n\right) = 0$$

can be reduced to the uniqueness result for linear equations with sufficiently smooth coefficients. Afterwards, much effort was concentrated on conditions for the uniqueness of solutions to the Cauchy problem for linear differential equations with non-holomorphic coefficients. For the history of these studies see, for example, Section I of L. Nirenberg's essay [III.295].

## 14.3    Memoir on the equilibrium of plates

In 1907 Hadamard submitted to the Academy for the *Prix Vaillant* a paper [I.145] on the equilibrium of homogeneous thin elastic plates clamped along their boundary. The mathematical formulation of the problem is as follows. Let $\Omega$ be a domain which is the mid-plane of a thin plate, and let $\partial\Omega$ be the contour bounding $\Omega$ which is assumed to be smooth. One looks for a normal displacement $w(x, y)$ at the point $(x, y) \in \Omega$ under the load $q(x, y)$. The function $w$ satisfies the equation

$$\Delta^2 w(x, y) = q(x, y) \qquad \text{in} \quad \Omega, \qquad (14.11)$$

first obtained by Sophie Germain in 1811 (we omit a constant factor before $q(x, y)$ which depends on the thickness of the plate and the material of which it is made). If the edge of the plate is clamped, then the following boundary conditions hold on $\partial\Omega$

$$w(x, y) = 0, \quad \frac{\partial w(x, y)}{\partial \nu} = 0, \qquad (14.12)$$

where $\nu$ is the outward normal to $\partial\Omega$.

The French Academy of Sciences proposed the following problem for the competition: to improve upon the analysis of the boundary value problem (14.11), (14.12), and especially study the case of the rectangular plate. One should remember that boundary value problems describing the equilibrium of plates very rarely admit explicit solutions. This happens in the case of a circular plate, considered by Mathieu [III.270] and by M. Lévy [III.238], who found series solutions.

*Giuseppe Lauricella (1867-1913)*

In 1896 Almansi [III.9] and Lauricella [III.228] simultaneously obtained the solution to a problem of a clamped circular plate in equilibrium in the form of an integral. They noticed that the solution of the equation $\Delta^2 w = 0$ in the disc $x^2 + y^2 < R^2$ admits the representation

$$w(x,y) = U(x,y)(x^2 + y^2 - R^2) + V(x,y),$$

where $U$ and $V$ are harmonic functions, and thus they reduced the problem concerning the circular clamped plate to two Dirichlet problems for the Laplace equation. Two other classical problems on the equilibrium of circular plates are those of a free boundary and a freely supported boundary. Both were solved by Hadamard in his first paper on the theory of plates, published in 1901 [I.85]. He showed that, by means of quadratures, both problems can be reduced to the Dirichlet problem for harmonic functions, despite the fact that the conditions on the boundary are more complicated in comparison with a clamped plate. This result was new, and its proof was technically interesting, but this was only Hadamard's debut in a new field. He demonstrated his true strength six years later in the prize memoir mentioned above, where he considered the problem (14.11), (14.12) for plates of arbitrary shape.

The jury for the *Prix Vaillant*, which consisted of Jordan, Appell, Humbert, Maurice Lévy, Darboux, and Boussinesq, considered twelve memoirs. Poincaré, Picard, and Painlevé were given the task of judging them. The verdict of the jury was to award Hadamard three quarters of the prize, and the remainder to be divided amongst Lauricella, Korn, and Boggio. In the report of Painlevé on Hadamard's work it was said that although it is rich in results, it is even more remarkable because of the corollaries which one could expect to derive from them [III.300]. It is still interesting, after all these years, to understand these results and to see if these expectations of the jury were justified.

*Arthur Korn (1870-1945)*

The main purpose of Hadamard's rather voluminous paper (it is 128 pages long!) is the investigation of Green's function of the boundary value problem (14.11), (14.12), that is, the solution of this problem corresponding to a normal concentrated load applied to a plate at an arbitrary point of $\Omega$. Using contemporary language, one can say that Green's function $G(P,Q)$, where $P$ and $Q$ are points of $\Omega$, is the solution for which $q(x,y) = \delta_Q(x,y)$, where $(x,y)$ are the cartesian coordinates of $P$ and $\delta_Q$ is the so-called Dirac function at $Q$. Of course, Hadamard used another definition of Green's function, as distributions only appeared thirty years later (see the definition of Green's function of the Dirichlet problem for the Laplace operator given in Sec. 14.1). The unique solvability of the problem (14.11), (14.12) can be established in different ways; some of them were already known at the beginning of this century (for instance, the method of boundary integral equations used by Hadamard).

Using Green's function, one can express the deflection of the plate for any load $q$ as

$$w(P) = \int_\Omega G(P,Q)q(Q)dQ,$$

thus reducing the solution of the boundary value problem to the calculation of the integral over the domain. Moreover, one can find an explicit expression for Green's function for special domains. For instance, if $\Omega$ is a disc of radius 1 we have

$$G(P,Q) = \frac{1}{8\pi}|P-Q|^2 \log\left|\frac{P-Q}{1-\overline{Q}P}\right|$$
$$-\frac{1}{16\pi}(|P-Q|^2 - |1-\overline{Q}P|^2),$$

where $P$ and $Q$ are now interpreted as complex numbers, and the bar denotes complex conjugation. It is not possible to give Green's function explicitly for an arbitrary domain, and even now, to find a good approximation for $G(P,Q)$ using computers is a difficult, albeit solvable, problem.

In Hadamard's memoir the dependence of Green's function on the boundary of the domain is investigated. It would be more correct to say three Green's functions, since, as well as $G(P,Q)$, Hadamard considered Green's functions for the Dirichlet and Neumann boundary conditions for the Poisson equation

$$-\Delta u(x,y) = q(x,y) \quad \text{in} \quad \Omega.$$

The paper begins with the derivation of variational formulas, which were later named after him. These formulas express the principal part of the variation of Green's functions under a small deformation of the domain. The proof is not difficult, and we give it here, omitting some insignificant technical details. We consider only Green's function $g(P,Q)$ of the Dirichlet problem for the Laplace operator which Hadamard had already studied in his article [I.110] in 1903.

Hadamard assumes that there is another domain $\Omega^*$ inside $\Omega$, bounded by a smooth contour $\partial\Omega^*$. This contour depends on a small parameter $\varepsilon > 0$ and approximates $\partial\Omega$ in such a way that the angle between the two outward normals at nearby points, is small. We denote by $g(P,Q)$ and $g^*(P,Q)$ Green's functions for $\Omega$ and $\Omega^*$ respectively, where $P, Q$ are fixed points of $\Omega^*$. Since the operator $-\Delta$ gives the Dirac delta function when applied to Green's function, we have the identity

$$g(P,Q) - g^*(P,Q)$$

$$= \int_{\Omega^*} (g^*(P,R)\Delta_R g(Q,R) - g(P,R)\Delta_R g^*(Q,R))dR.$$

Integrating by parts on the right and taking into account the zero boundary condition for the function $g^*$ on $\partial\Omega^*$, we obtain (see formula (14.9))

$$g(P,Q) - g^*(P,Q) = \int_{\partial\Omega^*} g(P,R)\frac{\partial g^*}{\partial \nu_R^*}(Q,R)ds_R^*,$$

where $ds_R^*$ is the line element of the contour $\partial\Omega^*$ at the point $R$, and $\nu_R^*$ is an inward normal to $\partial\Omega^*$ at the same point. The integral is approximately equal to

$$\int_{\partial\Omega} g(P,R^*)\frac{\partial g}{\partial \nu_R}(Q,R)ds_R,$$

the notation $ds_R$ and $\nu_R$ have the same meaning as before, and $R^*$ is the nearest point of $\partial\Omega^*$ to $R$. Now, $g(P,R) = 0$ for $R \in \partial\Omega$, so that

$$g(P,R^*) \simeq \frac{\partial g}{\partial \nu_R}(P,R)\delta\nu_R,$$

where $\delta\nu_R$ is the distance from $R$ to $\partial\Omega^*$, which is assumed to depend linearly on $\varepsilon$. Thus we obtain

$$g(P,Q) - g^*(P,Q) \simeq \int_{\partial\Omega} \frac{\partial g}{\partial \nu_R}(P,R)\frac{\partial g}{\partial \nu_R}(Q,R)\delta\nu_R ds_R.$$

Taking the principal part of the difference $g - g^*$, we finally obtain the Hadamard variational formula

$$\delta g(P,Q) = \int_{\partial\Omega} \frac{\partial g}{\partial \nu_R}(P,R)\frac{\partial g}{\partial \nu_R}(Q,R)\delta\nu_R ds_R.$$

We recall that $g$ is Green's function of the Dirichlet problem for the Laplace operator. In the same memoir Hadamard obtained a similar variational formula for Green's function $\gamma(P,Q)$ of the Neumann problem for the Laplacian as well as for Green's function $G(P,Q)$ for the plate problem. In the last case, by a slightly more difficult proof, Hadamard arrives at the formula

$$\delta G(P,Q) = \int_{\partial\Omega} \Delta_R G(P,R)\Delta_R G(Q,R)\delta\nu_R ds_R. \tag{14.13}$$

From each of the variational formulas for $g$, $\gamma$ and $G$, Hadamard easily obtains non-linear equations in terms of functional derivatives, all having the same form

$$\delta\psi(P,Q) = \int_{\partial\Omega} \psi(P,R)\psi(Q,R)\delta\nu_R ds_R, \tag{14.14}$$

where $\psi$ is one of the functions

$$\psi_1(P,Q) = \frac{\partial^2 g(P,Q)}{\partial\nu_P\partial\nu_Q}, \quad \psi_2(P,Q) = \frac{\partial^2\gamma(P,Q)}{\partial s_P\partial s_Q},$$
$$\psi_3(P,Q) = \Delta_P\Delta_Q G(P,Q).$$

The integration of (14.14) enables one, in principle, to construct the function $\psi$ by the method of successive approximations for the curve $\partial\Omega$ under the condition that the function is given for some initial contour. However, the question arose as to whether the solution depends on the way in which the surface is deformed. This was the problem of the complete integrability of (14.14) given by Hadamard to the young Paul Lévy (see Section 5.4).

In Chapters 2 and 3 of his book [III.239], P. Lévy called equation (14.14) "Hadamard's equation with functional derivatives", and studied it in detail. He showed that the appearance of this equation in the analysis of three different Green's functions was not a chance occurrence. Indeed, in some sense, (14.14) is the only completely integrable equation of the form

$$\delta\psi(P,Q) = \int_{\partial\Omega} f[\psi(P,R),\ \psi(Q,R),\psi(P,Q),\ P,\ Q,\ R]\delta\nu_R ds_R.$$

More precisely, if this general equation is completely integrable then, by a change of the unknown function, it can be reduced to Hadamard's equation (14.14).

This result explains the appearance of equation (14.14) in the variational formulas for Green's functions in quite different boundary value problems. Block [III.42] showed that the function

$$\psi(P,Q) = \frac{1}{2\sqrt{\pi}} \frac{\partial^2}{\partial x\partial\xi} G(x,t,\xi,\tau)$$

satisfies equation (14.14), where $P = (x,t)$, $Q = (\xi,\tau)$ and $G$ is Green's function of the Dirichlet problem for the parabolic equation

$$\frac{\partial^2 u}{\partial x^2} = a\frac{\partial u}{\partial x} + b\frac{\partial u}{\partial t} + cu$$

with variable coefficients. In the paper [I.204], Hadamard obtained a similar result for the perturbed wave equation

$$\frac{\partial^2 u}{\partial x^2} - \frac{\partial^2 u}{\partial y^2} + a\frac{\partial u}{\partial x} + b\frac{\partial u}{\partial y} + cu = 0.$$

In three articles [III.201]–[III.203] published in 1921, Julia found several variational formulas for the conformal mapping of a plane domain $\Omega$ onto a unit disc: $\sigma \to f(\sigma)$, $f(0) = 0$, which were closely related to Hadamard's equation (14.14). For example, he showed that

$$\psi(\sigma, \tau) = \frac{-1}{\pi i}\frac{f'(\sigma)f'(\tau)}{(f(\sigma) + f(\tau))^2}$$

satisfies an analogue of (14.14):

$$\delta\psi(\sigma, \tau) = \int_{\partial\Omega} \psi(\sigma, z)\psi(z, \tau)\delta z dz,$$

where the infinitesimal quantities $\delta z$, $dz$ are expressed in terms of $ds$, $\delta n$.

We also mention some consequences of formula (14.13) derived by Hadamard in [I.145]. From the obvious corollary

$$\delta G(P, P) = \int_{\partial\Omega} \left(\Delta G(P, R)\right)^2 \delta\nu_R ds_R$$

of (14.13), he concludes that $G(P, P)$ increases as the domain becomes larger. Hence, comparing $G(P, P)$ for a domain with that for a circle inscribed in the domain, Hadamard arrives at the inequality $G(P, P) > 0$ for any smooth domain, which is not clear *a priori* for Green's function of a fourth order operator.

We now pass to the account of two conjectures connected with Green's function of the clamped plate. Already in 1901, Tommaso Boggio had assumed that the function $G(P, Q)$ is always positive inside the domain [III.44].[5] This conjecture is true for the unit disc: this follows from the

---

[5]Since then, many authors who discussed this conjecture, called it the Hadamard hypothesis (except Hadamard himself, who referred to Boggio).

above formula for Green's function of the unit disc and from the obvious inequality $\log t \geq 1 - t^{-1}$ for $0 < t < 1$ which implies

$$\log \left| \frac{P - Q}{1 - \overline{Q}P} \right|^2 \geq 1 - \left| \frac{1 - \overline{Q}P}{P - Q} \right|^2 .$$

In [I.145] Hadamard added another "physically evident" conjecture that $G(P, Q)$ increases with the domain. A year later, he discussed Boggio's conjecture in his talk *Sur certains cas intéressants du problème biharmonique* [I.151], given to the International Congress of Mathematicians in Rome. "Mr. Boggio, who was the first to have noted the physical significance of $\Gamma_B^A$, deduced from this the hypothesis that $\Gamma_B^A$ is always positive. In spite of the absence of a rigorous proof, the validity of this hypothesis is not in doubt for convex domains" [I.406, p. 1299].

Hadamard also considers the non-convex domain bounded by Pascal's snail, which is obtained from the unit disc by the conformal mapping $\zeta = (z + \alpha)^2$, and he gives an outline of the proof of positivity of $G(P, Q)$ for $\alpha \geq 1$. In this talk at the Congress, Hadamard referred to the necessity of some additional assumptions on the domain, since the function $G(P, Q)$ has alternating sign for the case of an annulus with a large ratio between the external and internal radii. In 1995 Engliš and Peetre showed that $G(P, Q)$ is not positive for an arbitrary annulus [III.123].

The reader will possibly be interested to know that Boggio's conjecture has been disproved for convex domains as well. This was in 1949 when Duffin showed that Green's function changes sign in the case of a clamped plate whose shape is a long rectangle [III.112]. His counterexample is based on the fact that a biharmonic function in the strip $\{(x, y) : x > 0, |y| < 1\}$ satisfying zero Dirichlet boundary conditions for $y = \pm 1$ and for large $x$, has the asymptotics on the line $y = 0$

$$C \exp(-\sigma x) \cos(\tau x - \varphi)$$

as $x \to \infty$, where $\sigma$, $\tau$, $\varphi$ and $C$ are real constants. Consequently, it has infinitely many changes of sign. In the introduction to his paper, Duffin, in his turn, formulates two conjectures. It appeared to him that the alternation of sign of Green's function occurs for rectangles whose ratio of sides is greater than four. On the other hand, he supposed that for a square plate, the answer to Hadamard's question was affirmative. In favour of this assumption is the fact that the answer is affirmative for a circular plate.

In 1951 Garabedian [III.143] showed that Green's function changes sign even for the ellipse $x^2 + \left(\frac{5}{3}y\right)^2 < 1$. Let $\Omega$ be such an ellipse, and let

$G(P, Q) < 0$ for some points $P, Q$. We inscribe a domain $\Omega^*$ inside $\Omega$ such that $P \in \Omega^*$ and $Q \in \partial\Omega^*$. Then $0 = G^*(P, Q) > G(P, Q)$ which contradicts Hadamard's assumption on the monotonicity of $G$. Two years later, Loewner [III.247] and Szegö [III.384] gave several examples of domains bounded by analytic curves, for which Green's function changes sign.

Shapiro and Tegmark [III.362] in 1994 remarked that non-positivity of $G(P, Q)$ for the ellipse $\Omega = \{(x, y) : x^2 + 25y^2 < 1\}$ can be verified directly by considering the polynomial

$$\Pi(x, y) = (x^2 + 25y^2 - 1)(1 - x)^2(4 - 3x),$$

which satisfies the zero Dirichlet conditions for $\Delta^2$ on $\partial\Omega$. One shows that $\Delta^2\Pi > 0$ on $\Omega$, and then, assuming $G \geq 0$ on $\Omega$, one arrives at $\Pi \geq 0$ on $\Omega$, which is obviously wrong.

Hedenmalm [III.172] studied conditions for the positivity of the biharmonic Green's function, by using Hadamard's idea of considering a continuous movement of the boundary and of calculating the change of $G$ along this movement. In order to formulate his condition, one needs the function

$$H(P, Q) = \Delta_P G(P, Q) - g(P, Q),$$

where $g$ is the harmonic Green's function for the Dirichlet problem. The result is as follows: if $\Omega$ is star-shaped with real-analytic boundary, then $G(P, Q) \geq 0$ on $\Omega$ if and only if $H(P, Q) \geq 0$ on $\partial\Omega \times \Omega$. Hedenmalm's result means that a clamped plate bends in the direction of a point load everywhere, irrespective of where the load is applied; one merely needs to ascertain that this is so when the load is applied near the boundary.

Hayman and Korenblum [III.170] proved that Green's function of the Dirichlet problem for the polyharmonic operator on the unit $n$-dimensional ball is positive (see also [III.172], where a simpler proof is given). The same property of Green's function is preserved under small perturbations of the polyharmonic operator [III.159].

In [I.145] Hadamard also observed the following corollary of the inequality $G(P, P) > 0$

$$G(P, Q)^2 \leq G(P, P)G(Q, Q). \tag{14.15}$$

Indeed, note that any solution of (14.11)-(14.12) satisfies

$$\int_\Omega w(R)q(R)dR = \int_\Omega |\Delta w(R)|^2 dR \geq 0.$$

By setting here $q(R) = \alpha \delta_P(R) + \beta \delta_Q(R)$, so that $w(R) = \alpha G(R, P) + \beta G(R, Q)$ with arbitrary real $\alpha$, $\beta$, we obtain the inequality

$$\alpha^2 G(P, P) + 2\alpha\beta G(P, Q) + \beta^2 G(Q, Q) \geq 0$$

which gives the result.

In connection with Hadamard's inequality (14.15), let us remark that a theorem of Malyshev [III.262] states that the inequality

$$G(P, Q) \geq -\epsilon(\Omega) G(P, P)^{1/2} G(Q, Q)^{1/2}$$

is valid for any bounded domain $\Omega \subset \mathbf{R}^n$ with smooth boundary, where $\epsilon(\Omega)$ is some constant less than one. This estimate holds not only for $\Delta^2$ but also for any operator of the form $LL^*$, where $L$ is an elliptic differential operator of order $l > n/2$, with constant real coefficients and $L^*$ is the adjoint of $L$.

After deriving the variational formulas for Green's functions, and the corollaries of these formulas, Hadamard proceeded, in his memoir, to the study of variational properties of Green's function for the clamped plate. In the introduction to Chapter 3, he wrote:

> In generalizing the theory of maxima and minima to the case of unknown functions, the calculus of variations has until now limited itself to the extrema of either definite integrals, or (and this is what Kneser, in his recent treatise, still called 'the most general problem of the calculus of variations for the case of one independent variable') of solutions of ordinary differential equations. The study of the physical world, however, presents many more measurable phenomena which depend on a line or an arbitrary surface, and whose extrema must, consequently, be examined; and there is no reason for the particular category treated until now, to be the most important.

One of the best-known such facts is the isoperimetric property of the circle, which was discovered in ancient times: among all plane figures whose perimeter has a fixed length, the circle has the largest area. Similarly, the ball is the solution of the problem of the maximum volume of a body with a fixed surface area. Such beautiful regularity is not the prerogative of geometrical properties of bodies. In 1856, Saint-Venant, in his *Mémoire sur la torsion des prismes* [III.353], conjectured that, among all elastic cylinders with the same area of cross-section, the circular cylinder has the largest torsional rigidity. This assertion was supported by physical arguments and some numerical calculations, but he gave no rigorous mathematical proof: this was supplied by Pólya in 1948.

Another example, which probably influenced Hadamard even more, is the conjecture stated by Lord Rayleigh in his book *The theory of sound* [III.338]: among all plates with the same area, the circular plate has the smallest basic frequency. We mean here the first eigenvalue of the boundary value problem

$$\Delta^2 u = \lambda u \quad \text{in} \quad \Omega, \quad u = \frac{\partial u}{\partial \nu} = 0 \quad \text{on} \quad \partial\Omega. \tag{14.16}$$

Until recently Rayleigh's conjecture was an open problem. It was proved under the additional assumption of positivity of the first eigenfunction of the boundary problem (14.16) by Szegö [III.384] in 1950, and included in the book *Isoperimetric inequalities in mathematical physics* by Pólya and Szegö [III.325]. The positivity of the first eigenfunction is a corollary of the positivity of Green's function. It is a particular case of the Jentsch theorem on integral operators with positive kernels [III.197]. However, the first eigenfunction may also be positive if Green's function changes sign.

In 1982, Coffman showed [III.84] that the first eigenfunction of the problem (14.16) for a rectangle has infinitely many changes of sign near each of the vertices, and thus disproved Duffin's conjecture (see also the paper by Kozlov, Kondratiev, and Maz'ya [III.218], where more general results of this type are presented and the existence of convex domains with smooth boundaries for which the first eigenfunction changes sign, is noted). The alternation in sign of the first eigenfunction of problem (14.16) in the case of the annulus $\epsilon^2 < x^2 + y^2 < 1$ with sufficiently small $\epsilon$ was proved by Coffman and Duffin [III.85]. Only in 1995 was Rayleigh's conjecture justified by Nadirashvili [III.289].

There are many similar problems, both solved and unsolved. The traditional tools of the calculus of variations are insufficient for their solution, and Hadamard remarked upon this when he discussed his isoperimetric conjecture. This conjecture says that the maximum value of the functional $G(P, P)$ considered on the set of all domains $\Omega$ containing $P$ and having a prescribed perimeter, is attained when $\Omega$ is a disc with centre at $P$ (by $G$ we mean Green's function for the clamped plate). Hadamard calculated $G(P, P)$ for a contour which is "nearly a circumference" and showed that the variation of the functional is negative. The method he used is based on an application of the formula (14.14), which in principle can only give conditions for a relative extremum.

In Sections 12.2 and 13.7 we have already discussed Hadamard's method in the calculus of variations and a multiplicative inequality for differentiable functions which were included into the fourth part of his memoir.

This concludes our description of Hadamard's paper, although we have not exhausted its contents. We hope that we have convinced the reader that this memoir is deep and rich in results. However, as is the case with Hadamard's ideas in general, their power and value would become apparent only a long time afterwards. The main applications of Hadamard's variational formula concern the theory of functions of complex variables, and planar problems with free boundaries arising in hydrodynamics and the physics of plasmas. Its importance for the solution of extremal problems in complex domains and for various problems connected with conformal mappings is manifested in the works of Julia, Schiffer, Lavrentiev, Garabedian, Bergman, and many others.

An extension of this formula to fundamental solutions of general elliptic boundary value problems in arbitrary $n$-dimensional domains with smooth boundaries, has been obtained in [III.140]. It is also interesting to note that the right-hand side of Hadamard's formula can be considered as the main term of the asymptotic decomposition of Green's function in powers of a small parameter $\epsilon$, which characterizes the variation of the boundary. Thus one can consider Hadamard's memoir as the first in a series of works concerning the asymptotic representations for solutions of elliptic boundary value problems in regularly or singularly perturbed domains. This direction of research, important for applications as well as for pure mathematics, has been developed intensively during recent years [III.273].

## 14.4 Hadamard's equation for surface waves

Hadamard repeatedly addressed problems of hydrodynamics. We have already seen in Section 14.2 how, in *Leçons sur la propagation des ondes*, he systematically expounded the theory of the Neumann boundary value problem, in particular describing the irrotational flow of an ideal fluid in a closed vessel or past an obstacle. In the present section we will be concerned with Hadamard's contribution to the theory of surface waves.

Already in his 1907 memoir on the equilibrium of plates, in showing that the nonlinear integro-differential equation for the variation of Green's function "is in no way an exception in mathematical physics", Hadamard referred to an analogous linear equation which occurs in the theory of small-amplitude surface water-waves. In the note on his scientific works, written in 1909, he remarks that "these same mixed integral equations govern one of the most interesting, and least known, problems of hydrodynamics: the propagation of waves on the surface of a fluid"[I.158, p. 31]. Such equations were the subject of three short papers by Hadamard, with the same title *Sur les ondes liquides* [I.160], [I.161], [I.195].

The problem concerns the hydrodynamics of the ideal (inviscid) incompressible fluid, and deals only with irrotational flows. As is well known, this last condition implies the existence of the velocity potential $\Phi(x, y, z, t)$. In other words, the velocity vector of the fluid particle, $\mathbf{v} = (v_x, v_y, v_z)$, coincides with the gradient of the function $\Phi$ with respect to the spatial variables. From the equation of continuity div $\mathbf{v} = 0$ it follows that the velocity potential satisfies the Laplace equation. At the bottom the normal component of the velocity is zero (the condition of impenetrability), which in terms of the velocity potential means that the normal derivative $\partial\Phi/\partial n$ vanishes for each $t > 0$. The difficulty of this problems lies in the *a priori* unknown form of the fluid surface and in the nonlinearity of the boundary conditions on it. In the initial "exact" formulation, one looks for not only the velocity potential, but also for the domain occupied by the liquid. This domain is where the velocity potential is defined, and it depends on time. This problem is extremely difficult and has not been studied much up to this time. In order to solve the problem, it is usual to introduce some extra physical assumptions. The most common such condition is that the wave amplitude and speed are small, which enables one to linearize the problem. With this hypothesis, we can look for the potential in a fixed domain $\Omega$, bounded below by the bottom, and bounded above by the plane $z = 0$ (see Figure 7).

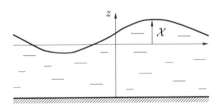

**Figure 7.**

The deviation $\chi(x, y, t)$ of the fluid surface from the plane $z = 0$ is connected with the velocity potential by the two conditions:

$$\chi = -\frac{1}{g}\frac{\partial\Phi}{\partial t}\Big|_{z=0},$$
$$\frac{\partial\chi}{\partial t} = \frac{\partial\Phi}{\partial z}\Big|_{z=0},$$

where $g$ is the gravity constant. The first (dynamic) equation follows from the vanishing of the pressure at the free surface, and the second (kinematic) one means that a fluid particle on the surface does not sink during motion. One should remember that $\Phi$ is a harmonic function satisfying the homogeneous Neumann condition at the bottom, and that at the initial time the potential $\Phi$ and the function $\chi$ are given. This boundary value problem is called the Cauchy-Poisson problem. The study of surface waves has a rich history, but in order to pass without delay to the discussion of Hadamard's equation, we limit ourselves to mentioning the names of Lagrange, Stokes, Kelvin, Rayleigh, Lamb, and Boussinesq, which are connected with the development of the theory.

Hadamard's intention was to find an equation for the function $\chi$ which does not contain the velocity potential. He introduced an auxiliary harmonic function $\psi = \partial\Phi/\partial t$, which obviously satisfies the boundary conditions:

$$\frac{\partial \psi}{\partial n} = 0 \quad \text{at the bottom,} \quad \psi = -g\chi \quad \text{for} \quad z = 0.$$

The function $\chi$ is not given, but this has no significance for Hadamard's argument. He calls the problem of defining a harmonic function by these conditions the mixed problem and says that the problem is solved if the corresponding Green's function $G(M, P)$ is known.

Thus, with the help of $G(M, P)$, the function $\psi$ and its derivative $\partial\psi/\partial z$ on the plane $z = 0$ are defined in terms of $\chi$. Differentiating with respect to time the dynamic boundary condition, one concludes that $\partial\psi/\partial z$ coincides with $\partial^2\chi/\partial t^2$ for $z = 0$ and, consequently, the required relation is obtained. The final form of Hadamard's equation is:

$$\frac{4\pi}{g} \frac{\partial^2 \chi}{\partial t^2} = 2\left(\frac{\partial^2}{\partial x^2} + \frac{\partial^2}{\partial y^2}\right) \int\int \chi \frac{d\xi d\eta}{r} + \int\int \chi \frac{\partial^2 H}{\partial z \partial \zeta} d\xi d\eta. \qquad (14.17)$$

Here $r$ is the distance between the points $(x, y, z)$ and $(\xi, \eta, \zeta)$, and integration is over the plane $\zeta = 0$. The function $H$ is connected with Green's function by the equality

$$G(x, y, z; \xi, \eta, \zeta) = \frac{1}{r} - \frac{1}{r'} + H(x, y, z; \xi, \eta, \zeta),$$

where $r'$ is the distance between $(x, y, z)$ and $(\xi, \eta, -\zeta)$.

However, Hadamard does not restrict himself to the derivation of an integro-differential equation for $\chi$. In the article [I.195], published in 1916, he derives from (14.17) (albeit not rigorously) the so-called shallow water equation in the acoustic approximation. By this the usual wave equation is meant:

$$\frac{\partial^2 \chi}{\partial t^2} = gh\Delta\chi, \tag{14.18}$$

where $h$ is the small constant depth of the fluid in its equilibrium state. This equation for $\chi$ was first obtained by Lagrange from purely physical considerations.

If the liquid fills the half-space $z < 0$ then from (14.17) Hadamard derives the following equation for the elevation of the free surface

$$\frac{\partial^4 \chi}{\partial t^4} + g^2\left(\frac{\partial^2 \chi}{\partial x^2} + \frac{\partial^2 \chi}{\partial y^2}\right) = 0 \tag{14.19}$$

which, as he himself remarked, is not equivalent to (14.17). Equation (14.19) was first found by Cauchy for a liquid of infinite depth which extends infinitely horizontally.

In the case of the liquid in a vessel, Hadamard also derived an equation similar to (14.17), and showed that it can be transformed to the following generalization of the Cauchy equation (14.19):

$$\frac{\partial^4 \chi}{\partial t^4} + g^2\left(\frac{\partial^2 \chi}{\partial x^2} + \frac{\partial^2 \chi}{\partial y^2}\right) = \int_\omega \chi(\xi, \eta)K(x, y; \xi, \eta)d\xi d\eta, \tag{14.20}$$

where $\omega$ is the surface of the liquid, and the kernel $K$ is defined in terms of Green's function of the corresponding mixed problem.

In constructing the above-mentioned Green's function, Hadamard encountered a difficulty the clarification of which is one of the achievements of the modern theory of elliptic equations. This was the problem of singularities of solutions of the mixed problem, which arise near the line $C$ of contact of the walls of the vessel and the surface of the liquid. Hadamard noticed that in the case of a liquid in a hemispherical vessel, the derivative $\partial^2 \chi/\partial t^2$ has a logarithmic singularity on the circle $C$ even when $\chi$ and $\partial\chi/\partial t$ are finite. Calling attention to this paradox – the unboundedness of the acceleration for small displacements and velocities – Hadamard poses the problem of the study of singularities of solutions of equation (14.19) on the line $C$. He is able to derive equation (14.20) only under the *a priori* assumption of regularity of the function $\partial^2 \chi/\partial t^2$ on the contour $C$.

In his doctoral thesis, Bouligand showed that the right-hand side of equation (14.20) vanishes identically in the case when the vessel is a vertical cylinder. The integro-differential equation (14.20) then becomes the differential equation (14.19) in the domain $\omega$. Bouligand summarized his studies in his book *Sur divers problèmes de la dynamique des liquides* [III.56], published in 1930. He begins with the words: "The substance of the present exposition comes from the work of Mr. Jacques Hadamard on theoretical hydrodynamics (ideal fluids)". Further on, he writes:

> [...] there opens up a vast field of research, whose development is of decisive importance. It is this which Mr. Hadamard has demonstrated, in various articles, in studying the theory of liquid waves. The possibility to associate it with a mixed problem allowed him to present this question in a new and systematic form, in which he was able to place the earlier results of Lagrange and Cauchy. But in science, each small corner often constitutes a new domain of research which is as fertile as those areas which contain it.

We conclude with some general remarks about equation (14.17). Some of its important properties, for instance the unique solvability of the Cauchy problem, are not connected with the particular form of the operator on the right-hand side. What is important is the structure of the equation:

$$\frac{d^2\chi}{dt^2} + A\chi = 0,$$

where $A$ is a $t$-independent, positive, self-adjoint operator in a Hilbert space. Equations of this form are easily studied within the framework of abstract operator theory. If the form of the bottom of the vessel varies with time, or if there are moving bodies in the liquid, the derivation of equation (14.17) remains valid, but the operator $A$ becomes a function of $t$. The uniqueness and existence theorem in this more difficult case was proved by Garipov in 1967 [III.144].

It appears that, until recently, Hadamard's equation was not used in the effective solution of hydrodynamic problems. It is quite a complicated equation: it is integro-differential and an explicit description of its right-hand side requires the knowledge of Green's function. However, in view of the practical importance of this equation for hydrodynamics, one should not disregard it. In the same article of 1916, Hadamard remarks on the possibility of the synthesis of equation (14.17) and his variational formula for Green's function. This would lead to some approximate equations of type (14.17) for some cases when the bottom does not have a simple shape. In general, the introduction of some small parameters may considerably simplify equation

(14.17). During recent years, works have appeared concerning the application of this equation to the problem of tsunami generated by instantaneous shifts of the bottom. Finally, one can hope that the ever-growing capabilities of computers and numerical methods will give satisfactory numerical solutions of Hadamard's equation.

## 14.5   Motion of a droplet in a viscous liquid

*George Gabriel Stokes (1819-1903)*

In 1911 two independent papers by Rybczynski [III.350] and Hadamard [I.168] appeared in which the steady axisymmetric motion of a viscous droplet in a viscous liquid under the action of gravity was discussed. Hadamard writes: "This question arises in investigations leading to the determination of the size of atoms, which have led to a series of important experiments..." [I.406, p. 1311]. These studies were based on the analysis of the problem on the descent of a rigid sphere in a viscous liquid by Stokes, in 1851. However, as physicists noted, there was nothing to suggest the validity of the assumption that an atom can be considered as a rigid sphere. In connection with this, Hadamard tried to find an analogue of Stokes' solution, assuming that the falling sphere is also liquid.

In general, the stationary motion of a liquid is described by the Navier-Stokes system, of which three equations are non-linear. Under the condition that the Reynolds number is small, the non-linear terms can be neglected, as was done by Stokes. Using the linearization of Stokes, Hadamard found an explicit solution for the problem of a falling drop, taking into account both external and internal motion. We shall not describe his arguments, but refer the reader to [III.221]. The final formula for the velocity of a falling droplet of radius $R$ with coefficient of viscosity $\eta'$ and density $\rho'$ is

$$u = \frac{2R^2 g(\rho' - \rho)(\eta + \eta')}{3\eta(2\eta + 3\eta')},$$

where $\eta$ and $\rho$ are the coefficient of viscosity and the density of the surrounding liquid respectively, and $g$ is the acceleration due to gravity. As $\eta' \to \infty$ the above expression becomes the formula obtained by Stokes for the rigid sphere, and for $\eta' = 0$ it gives the velocity of ascent of a gaseous bubble. The change in the form of the droplet is neglected in the linear model of Rybczynski and Hadamard, which in reality corresponds to a high surface tension of the droplet. Hadamard understood that his solution has a narrow domain of practical application. Concluding the article, he writes that his formula for the velocity "reveals substantial deviation from the experimental results obtained so far (and as yet unpublished). It seems, therefore, that until further notice, in the cases which have been studied, the classical hypotheses we began with should be modified".

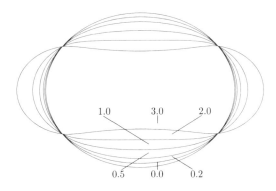

**Figure 8.**

A more realistic solution of the problem, taking into account the nonlinearity of the Navier-Stokes system as well as the change in the shape of the droplet, is more difficult. Here we have a boundary value problem with *a priori* unknown boundary, or in modern terms, with a free boundary. The equation for the surface of the droplet has to be found together with the velocity vector, both outside and inside the droplet. On the surface of the droplet one has to deal with the condition of equality of the normal and capillary stresses, as well as the impenetrability condition and the continuity of the tangential stresses and velocity components. The first boundary condition contains auxiliary information defining the shape of the droplet. However, obtaining this information was impossible at the beginning of this century. The problem was solved recently with the aid of a fast computer.

Figure 8 shows the form of a drop for various values of the Weber number,

which characterizes its surface tension, and for the Reynolds number equal
to 30. This can be found in [III.346]. The solution due to Rybczynski and
Hadamard was taken as a zero-approximation in the iteration process. For
pure liquids the numerical results give good agreement with experiment.[6]

---

[6]For a rich survey on the modern drop deformation literature, see H.A. Stone's paper
[III.380].

# Chapter 15

# Partial Differential Equations

## 15.1 Well-posedness in the sense of Hadamard

A boundary value problem in mathematical physics is said to be well-posed in the sense of Hadamard if it satisfies the following three conditions:

1. the solution exists;
2. the solution is unique;
3. the solution depends continuously on the data.

Each of these conditions is determined by the mechanical or physical origin of the problem. The first condition expresses the consistency of the mathematical model; the second one reflects the definiteness of the real situation; and the third condition corresponds to an intuitive belief on the part of the observer, that small errors in the initial conditions or in the source terms lead to small deviations in the solution.

The conditions formulated above only represent some general principles for the mathematical formulation of physical problems, and allow modifications in concrete problems. For instance, solvability may occur under certain restrictions on the right-hand sides. Some degree of arbitrariness is possible when we choose the solution. This is, for example, the case with eigenvalue problems. The notion of stability of the solution under the variation of the data requires, on the one hand, some more precision, and on the other hand, admits exceptions, since there exist discontinuous or rapidly-varying processes. In general, however, the above three conditions reflect general laws which have a deep mathematical meaning.

The concept of well-posedness of boundary value problems appeared in Hadamard's early works on partial differential equations. In the beginning of his article *Sur les problèmes aux dérivées partielles et leur signification physique* of 1902 [I.89] he wrote that, to the varied problems of mathematical physics there correspond two general types of boundary conditions for partial differential equations: the Dirichlet condition and its analogues like Neumann condition, on the one hand, and the Cauchy problem, on the other. Each of these may be well-posed (*parfaitement bien posé*), that is, admissible and defined (*possible et déterminé*). Hadamard relates the last two properties (solvability and uniqueness of the solution) to the physical origin of the problem.

As an illustration of his idea, Hadamard uses the Laplace equation

$$\frac{\partial^2 u}{\partial x^2} + \frac{\partial^2 u}{\partial y^2} + \frac{\partial^2 u}{\partial z^2} = 0$$

in the half-space $x \geq 0$, for which the Dirichlet problem with the data on the plane $x = 0$ is uniquely solvable in the class of bounded functions, and has a clear physical meaning. Then he considers the Laplace equation with Cauchy data, still on the plane $x = 0$ "without any physical meaning":

$$u = u_0, \quad \frac{\partial u}{\partial x} = u_0' \tag{15.1}$$

and shows that this problem is not always solvable.

A second example is the wave equation (14.4) for which the Cauchy problem with initial data at $t = 0$ (arising naturally in the theory of oscillations) is uniquely solvable. However, "one must beware of formulating the conclusion as: 'the Cauchy problem for equation (14.4) is admissible and defined'", writes Hadamard [I.89]. The point is that, if we adjoin condition (15.1) for $x = 0$ in place of (14.6) to this equation on the set

$$\{(x, y, z, t) : x > 0, (y, z) \in R^2, t > 0\},$$

we get an ill-posed problem. Indeed, suppose $u_0$ and $u_0'$ are independent of $t$. One can then show that the solution (if it is unique) is also independent of $t$. As a result, equation (14.4) becomes the Laplace equation, for which Cauchy problem is unsolvable, as we have already mentioned.

Hadamard often returned to this concept of well-posedness. Mandelbrojt and Schwartz comment on this as follows:

> He repeated this idea constantly so that, even now, a problem is called "well-posed in the sense of Hadamard" if it has the property of continuity of the solution with respect to the data. This idea was even more fruitful than he himself imagined; for the

analysts were then obliged to examine, as he says, the 'different types of neighbourhoods and continuity', which led unavoidably to functional spaces, general topology and functional analysis; it is surely one of the sources of functional analysis, and it is still now one of the best fields for applications of functional analysis. The modern ways for solving partial differential equations use '*a priori* estimates', which means that one actually proves the existence and the uniqueness of a solution by proving, first, its continuity with respect to the data; functional analysis (essentially Banach's and F. Riesz' theorems) yields then the result [II.43, p. 114].

The notion of well-posedness of a boundary value problem, which is so familiar nowadays, was far from being natural at the beginning of this century. In particular, many authors thought that the solution of the Cauchy problem had been dealt with once and for all by the Cauchy-Kovalevskaya theorem. This theorem (we restrict ourselves to the case of one equation, but an analogous result holds for systems of equations) concerns the equation

$$\frac{\partial^k u}{\partial t^k} = f(t, y_1, \ldots, y_n, \frac{\partial u}{\partial t}, \frac{\partial u}{\partial y_1}, \ldots, \frac{\partial^k u}{\partial y_n^k}),$$

where $f$ is an analytic function of its arguments which vary in some neighbourhood of the origin. In other words, the function $f$ can be written as a power series in the variables $t, y_1, \ldots, \partial^k u/\partial y_n^k$. Analytic functions $\varphi_j(y_1, \ldots, y_n)$, $0 \le j \le k - 1$, are given near the point $(y_1, \ldots, y_n, t) = 0$ in the hyperplane $t = 0$. The Cauchy problem consists in finding a solution of the above equation which satisfies the initial conditions

$$u(0, y_1, \ldots, y_n) = \varphi_0(y_1, \ldots, y_n),$$

$$\ldots \ldots \ldots \ldots$$

$$\frac{\partial^{k-1} u}{\partial t^{k-1}}(0, y_1, \ldots, y_n) = \varphi_{k-1}(y_1, \ldots, y_n).$$

The Cauchy-Kovalevskaya theorem says that the problem has a unique solution which is analytic near the origin. For the history of this theorem see the book by R. Cooke [III.87].

*Sofia Kovalevskaya (1850-1891)*

Hadamard arrived at the definition of well-posedness by understanding the restrictiveness of this result. Here is what he writes in his book *Lectures on the Cauchy problem for linear partial differential equations* (which is based on lectures he gave at Yale University in 1920) [I.223, p. 23]:

> The reader will probably wonder at our systematically employing a conditional form and seeming to consider as doubtful one of the most classic and well-known demonstrations of analysis. The fact is that things are not so simple as would be suggested by the above arguments [the Cauchy-Kovalevskaya theorem]. Indeed, the circumstances which we shall meet with will appear as quite paradoxical from the purely mathematical point of view and could only be forseen by physical hints. No question offers a more striking illustration of the ideas which Poincaré developed at the first International Mathematical Congress at Zürich, 1897 (see also *La Valeur de la Science*, pp. 137-155), viz. that its physical applications which show us the important problems we have to set, and that again Physics foreshadows the solutions.
>
> The reasonings of Cauchy, S. Kowalewsky and Darboux, the equivalent of which has been given above, are perfectly rigorous; nevertheless, their conclusion must not be considered as an entirely general one. The reason for this lies in the hypothesis, made above, that Cauchy's data, as well as the coefficients of

the equations, are expressed by *analytic* functions; and the theorem is very often likely to be false when this hypothesis is not satisfied.

Here, Hadamard has in mind the fact that not all boundary value problems are well-posed, but those which correspond to mathematical models of physical phenomena are, and he went against the tendency of that time to consider the Cauchy-Kovalevskaya theorem as being valid for both analytic and smooth functions. The tempting idea of approximating continuous functions by polynomials, which goes back to Weierstrass, seemed to exhaust the problem of solvability especially after the well-known result of Holmgren of 1901 [III.178], saying that the uniqueness of the solution of a linear differential equation holds in the class of sufficiently smooth functions, under the conditions of the Cauchy-Kovalevskaya theorem.

Let us cite Hadamard's arguments refuting similar views and calling attention to various properties of problems with analytic and non-analytic data (even infinitely differentiable). This excerpt from his *Lectures on the Cauchy problem* [I.223] contains, in particular, a well-known example demonstrating the ill-posedness of the Cauchy problem for the Laplace equation, which Hadamard gave in 1917 to the Swiss Mathematical Society, in Zürich:

> I have often maintained, against different geometers, the importance of this distinction. Some of them indeed argued that you may always consider any function as analytic, as, in the contrary case, they could be approximated with any required precision by analytic ones. But, in my opinion, this objection would not apply, the question not being whether such an approximation would alter the data very little, but whether it would alter the solution very little. It is easy to see that, in the case we are dealing with, the two are not at all equivalent. Let us take the classic equation of two-dimensional potentials
>
> $$\frac{\partial^2 u}{\partial x^2} + \frac{\partial^2 u}{\partial y^2} = 0$$
>
> with the following Cauchy data
>
> $$\begin{aligned} u(0, y) &= 0, \\ \frac{\partial u}{\partial x}(0, y) &= u_1(y) = A_n \sin(ny), \end{aligned}$$
>
> $n$ being a very large number, but $A_n$ a function of $n$ assumed to be very small as $n$ grows very large (for instance $A_n = 1/n^p$).

These data differ from zero as little as can be wished. Nevertheless, such a Cauchy problem has for its solution

$$u = \frac{A_n}{n} \sin(ny) \sinh(nx),$$

which, if $A_n = 1/n, 1/n^p, e^{-\sqrt{n}}$, is very large for any determinate
value of $x$ different from zero on account of the mode of growth
of $\exp nx$ and consequently $\sinh(nx)$.

In this case, the presence of the factor $\sin(ny)$ produces a
"fluting" of the surface, and we see that this fluting, however
imperceptible in the immediate neighbourhood of the $y$-axis, becomes enormous at any given distance of it, however small, provided the fluting be taken sufficiently thin by taking $n$ sufficiently
great [I.223, p. 33-34].

In modern terms, Hadamard's example shows that the solution of the
Cauchy problem for the Laplace equation does not necessarily depend continuously on the Cauchy data, even when they converge in the $C^\infty$ topology
(to be the case when $A_n = e^{-\sqrt{n}}$), and the solution is understood in a very
weak topology. However, the Cauchy problem for the equation of a vibrating
string

$$\frac{\partial^2 u}{\partial x^2} - \frac{\partial^2 u}{\partial y^2} = 0, \qquad x > 0,$$

$$u(0, y) = \varphi(y), \qquad \frac{\partial u}{\partial x}(0, y) = 0, \tag{15.2}$$

where $\varphi$ is a bounded continuous function on the real axis, is well-posed in
the class of bounded continuous functions in the half-plane $x \geq 0$. This is
an immediate consequence of the formula which gives the unique solution

$$u(x, y) = [\varphi(x + y) + \varphi(y - x)]/2.$$

Here is a manifestation of "one of the most curious facts in this theory,"
Hadamard writes, "that apparently very slightly different equations behave
in quite opposite ways in this matter" [I.223, p. 23].

Hadamard's paper [I.250], published in 1926, concerns the unsolvability
of the Cauchy problem for the Laplace equation on the half-plane $\{(x, y) :
x > 0\}$ with non-analytic data

$$u(0, y) = f(y), \qquad \frac{\partial u}{\partial x}(0, y) = g(y).$$

Hadamard considers two problems: the first one when the solution is required for all the $x$-axis, and the second one when one looks for the solution on half of this axis. He proves the impossibility of solving the first problem, using the results of Painlevé's article *Sur les lignes singulières des fonctions analytiques* [III.299], and shows that the second problem is admissible even in the case of non-analytic initial data.

At first, Hadamard defined well-posedness as solvability and uniqueness, insisting that the continuous dependence on the initial data is important only for the Cauchy problem. He writes in his book *La théorie des équations aux dérivées partielles*: "This third condition, which we included in our *Leçons sur le problème de Cauchy*, but without considering it as belonging to the definition of the well-posedness of a problem, has been added quite correctly by Hilbert and Courant (*Methods of Mathematical Physics, v. 2*). We adopt their point of view here" [I.405, p. 20].

The necessity of including the condition of continuous dependence of solutions on the data seems to be quite a delicate matter. The point is that, according to Banach's closed graph theorem [III.57], the unique solvability of the linear problem entails the boundedness of the inverse operator, and therefore continuous dependence of the solution on the right-hand sides. Thus, at first sight, the initial definition of well-posedness seems to be sufficient for a large class of problems. However, one can argue that the data of the problem include the coefficients of the differential operators and the domains of the functions, as well as the right-hand sides. A problem can have a unique solution for any right-hand side, but one cannot exclude the possibility that the solution varies by a large amount under a negligible variation of the boundary. Such a situation arose when Hadamard considered the Dirichlet problem for a hyperbolic equation. Before passing to this problem, we would like to remark that, since the above objection is valid, the final variant of the definition of a well-posed problem including all three conditions is preferable. It is also most desirable when the problem is non-linear.

In trying to understand the nature of well-posed problems, Hadamard turned in 1921 to the Dirichlet problem for the vibration of a string [I.209]

$$\frac{\partial^2 u}{\partial x \partial y} = 0 \quad \text{in} \quad \Omega, \qquad u = \varphi \quad \text{on} \quad \partial\Omega. \tag{15.3}$$

(Equations (15.2) and (15.3) are equivalent: one can be transformed into the other by a rotation of the coordinate axes through the angle $\pi/4$).

This problem is not as well-behaved as the Dirichlet problem for the Laplace equation, as is shown by the example of a rectangle with its sides parallel with the coordinate axes. One can easily verify, using the general solution of equation (15.3)

$$u(x, y) = f(x) + g(y), \tag{15.4}$$

that prescribing the solution on any two adjacent sides, uniquely defines it inside the rectangle, and consequently it is impossible to prescribe the Dirichlet data arbitrarily on the whole of the boundary.

In connection with this, it is worth reading Sommerfeld's 1904 review article *Randwertaufgaben in der Theorie der partiellen Differentialgleichungen* written for the *Encyklopädie der mathematischen Wissenschaften* [III.372]. When commenting on du Bois-Reymond's proposal to divide boundary conditions into 'one-directional' (the solution and its normal derivative are prescribed on an arc) and 'two-directional' (the values of the solution are given on two arcs, the case of a closed arc not being excluded), Sommerfeld optimistically claims the possibility of posing boundary value problems also for hyperbolic equations. "There is as yet no proof of this assumption", he remarks [III.372, p. 512-513].

In the paper [I.209] Hadamard considered the instructive example of an ellipse $\partial\Omega = \{(x, y) : x = a\cos t, \quad y = b\cos(t - h)\}$. We follow his reasoning. Let

$$\varphi = \sum(\alpha_n \cos nt + \beta_n \sin nt)$$

be the Fourier expansion of the Dirichlet data. Represent the functions $f$ and $g$ in (15.4) as the series

$$f(x) = \sum \lambda_n \cos nt, \qquad g(y) = \sum \mu_n \cos n(t - h)$$

with unknown coefficients $\lambda_n$ and $\mu_n$. Defining the coefficients using the boundary condition, one obtains the relations

$$\lambda_n + \mu_n \cos nh = \alpha_n, \qquad \mu_n \sin nh = \beta_n.$$

This implies, first of all, that in the case when $h$ and $\pi$ are commensurable quantities, the problem (15.3) has infinitely many linearly independent solutions. If they are incommensurable, then uniqueness holds, but one may encounter difficulties with existence of a solution since the series for $f$ and $g$ may diverge. In fact, let us write $h/\pi$ as the continued fraction

$$\frac{h}{\pi} = a_0 + \cfrac{1}{a_1 + \cfrac{1}{a_2 + \cfrac{1}{a_3 + \dots}}}$$

If the sequence $\{a_j\}$ is bounded, it is known that the fractional parts of the quantities $nh/\pi$, and consequently $\sin nh$, decrease no faster than $n^{-2}$. Therefore, it is sufficient to assume that the function $\varphi$ is three times continuously differentiable in order that the series defining the functions $f$ and $g$ converge (under this condition one has $\beta_n = O(n^{-4})$). In general, the class of admissible boundary values becomes smaller as the approximation of the number $h/\pi$ by rationals improves. Thus the problem under consideration becomes ill-posed in the sense of Hadamard for some values of $h$.

As well as this example, in the same article Hadamard studied the Dirichlet problem (15.3) for the domain formed by two arcs with the same endpoints $A$ and $B$, on each of which both coordinates $x$ and $y$ are monotone.

The next development in this area is due to Huber in 1932 [III.182], whose starting point was Hadamard's results. Huber assumed that the domain is convex with respect to the coordinate directions, that is that the contour intersects any straight line parallel with the coordinate axes in at most two points. Then both $x$ and $y$ have only one maximum and minimum on the contour. The points at which either $x$ or $y$ achieve their extremal values are called vertices by Huber. Clearly, there are only two, three, or four vertices, and the last case is the general one. One can verify that the Dirichlet problem is in general unsolvable in the first two cases, whereas in the third case the situation is not so simple, as is shown by the example of the ellipse. Huber, as well as Hadamard, who returned to this problem in 1936, obtained some partial results concerning the case of four vertices. An important role in this study is played by certain topological mappings of the contour into itself, connected with the movement along the characteristics of the equation for the vibrating string.

Hadamard's 1921 paper was the first in a long series of works by various authors concerning ill-posed problems for hyperbolic equations. In 1939, Bourgin and Duffin [III.59] studied the Dirichlet problem for the equation

$$\frac{\partial^2 u}{\partial y^2} - \frac{\partial^2 u}{\partial x^2} = 0$$

in the rectangle $0 \leq x \leq a, \quad 0 \leq y \leq b$. They proved that if $b/a$ is irrational, uniqueness holds in the class of continuously differentiable functions.

Another result in the same paper gives sufficient conditions for the solvability of the problem expressed in terms of the rate of convergence of the approximation of $b/a$ by rationals.

Let us also mention the deep work of F. John [III.198] published in 1941, where he used the special mappings of the boundary mentioned above. Exploiting the results of the general theory of topological mappings of closed curves into themselves (created by Poincaré in his study of first-order differential equations on the torus), John established the following alternative for an arbitrary contour $\partial\Omega$ which is convex with respect to the coordinate axes: the contour $\partial\Omega$ can either be divided into two parts in such a way that the values of the function $\varphi$ in one part are uniquely defined by the Dirichlet data in the other part, or there exists a mapping of the form $\xi = \alpha(x)$, $\eta = \beta(y)$, where $\alpha$ and $\beta$ are increasing, continuous functions, preserving the type of the equation (15.2) and transforming $\partial\Omega$ into a rectangle formed by lines with slopes $\pm 1$ and whose sides are not commensurable. Clearly, in the first case problem (15.3) is not solvable, whereas in the second case one can apply the results of Bourgin and Duffin.[1]

The Dirichlet problem for hyperbolic equations interested Hadamard for many years. Returning to it while in the United States during the Second World War, he showed that, unlike the equation of the vibrating string, new, unexpected effects appear in the general equation

$$\frac{\partial^2 u}{\partial x \partial y} + A(x,y)\frac{\partial u}{\partial x} + B(x,y)\frac{\partial u}{\partial y} + C(x,y)u = 0$$

(see [I.361]).

Many variations of the problem of well-posedness occur again and again in Hadamard's works throughout his long life. Already in 1906 he noticed that even the classical Dirichlet problem for the Laplace equation with continuous boundary data, which is uniquely solvable in the space of continuous functions in $\overline{\Omega}$, becomes ill-posed if its solution is sought in the class of functions with gradient belonging to $L^2(\Omega)$. This observation of Hadamard, which is of great significance, was discussed in Section 12.3 on the Dirichlet principle.

A series of Hadamard's works concerned the solvability and properties of solutions of the so-called mixed problems for second-order hyperbolic equations (see also his book [I.223]). This term, introduced by Hadamard and in general use for some time, is mostly displaced nowadays by the term

---

[1] Berezanskiĭ, studying generalized solutions of the Dirichlet problem with zero boundary data for the equation $\partial^2 u/\partial y^2 - \partial^2 u/\partial x^2 = f(x,y)$ described a class of domains for which solvability is stable under small variations of the contour [III.29, Ch. 4, Sec. 2].

"initial-boundary value problems". The equation is considered in a domain of cylindrical type (possibly curvilinear) whose base is space-oriented and the lateral boundary is time-oriented. It is complemented by the Cauchy data on the base and boundary conditions (for instance, Dirichlet or Neumann) on the lateral boundary.

How can one formulate well-posed problems for equations which do not belong to one of the three classical types? This question interested Hadamard deeply. He noticed, for instance, the impossibility of well-posed problems for the ultra-hyperbolic equation

$$\frac{\partial^2 u}{\partial x \partial y} = \frac{\partial^2 u}{\partial z \partial t},$$

which was most probably considered for the first time by Hamel [III.164] in his thesis, in connection with some geometrical problems.[2]

In his review paper [I.338] Hadamard mentioned the results of Tricomi who was the first to study the unique solvability of the boundary value problem for equations with variable coefficients which are elliptic in one part of the domain, hyperbolic in the other part, and parabolic on the line separating these parts. A typical example is the equation

$$y \frac{\partial^2 u}{\partial x^2} + \frac{\partial^2 u}{\partial y^2} = 0 \quad \text{in the plane} \quad (x, y) \in R^2.$$

In the paper [I.309], Hadamard turned to equations with constant coefficients which have both real and complex characteristics. The simplest example of such equations of "composite type", as Hadamard called them, is

$$\frac{\partial}{\partial x} \Delta u = 0, \tag{15.5}$$

where $u$ is a function of $x$, $y$. Hadamard proved that some boundary value problems are well-posed for this equation, as well as for

---

[2]In 1964, Blagoveščenskii [III.41] showed that the problem of finding $u$ satisfying the ultra-hyperbolic equation

$$\sum_{j=1}^{p} \frac{\partial^2 u}{\partial x_j^2} = \sum_{j=p+1}^{n} \frac{\partial^2 u}{\partial x_j^2}$$

with Dirichlet data on the characteristic surface

$$\sum_{j=1}^{p} x_j^2 = \sum_{j=p+1}^{n} x_j^2$$

is well-posed in a certain class of functions and obtained an explicit solution.

$$\frac{\partial^2}{\partial x \partial y} \Delta u = 0.$$

A boundary value problem which is well-posed for equation (15.5) with domain as in Figure 9 is the following. Suppose that the functions $u$ and $\partial^2 u/\partial x^2$ are given on the segment $AB$, and that $u$ is given on the arc $L$. Since $\Delta u$ is given on $AB$ by the data of $\partial^2 u/\partial x^2$ and of $u$ (which provides $\partial^2 u/\partial y^2$), equation (15.5) implies that $\Delta u = f(x,y)$ in the domain $ABL$ for some given $f$. It then remains to solve the Dirichlet problem for the equation $\Delta u = f(x,y)$ with the given values of $u$ on the closed contour.[3]

**Figure 9.**

In general, the problem of describing "good" boundary conditions for differential equations "without type" is still far from being solved. An important step in the direction initiated by Hadamard is the well-known theorem due to Hörmander [III.180] stating the existence of at least one well-posed boundary value problem for any partial differential operator with constant coefficients in a bounded domain. For partial differential operators with variable coefficients which do not belong to one of the three classical types, the situation turns out to be much more complicated. It is sufficient to quote the example due to Hans Lewy [III.243]: the equation

$$\frac{\partial u}{\partial x_1} + i\frac{\partial u}{\partial x_2} - 2i(x_1 + ix_2)\frac{\partial u}{\partial x_3} = f(x_3)$$

with infinitely differentiable right-hand side has no solution in the class of generalized (and *a fortiori* smooth) functions in a neighbourhood of the origin. In contrast, a vast theory of well-posed boundary value problems for

---

[3]Two papers by O. Sjöstrand [III.365], [III.366] appeared soon afterwards, in which the unique solvability of a more complicated problem for (15.5) was proved. In a particular case, that problem is formulated as follows: look for a solution of (15.5) in a disc with prescribed values on the boundary and on the diameter which is parallel to the $y$-axis.

equations and systems of elliptic, hyperbolic, and parabolic types, has now been constructed through the efforts of many mathematicians.

The notion of well-posedness has repeatedly shown its universal and deep character. For example, in investigating general linear elliptic boundary value problems one does not speak of unique solvability: this can only be decided in comparatively few cases. Well-posedness is identified with the so-called Fredholm property of the problem. This means that the operator in question has a closed domain, the homogeneous problem has a finite number of non-trivial solutions and the inhomogeneous problem is solvable under a finite number of orthogonality conditions for the right-hand sides.

Hadamard did not participate in the creation of this general theory, which required new tools from operator theory and new "local" analytic techniques. He played a pioneering role when, at the beginning of the twentieth century, he recognised and then developed and propagated the idea of well-posedness. Hadamard's thesis that only well-posed boundary value problems are important in practice could not be accepted in its absolute form. Applications, such as in gravimetrics, spectroscopy, radio astronomy, atmospheric soundings, modelling of optimal systems and constructions, and others, led to the formulation of mathematical problems whose solutions are not stable under small changes of initial data. Among such ill-posed problems we find the Cauchy problem for the Laplace equation, which occurs when one is looking for the extension of the gravitational potential (measured on the earth's surface) in the direction of the normal to the surface. Methods for the approximate solution of ill-posed problems have been developed during the past decades (A.N. Tikhonov, F. John, J.-L. Lions, and others).

## 15.2 From the Cauchy problem to the problem of quasi-analyticity

When studying relations between differentiablity and analyticity for functions of a complex variable, Borel showed that the unique continuation of a function given in a neighbourhood of a point is not an exclusive property of analytic functions (1894). In 1912 Hadamard approached the problem of unique continuation in a short note [I.178] (no longer than a page) which begins with the words: "Independent of the extensions to this notion which can be done in the ways indicated by Borel, there are others to which one is led by analogies suggested by the theory of partial differential equations".

Hadamard departed from his considerations on the role of analyticity of initial data in the Cauchy problem and a remark by Holmgren, and formulated the so-called problem of quasi-analyticity.

We shall follow Hadamard's reasoning which led to the formulation of the problem. Let $u(x, y)$ be a continuous function on the square $Q = \{(x, y) : |x| < 1, |y| < 1\}$, odd in $x$ and harmonic for $x \neq 0$. It can be shown that $u$ is harmonic in $Q$ and hence $\frac{\partial u}{\partial x}(0, y)$ is real analytic on the interval $(-1, 1)$. This argument, given in Hadamard's article [I.89] proves, in particular, the non-solvability of the Cauchy problem for the Laplace equation with data

$$u(0, y) = 0, \qquad \frac{\partial u}{\partial x}(0, y) = u_1(y), \tag{15.6}$$

where $u_1$ is non-analytic. (Here the Cauchy data are given on a line which is interior to the domain.) It is important that analyticity of $u_1$ is equivalent to the sequence of inequalities

$$\left| \frac{d^n u_1}{dy^n} \right| < \frac{n! \, M}{\rho^n}; \quad n = 0, 1, \ldots \tag{15.7}$$

where $M$ and $\rho$ may depend on $y \in (-1, 1)$ but not on $n$.

What happens if we replace the Laplace equation with the heat equation

$$\frac{\partial^2 u}{\partial x^2} - \frac{\partial u}{\partial y} = 0 \ ? \tag{15.8}$$

This question was answered by Holmgren in a remark on p. 324 of his paper [III.178]. Let us denote by $u(x, y)$ a continuous function on $Q$, odd in $x$, satisfying equation (15.8) for $x \neq 0$. Just as in the case of the Laplace operator, one obtains that $u$ is a solution of the heat equation (15.8) in $Q$. Moreover, one can show that this function is analytic in $x$. We represent $u$ as the power series

$$u(x, y) = \sum_{n=0}^{\infty} \frac{u_{2n+1}(y)}{(2n + 1)!} x^{2n+1}, \tag{15.9}$$

where

$$u_{2n+1} = \frac{\partial^{2n+1} u}{\partial x^{2n+1}} \Big|_{x=0}.$$

Using equation (15.8), one finds

$$\frac{\partial^{2n+1} u}{\partial x^{2n+1}} = \frac{\partial^n}{\partial y^n}\left(\frac{\partial u}{\partial x}\right)$$

and therefore $u_{2n+1} = d^n u_1/dy^n$. Hence and by the convergence of the series (15.9) we have

$$\left|\frac{\partial^n u_1}{\partial y^n}\right| < \frac{(2n+1)!\,M}{\rho^{2n+1}}; \quad n = 0, 1, \dots, \quad y \in (-1, 1), \tag{15.10}$$

with $M$ and $\rho$ independent on $n$. It is clear that the sequence of inequalities (15.10) is a less stringent requirement on the function $u_1$ than (15.7); the condition (15.10) is also satisfied by non-analytic functions, a simple example of which is the function $\exp(-1/|y|)$ on the interval $(-1, 1)$.

There is a fundamental difference between conditions (15.7) and (15.10). An analytic function, that is, one satisfying (15.7), is completely defined by its values and the values of all its derivatives at a given point. The class of functions satisfying the sequence of inequalities (15.10) does not have this property. Indeed, the function which is identically zero on the interval $[-1, 0]$ can be extended to the interval $(0, 1]$ by zero or by the function $\exp(-1/y)$.

In [I.178] Hadamard poses the problem "of investigating the conditions on the growth of the consecutive derivatives of a function of a real variable, which would determine the extension of that function...". We give a precise formulation of this problem. Let $\{M_n\}_{n\geq 1}$ be a sequence of positive numbers. The class $C\{M_n\}$ is the set of infinitely differentiable functions $f(x)$ on the interval $a \leq x \leq b$ for which there exists a constant $K$ such that for all $x \in [a, b]$ the inequalities

$$|f^{(n)}(x)| \leq K^n M_n$$

are satisfied for every $n \geq 1$. It is easy to see that $C\{n!\}$ is the class of analytic functions. The class $C\{M_n\}$ is called a class of quasi-analytic functions if any function of that class, that vanishes together with all its derivatives at some point $x_0 \in [a, b]$, is identically zero on $[a, b]$. Hadamard's problem can be now formulated as follows: for which sequences $\{M_n\}$ are the classes $C\{M_n\}$ quasi-analytic?

In 1921, Denjoy [III.100] proved the quasi-analyticity of the class $C\{M_n\}$ when the sequence $\{M_n\}$ satisfies some regularity restrictions and is such that

$$\sum_{n\geq 1} \frac{1}{\sqrt[n]{M_n}} = \infty.$$

In particular, Denjoy's conditions are valid for the sequences

$$M_n = n!\,(\log n)^n, \quad M_n = n!\,(\log n)^n(\log\log n)^n, \quad \ldots$$

Carleman [III.68] removed regularity restrictions to $\{M_n\}$ in Denjoy's theorem and in 1926 found a necessary and sufficient condition for quasi-analyticity:

$$\sum_{n\geq 1}\frac{1}{m_n} = \infty, \tag{15.11}$$

where $m_n = \inf_{k\geq n} \sqrt[k]{M_k}$ [III.69].

For instance, the class $C\{n!\,(\log n)^{n(1+\varepsilon)}\}, \varepsilon > 0$, is not quasi-analytic since by Stirling's formula

$$n! = \sqrt{2\pi n}(n/e)^n(1 + o(1))$$

one has $m_n \approx n(\log n)^{1+\varepsilon}/e$ so that the series $\sum m_n^{-1}$ converges.

Another condition, equivalent to (15.11) but different in form, was given in 1929 by Ostrowski [III.297]:

$$\int^{\infty} \log T(r)\frac{dr}{r^2} = \infty,$$

where $T(r) = \sup_{n\geq 1} r^n/M_n$.

After describing recent progress in the problem of quasi-analyticity, Hadamard and Mandelbrojt [I.253, p. 28] wrote: "It is quite untrue, as one can see, that the analytic functions are the most general ones which one needs to consider, and it would be a grave error to suppose that they are. But they are the simplest and most important of all, in the sense that they furnish *concrete, precise* examples of functions which, up to now, have been introduced from applications". This remark is accompanied by a footnote: "The recent works of Julia give a glimpse, through examples, such as the problem of iteration, that it may become otherwise in the future." In retrospect, one can see in these words a prophetic presentiment of the appearance of the theory of fractals some thirty years later in the work of Mandelbrojt's nephew Benoit Mandelbrot.

Although the problem of quasi-analyticity had been solved completely by 1926, the development of the ideas to which it gave birth, did not stop at this. Various modifications of the problem were formulated and studied (Bernstein, Mandelbrojt, Beurling and others), and various applications to Fourier series, the problem of moments, uniqueness theorems for analytic functions and other areas of analysis, were discovered. The reader can acquaint himself further with this area in Mandelbrojt's book [III.263].

## 15.3  Fundamental (elementary) solutions

Hadamard's article [I.114] on fundamental solutions and the integration of linear partial differential equations appeared in 1904 in the *Annales de l'École Normale Supérieure*. In this memoir, a procedure was developed for the construction of special solutions for equations of the general form

$$\sum_{i,j=1}^{m} A_{ij}(y)\frac{\partial^2 u}{\partial y_i \partial y_j} + \sum_{i=1}^{m} B_i(y)\frac{\partial u}{\partial y_i} + C(y)u = 0 \qquad (15.12)$$

with analytic coefficients, where $y = (y_1, \dots, y_m)$.

Nowadays, following Laurent Schwartz, fundamental solutions are functions (or distributions) which satisfy the equation with a $\delta$-function on the right-hand side. The knowledge of such solutions is important, since they represent the kernels of integral operators which are the inverses of the differential operators. Thus, with their aid, we can express the solution of a differential equation whose right-hand side is an arbitrary function. In Hadamard's time, mathematicians used another definition (equivalent, in the elliptic case) which singled out the fundamental solutions by their asymptotic behaviour in the neighbourhood of a singular point or surface. We have already spoken about one fundamental solution which Hadamard studied, namely Green's function for the clamped plate.

The Laplace operator in $\mathbf{R}^m$ has the fundamental solution

$$v(y) = \begin{cases} (2-m)^{-1} s_m |y|^{2-m} & \text{for } m > 2, \\ (2\pi)^{-1} \log|y| & \text{for } m = 2, \end{cases}$$

where $|y| = (y_1^2 + \dots + y_m^2)^{1/2}$ and $s_m$ is the $(m-1)$-dimensional area of the boundary of the unit $m$-dimensional ball. For equations with variable coefficients one can find explicit fundamental solutions only in rare cases. In 1891, Picard presented a method to obtain the fundamental solution (that is, one having a logarithmic singularity) for the equation

$$\frac{\partial^2 u}{\partial y_1^2} + \frac{\partial^2 u}{\partial y_2^2} + C(y)u = 0 \quad \text{in} \quad R^2.$$

Picard's algorithm leads to the representation of the solution as a convergent series. Later, Sommerfeld, Hilbert, and Hadamard solved the same problem for the general equation with analytic coefficients, when $m = 2$. Partial results for the multi-dimensional case were obtained by Fredholm and Holmgren. One of the results contained in the 1904 article by Hadamard was the existence, in the neighbourhood of a point, of a fundamental solution of the elliptic equation (15.12) with analytic coefficients depending on $m$ variables. The requirement of analyticity was later removed by E. Levi and Hilbert.

We shortly describe Hadamard's result. He begins with introducing a particular length of an arc as the following integral along this arc

$$l = \int \sqrt{\sum_{i=1}^{m}\sum_{j=1}^{m} B_{ij} dy_i dy_j},$$

where $[B_{ij}]$ is the inverse matrix of $[A_{ij}]$. In the case of elliptic equations with real coefficients, this integral is positive, and by minimizing it over all arcs joining two points, one arrives at the "distance" between these points. The arcs along which this integral takes its minimum value are called geodesics. By $\Gamma(y)$ Hadamard denoted the square of this new distance from the origin to the point $y \in \mathbf{R}^m$. He constructed the fundamental solution in the form

$$U = \Gamma^{\frac{2-m}{2}} \sum_{k=0}^{\infty} u_k \Gamma^k, \tag{15.13}$$

if $m$ is odd, and

$$U = \Gamma^{\frac{2-m}{2}} \sum_{k=0}^{\infty} u_k \Gamma^k + \log \Gamma \sum_{k=0}^{\infty} v_k \Gamma^k, \tag{15.14}$$

if $m$ is even.

In calculating the coefficients of the decompositions in (15.13), (15.14), Hadamard considered the variables $y_1, \dots, y_m$ to be complex. The functions $u_k$ and $v_k$ are defined as solutions of certain ordinary differential equations along the geodesics starting at the origin, and the condition of regularity of the $u_k$ and $v_k$ at the origin enables one to find the fundamental solution up to a constant factor.

By using the complex domain Hadamard could consider both elliptic and hyperbolic equations (15.12) from a unified standpoint. In the hyperbolic case, the integral $l$ may be non-real or zero and the "distance" between $y$ and the origin should be defined not as a minimum, but as a stationary value of the integral $l$. Thus, solutions (15.13) and (15.14) of hyperbolic equations have singularities not only at the origin, but also on the real surface $\{y \in \mathbf{R}^m : \Gamma(y) = 0\}$, which is called the characteristic conoid. This conoid coincides with the characteristic cone in the case of constant coefficients $A_{ij}, 1 \leq i, j \leq m$.

Strictly speaking, in the hyperbolic case, the solutions (15.13) and (15.14) constructed by Hadamard are not fundamental according to the terminology used in contemporary mathematics. In accordance with what we said above, by a fundamental solution of the Cauchy problem for the wave equation in $\mathbf{R}^n$, i.e., $(t, x) \in \mathbf{R}^{n+1}$,

$$u_{tt} - \omega^2 \Delta u = 0,$$

one means a distribution $v$ satisfying

$$v_{tt} - \omega^2 \Delta v = \delta(x - x_0)\delta(t - t_0), \quad v|_{t < t_0} = 0.$$

For instance, in the case of odd $n$ the following formula holds

$$v(x, t) = (2\pi^{(n-1)/2}\omega)^{-1}\delta^{(n-3)/2}(\omega^2(t - t_0)^2 - |x - x_0|^2),$$

where $t > t_0$ and $\delta^{(k)}$ is the derivative of order $k$ of the $\delta$-function of one variable. Naturally, Hadamard could not use this notion, his "fundamental solution" of the wave equation has the following form for odd $n$

$$w(x, t) = (\omega^2(t - t_0)^2 - |x - x_0|^2)^{(1-n)/2}.$$

In what follows, in order to avoid confusion with terminology, we shall call the solution constructed by Hadamard elementary, in keeping with his later work, in particular his book [I.223].

Hadamard's method turned out to be also suitable for the construction of solutions of hyperbolic equation (15.12) which are fundamental in the modern sense (see [III.90, p.734-737]). Later it was modified by decomposing the $\delta$-function into plane waves (Asgeirsson, John). Courant writes: "Hadamard's achievement was in the first place the construction of the fundamental solution; this construction proceeds directly without the benefit of the simplification [obtained] by first decomposing the $n$-dimensional $\delta$-function into plane waves and then integrating over a unit sphere". It is interesting to cite Hadamard's words in his book [I.372, p. 53]:

> I must close the enumeration of these failures with one which
> I can hardly explain: having found, for constructing conditions

of possibility for a problem in partial differential equations, a
method which gives the result in a very complicated and intricate
form, how did I fail to notice, in my own calculations, a feature
which enlightens the whole problem, and leave that discovery to
happier and better inspired successors? That is what is difficult
for me to conceive.

Hadamard is probably too hard on himself. Babich, in a letter to the
authors, expresses the following opinion: "I definitely think that in the case
of linear hyperbolic second order equations, decomposing the $\delta$-function into
plane waves is worse than Hadamard's method: the solution is obtained in
an unjustifiably complicated form."

Let us note, in conclusion, that Hadamard's construction of elementary
solutions was the first mathematical application of the so-called "space-time
ray method" for the asymptotic description of wave phenomena, which was
developed intensively during recent years (see [III.17]).

## 15.4　The Cauchy problem for hyperbolic equations

Hadamard's greatest achievement in the theory of partial differential equations was the complete solution of the Cauchy problem for general, linear,
second-order hyperbolic equations. In 1905, in the continuation of his memoir on fundamental solutions, which we have just described, he obtained his
results for the case of three space variables [I.119]. The case of arbitrary
dimensions was studied in 1908 [I.148]. These researches were summarized
in his lectures given at Yale University in 1920 and published as a book in
1923 in English [I.223]. The French edition appeared in 1932 [I.305]. Mandelbrojt and Schwartz write in their essay on Hadamard's life and scientific
achievements [II.43, p. 115]: "This book is a real masterpiece and, by its
content, its clarity, and the abundance of its ideas, it has inspired all the
investigators on partial differential equations of the following generation."

Hadamard's direct predecessors were Kirchhoff, Beltrami, and Volterra.
In their papers from the end of the last century they developed a mathematical theory of light (or acoustic) waves described by the Cauchy problem for
the wave equation. Hadamard wrote in [I.239, p. 66]: "Being struck, as were
all geometers, by the beauty of Volterra's results, I set out to extend them to
linear partial differential equations with variable coefficients, in other words
to wave propagation in heterogeneous media".

*Eugenio Beltrami (1835-1900)*

In order to make the presentation of Hadamard's results clearer, we recall some classical facts about this problem. Consider the wave equation (14.4), where $\omega$ is the constant velocity of light or sound, and $\Delta$ is the Laplace operator with respect to the space variables $(x_1, \ldots, x_n) = x$. Let the solution $u$ satisfy the Cauchy data (14.6).

In the case $n = 3$ the solution of the problem (14.14), (14.16) is given by the Poisson formula

$$u(x, t) = t M_{x,\omega t}(g) + \frac{\partial}{\partial t}(t M_{x,\omega t}(f)), \tag{15.15}$$

where $M_{x,r}(f)$ is the mean value of the function $f$ on the sphere with centre $x$ and radius $r$. The solution of the problem (14.14), (14.16) for $n = 2$ can be obtained from (15.15) in the following way. Suppose that in (14.16) the functions $f$ and $g$ do not depend on the third space variable. After simple transformations of the right-hand side in the Poisson formula, one arrives at the following representation for the solution

$$u(x, t) = \frac{1}{2\pi\omega}[\mu_{x,\omega t}(g) + \frac{\partial}{\partial t}\mu_{x,\omega t}(f)], \tag{15.16}$$

where

$$\mu_{x,r}(f) = \int\int_{|x-z|<r} \frac{f(z)dz}{(r^2 - |x-z|^2)^{1/2}}.$$

Hadamard wrote in his book [I.223, p. 49]:

> We thus have a first example of what I shall call a 'method of descent'. Creating a phrase for an idea which is merely childish and has been used since the very first steps of the theory is, I must confess, rather ambitious; but we shall come across it rather frequently, so that it will be convenient to have a word to denote it. It consists in noticing that he who can do more can do less: if we can integrate equations with $m$ variables, we can do the same for equations with $(m-1)$ variables.

We shall see later that the method of descent used by Hadamard in a rather non-trivial case, enabled him to achieve a fundamental simplification of the proof.

Now let us return to (14.14). One is looking for a representation of the solution of the more general Cauchy problem

$$u|_S = f, \qquad \frac{\partial u}{\partial \nu}\Big|_S = g, \tag{15.17}$$

where $f$ and $g$ are sufficiently smooth functions given on a surface $S$ and $\nu$ is the normal to $S$. The surface must be space-like, i.e., at all its points the cosine of the angle between the normal $\nu$ and time direction should satisfy the inequality

$$|\cos(\nu, t)| > \omega(1+\omega^2)^{-1/2}.$$

If this condition is violated, the Cauchy problem is, generally speaking, unsolvable.

Let $P_0 = (x_0, t_0)$ be a point, at which we shall find the solution, and let the characteristic cone

$$|\cos(\nu, t)| = \omega(1+\omega^2)^{-1/2}$$

with vertex $P_0$ cut $S$ along the surface $\sigma(P_0)$. The domain in $\mathbf{R}^{n+1}$ which is bounded by the characteristic cone and $\sigma(P_0)$ is denoted by $\tau(P_0)$ (see Figure 10).

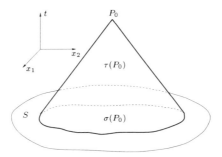

**Figure 10.**

In order to express $u(P_0)$ in terms of integrals, we can apply the following Green's formula, which is easily verified by integration by parts

$$\int_{\tau(P_0)} uLvd\tau = \int_{\sigma(P_0)} (uNv - vNu)d\sigma + \int_{\tau(P_0)} vLud\tau,$$

where $L = \partial^2/\partial t^2 - \omega^2 \Delta$ and $N$ is the differential operator defined on $S$ by

$$N = \cos(\nu, t)\frac{\partial}{\partial t} - \omega^2 \sum_{i=1}^{n} \cos(\nu, x_i)\frac{\partial}{\partial x_i}.$$

The functions $u$ and $v$ in Green's formula are arbitrary, and if we take $u$ as a solution of the problem (14.4), (14.6), and take $v$ as the fundamental solution of the Cauchy problem, which we discussed in the previous section, then on the left-hand side we obtain the sought-for value of the function, and on the right-hand side some expression which does not contain the unknown functions. Of course, one must verify that the formula for $u(P_0)$ which we obtain is actually the solution of the problem.

The above procedure, which now seems to be quite simple in terms of ideas and technique, was, in Hadamard's time (and before him), impossible due to the lack of the notion of distribution. However, the Cauchy problem for the wave equation and for some of its generalizations, was solved in the works of Riemann, Kirchhoff, and Volterra, and later in the works of Tedone, Coulon, and d'Adhémar. All these authors used a modification of Green's formula. The appearance of divergent integrals was overcome by replacing the solution $u(x, t), x \in \mathbf{R}^n, t > 0$ with the integral

$$\int_0^t (t - \tau)^{n-2} u(x, \tau)d\tau,$$

after which the solution was obtained by differentiation. This device worked in the cases it was applied to, but could not be extended to general linear second-order hyperbolic equations with variable coefficients.

*Gustav Robert Kirchhoff (1824-1887)*

Yet another characteristic feature in the methods of Hadamard's predecessors can be seen in the example of the auxiliary solution

$$v(x, t; x_0, t_0) = \log \frac{\omega(t_0 - t) + \sqrt{\omega^2(t_0 - t)^2 - |x_0 - x|^2}}{|x_0 - x|}$$

used by Volterra in the derivation of his representation of cylindrical waves. This function $v$ tends to infinity not only at the boundary of the characteristic cone, but also on its axis. Volterra's solution corresponds physically to continuously acting sources of vibrations.[4] In the general case, such solutions are difficult to construct. Volterra's solutions are most probably unsuitable as a starting point for the theory of more or less general hyperbolic equations.

Hadamard sets himself the goal of finding a representation for the solution of the Cauchy problem for an arbitrary linear hyperbolic equation of order two

---

[4]Hadamard's elementary solution for an even number of space variables corresponds to a source of vibrations of an instantaneous impulsive character.

$$Lu := \sum_{i,j=1}^{n+1} a_{ij} \frac{\partial^2 u}{\partial x_i \partial x_j} + \sum_{i=1}^{n+1} b_i \frac{\partial u}{\partial x_i} + cu = 0 \qquad (15.18)$$

with variable coefficients. In the preface to his lectures, Hadamard writes:

> The origin of the following investigations is to be found in Riemann, Kirchhoff and still more Volterra's fundamental Memoirs on spherical and cylindrical waves. My endeavour has been to pursue the work of the Italian geometer, and so to improve and extend it that it may become applicable to all (normal) hyperbolic equations, instead of only to one of them [I.223, p. 7].

Let us recall one of the basic notions of the theory of hyperbolic equations of the form (15.18). Consider the conic surface

$$C(P_0) = \{P = (x_1, \dots, x_{n+1}) : \sum_{i,j=1}^{n+1} h_{ij}(x_i - x_i^0)(x_j - x_j^0) = 0\}$$

with vertex $P_0 = (x_1^0, \dots, x_{n+1}^0)$, where $[h_{ij}]_{i,j=1}^{n+1}$ is the inverse matrix of $[a_{ij}]_{i,j=1}^{n+1}$. Due to the hyperbolicity of the equation, the surface has two sheets and divides the $n + 1$-dimensional space into three non-overlapping regions: two "internal" and one "external". A surface $S$ is said to be space-like (or, a surface of space type) if and only if the normal to $S$ at any point $P_0$ in $S$ is directed strictly inside one of the internal zones bounded by the surface $C(P_0)$ (see Figure 11). One can easily verify that this definition agrees with the definition of a space-like surface given previously for the wave equation.

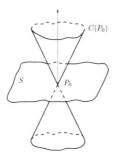

**Figure 11.**

As in the case of the wave equation, the Cauchy data for equation (15.18) are given on space-like surfaces. It is precisely in this way that Hadamard solves the Cauchy problem (15.18), (15.17). We briefly describe his results, following the book [I.223]. The first step, the construction of the elementary solution of the adjoint equation

$$L^*(v) := \sum_{i,j=1}^{n+1} \frac{\partial^2}{\partial x_i \partial x_j}(a_{ij}v) - \sum_{i=1}^{n+1} \frac{\partial}{\partial x_i}(b_i v) + cv = 0 \qquad (15.19)$$

has already been described in Section 15.3. As a result, Hadamard obtains a function $v$ of the points $P$ and $P_0$ of the form

$$v = \begin{cases} V\Gamma^{-(n-1)/2} & \text{if } n \text{ is even,} \\ V\Gamma^{-(n-1)/2} + W \log \Gamma & \text{if } n \text{ is odd,} \end{cases}$$

where $V$ and $W$ are holomorphic functions of $2n + 2$ coordinates, and $\Gamma$ is the square of the geodesic distance between the points $P$ and $P_0$.

If Hadamard had been able to substitute the fundamental solution of the Cauchy problem (in our sense of the notion) into Green's formula for the operators $L$ and $L^*$, then the problem would have been solved in the same way as for the wave equation. However, he could only operate with the elementary solution $v(P, P_0)$, and a direct substitution of this in Green's formula leads to divergent integrals. To overcome this difficulty, Hadamard introduced a new notion for divergent improper integrals of functions of one or several variables. This was the concept of the "finite part" (*partie finie*) of a divergent integral.

Consider for instance the integral

$$I(\varepsilon) = \int_\varepsilon^h \frac{f(x)}{x^{r+1}} dx,$$

where $r$ is a positive non-integer number. If the function $f$ has a sufficient number of derivatives at the point 0, then there exist coefficients $c_k$ ($0 \leq k < r$, $k$ is an integer) such that the difference

$$I(\varepsilon) - \sum_{k=0}^{[r]} c_k \varepsilon^{k-r}$$

has a finite limit as $\varepsilon \to 0+$. This limit is what Hadamard called the finite part of the integral $I(0)$ and denoted by the symbol

$$\boxed{\int_0^h \frac{f(x)}{x^{r+1}}\,dx}\;.\tag{15.20}$$

He extended the validity of the usual rules of change of variable, integration by parts and differentiation with respect to the upper limit of integration to this new object. Then Hadamard passed to multiple integrals and, by reducing them to one-dimensional integrals in the usual way, he defined and studied the expression

$$\boxed{\int_\Omega \frac{f(x)}{G(x)^{r+1}}\,dx},$$

where $\Omega$ is an $n$-dimensional domain, a part of whose boundary is formed by a smooth surface $G(x) = 0$.

Many years later, when analysing the connections between the intuitive and logical ways of mathematical discovery, Hadamard wrote:

> I have been asked by what kind of guessing I thought of the device of the "finite part of infinite integral", which I have used for the integration of partial differential equations. Certainly, considering in itself, it looks typically like "thinking aside". But, in fact, for a long while my mind refused to conceive that idea until positively compelled to. I was led to it step by step as the mathematical reader will easily verify if he takes the trouble to consult my researches on the subject, especially my *recherches sur les solutions fondamentales et l'intégration des équations linéaires aux dérivées partielles*, 2nd Memoir, especially p. 121 (*Annales Scientifiques de l'École Normale Supérieure*, Vol.**XXII**, 1905). I could not avoid it any more than the prisoner in Poe's tale *The Pit and the Pendulum* could avoid the hole at the center of his cell [I.372, p. 110].

It is not well known that the term "finite part of a divergent integral" was introduced by d'Adhémar in his thesis presented at the Sorbonne in December 1903, and defended in April 1904. Referring to Hadamard's article [I.110] of December 7, 1903, d'Adhémar writes "Independently of each other, we understood the role of these *finite parts*" [III.4, p. 371]. In d'Adhémar's thesis this notion was applied to the construction of solutions of equations for cylindrical waves, whereas Hadamard used finite parts for the solution of the Cauchy problem for second order equations with variable coefficients and an arbitrary number of independent variables.

The calculation of the finite part of a divergent integral is, in modern terms, one of the ways of regularizing distributions. At present the regularization of divergent integrals, which consists in replacing them with their finite part in the sense of Hadamard, has mostly historical value: the regularization by analytic extension of integrals with respect to a parameter is more convenient and is better developed.

However, in the theory of one-dimensional hypersingular integral equations, arising, in particular, in crack problems in elasticity and in thin airfoil theory, the notion of finite part of a divergent integral in the sense of Hadamard is often used (see [III.79] for references).

*Cauchy's paper of 1826 on the extraordinary integrals. In his 1844 memoir [III.74] he wrote that the work on the* intégrale extraordinaire *had been presented in its initial form to the* Académie des Sciences *on January 2, 1815.*

Surprisingly, the regularization of the divergent integral

$$\int_0^h \frac{f(x)}{x^{r+1}}\,dx,$$

where $r$ is any positive non-integer number, had already been done by Cauchy (see [III.75], which appeared in 1826). He called this expression *l'intégrale extraordinaire* and by this he meant the same as we do, namely

$$\int_0^h \frac{f(x) - F(x)}{x^{r+1}}\,dx,$$

where $F(x)$ is the Taylor polynomial

$$F(x) = \sum_{k=0}^{[r]} f^{(k)}(0)\frac{x^k}{k!}.$$

An elementary calculation shows that the difference between the finite part (15.20) and Cauchy's *intégrale extraordinaire* equals

$$\sum_{k=0}^{[r]} f^{(k)}(0)\frac{h^{k-r}}{k!(k-r)}.$$

Cauchy proved that the usual rules for differentiation and integration with respect to a parameter are valid for extraordinary integrals.[5]

Let us now return to the solution of the Cauchy problem (15.18), (15.17), which was obtained by Hadamard, at first for even values of $n$. This restriction, which we assume to hold for the time being, was caused by the difference in asymptotic representations for elementary solutions of the equation (15.19) for even and odd $n$.

Hadamard wrote Green's formula for the operators $L$ and $L^*$ in terms of finite parts of integrals. This new Green's formula in which $u$ is the solution of the Cauchy problem, and $v$ is the elementary solution of the equation (15.18), has the form

$$\overline{\left| \int_{\Sigma(P_0)} (uN(v) - vN(u) + Muv)\, d\Sigma \right.}$$

$$= \overline{\left| \int_{\sigma(P_0)} (uN(v) - vN(u) + Muv)\, d\sigma. \right.} \tag{15.21}$$

Here, $N$ is the differential operator given by

$$N(v) = \sum_{i,j=1}^{n+1} a_{ij} \cos(\nu, x_j)\frac{\partial v}{\partial x_i},$$

$M$ is a smooth function, the set $\sigma(P_0)$ is defined in the same way as in the case of the wave operator, but the role of the characteristic cone passes to the characteristic conoid $\{P : \Gamma(P, P_0) = 0\}$ with vertex at the point $P_0$. This conoid represents the characteristic surface of the equation (15.18), which has a conic singularity at $P_0$ or, in other words, is tangent to the cone

---

[5] For the history of the notion of the finite part of the divergent integral see F.J. Bureau's paper [III.63].

$C(P_0)$ at the point $P_0$. By $\Sigma(P_0)$ we denote a part of the surface of the $(n+1)$-dimensional ball of small radius $\delta$ centered at $P_0$ which is contained inside the same conoid (Figure 12).

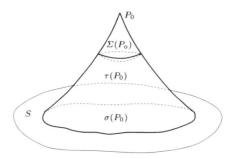

**Figure 12.**

Let us stress that (15.21) has no meaning without the finite part sign, because when the point $P$ approaches the edges of the surfaces $\Sigma(P_0)$ and $\sigma(P_0)$ on the conoid $\Gamma = 0$ the function $v$ tends to infinity as $\Gamma^{(1-n)/2}$. Also important is the fact that the right-hand side of the equation is given in terms of the Cauchy data, and is therefore known. As to the left-hand side, letting the radius $\delta$ go to zero, Hadamard showed that it tends to $u(P_0)$, up to a constant factor. Thus, for even $n$, he obtained the formula for $u(P_0)$. Hadamard then verified that the right-hand side of (15.21) is indeed a solution of the Cauchy problem.

The case of odd $n$ needed a special approach, in the first place because it is not possible to define the finite part of the integrals which arise. Remember that in the definition of the finite part, the number $r$ was non-integer. Now the exponent of $\Gamma$ in the denominator of the integrand is $(n-1)/2$, which is an integer. Thus Hadamard arrives at his goal using, with great virtuosity, "descent" from even $n$ to odd $n$. We have already described it when speaking about the wave equation. Later, in 1924, he gave a direct construction of the solution for odd $n$, but for this he used the so-called logarithmic part of a divergent integral, instead of a finite part. The representations Hadamard obtained for solutions of the Cauchy problem for even and odd $n$ led him to interesting conclusions about the so-called Huygens' principle, which we shall discuss in the next section.

The difficulties Hadamard met in the case of odd $n$ are connected with the use of elementary solutions in the form

$$V\Gamma^{(1-n)/2} + W \log \Gamma,$$

which is inconvenient for calculations. Nowadays, using Hadamard's methods, one can construct solutions of the form

$$V_1 \delta^{(n-3)/2}(\Gamma) + W_1 \theta(\Gamma),$$

where $\theta$ is the Heaviside function, and $\delta^{(l)}$ is the derivative of order $l$ of the $\delta$-function of one variable (see [III.90]). Substituting the latter solution into Green's formula, one can find the solution of the Cauchy problem in a similar way to that of Hadamard, for even $n$. Obviously, Hadamard had no derivatives of the $\delta$-function at his disposal at the beginning of the century.

Other ideas, generated by the representations of solutions which we discussed, were outlined by Hadamard, and are connected with potential theory for hyperbolic equations. Integral operators, of simple- and double-layer potential type, for the wave equation with two space variables, were first defined by Volterra [III.410]. Hadamard introduced potentials in the multidimensional case and stressed their analogy with the usual harmonic potentials. Comparatively recently, the theory of hyperbolic potentials was developed and applied to the solving of the mixed problem for the wave equation [III.18].

Hadamard finished his book on the Cauchy problem with an investigation of hyperbolic equations with sufficiently smooth, non-analytic coefficients. His starting point is the work of Hilbert and that of a young Eugenio Levi, who perished in the First World War. In studying elliptic equations, both started from an explicit first approximation to the fundamental solution. This approximation, called the parametrix by Hilbert, gives, on substitution into the equation, a singularity of low order in the right-hand side. In particular, Levi was able to construct the fundamental solution using the parametrix. However, in the hyperbolic case, the method of Hilbert and Levi encountered serious difficulties because of the singularity of the solution on the characteristic conoid. Hadamard reduced the Cauchy problem to an integral equation of Volterra type using an approximation of the elementary solution of the hyperbolic equation with sufficiently smooth coefficients. This integral equation can be solved by the method of successive approximations.

Hadamard's study of the Cauchy problem, which he carried out at the beginning of the century, became a starting point for further progress in the theory of hyperbolic equations. Inspired directly by Hadamard's book, the works of Mathisson [III.271], Schauder [III.356], and Sobolev [III.368], [III.370] appeared in the 1930s. They were concerned with other approaches to the solution of the Cauchy problem for the hyperbolic equation (15.18). Schauder used the energy inequality to extend a local Cauchy-Kovalevskaya solution and thus constructed the global solution of the Cauchy problem. Simultaneously, Sobolev published his paper *Le problème de Cauchy dans l'espace des fonctionnelles* [III.369], in which, for the first time, he introduced distributions as continuous functionals on the space of *s*-times differentiable functions, vanishing outside some characteristic conoid. The apparatus of distributions, which several years later was developed by Schwartz, revolutionized the theory of differential equations.

Let us note that Sobolev arrived at his embedding theorems also in connection with the Cauchy problem [III.370]. Here is what Sobolev wrote in this article: "In some particular cases, for instance in the theory of quasilinear, hyperbolic equations considered by Schauder, the use of delicate estimates enables one to find the necessary number of continuous derivatives of the initial conditions for such an equation". In a conversation with the authors, Sobolev recalled that the reason why he wrote this article was a bet with the mathematician and aerodynamicist F.I. Frankl about the least number of derivatives of the Cauchy data which ensures that the generalized solution coincides with the classical one. Later, embedding theorems became a large domain of functional analysis.

After constructing the theory of the Cauchy problem for second-order hyperbolic equations, Hadamard was naturally interested in the possible extension of his method to more general equations and systems. He discussed this in his book [I.223]. Many years later, it turned out that, proceeding from Hadamard's construction, one can write an asymptotic decomposition for the fundamental solution of ultra-hyperbolic equations, for the high-frequency point-source oscillations in a non-homogeneous medium, for the fundamental solution of the Cauchy problem for second order parabolic equations, and so on (a survey of such studies is given in [III.16]). However, attempts at direct extension to the general case did not succeed. The construction of the fundamental solution of the Cauchy problem for general hyperbolic systems was done only with the aid of Radon and Fourier formulas, which represent the $\delta$-function as a superposition of plane waves. The fundamental solutions for elliptic equations were constructed in a similar way by John, and for hyperbolic equations by Asgeirsson, John, Courant and Lax, Babich (using the Radon formula), and Lax (using the Fourier formula). The theory of the Cauchy problem continues to develop intensively. A far-from-complete bibliography on this topic can be found in the book [III.129], and takes up more than one hundred pages.

In conclusion we add that Petrovskiĭ introduced the notion of a hyperbolic system. By generalizing the above-mentioned Schauder's approach he developed deep methods for the study of the Cauchy problem for such systems [III.307], [III.308]. Interestingly, Petrovskiĭ's study during 1930s was not a result of directly following Hadamard's works on second order hyperbolic equations. Petrovskiĭ once said, that if he had fallen under Hadamard's spell and followed Hadamard's way he would never have arrived at his own results. In the 1950s, Petrovskiĭ's construction was simplified and "modernized" in the works of Leray and Gårding. The solution of the Cauchy problem for hyperbolic and more general equations was constructed with the aid of new tools: the so-called Fourier integral operators, by Maslov, Hörmander, and others.

## 15.5 Huygens' principle

*Christian Huygens (1629-1695)*

Around 1900, Hadamard's attention was attracted by "a remarkable circumstance which arises in the integration of certain partial differential equations of second order" (cf. [I.87, p. 19]). Hadamard means the Huygens' principle, a property of solutions of hyperbolic equations which is a characteristic of the propagation of light or acoustic waves. In his memoir on light, published in 1690 in the Hague, the famous Dutch scientist Christian Huygens, considered light as a perturbation of the ether which was supposed to fill all space and proposed a geometrical law for the propagation of this perturbation, later named after him. In its initial form, Huygens' principle consisted in the following: a perturbation produced by a source of light placed at a point $O$ at the initial moment $t = 0$, propagates as a spherical wave with constant speed. Each point at which the light arrives becomes, in its turn, a source of light and also produces a spherical wave. The total effect at time $t = t_0$ is the envelope of the secondary spherical waves arising at $t = t'$ for every $t'$ such that $0 \le t' < t_0$. Thus, Huygens' principle first consisted in the construction of wave fronts.

In his lectures on the Cauchy problem, Hadamard gave a logical analysis of the above principle and expressed it in the form of the following syllogism [I.223, p. 53-54]:

(A)(major premise). The action of phenomena produced at the instant $t = 0$ on the state of matter at the later time $t = t_0$

takes place by the mediation of every intermediate instant $t = t'$, i.e. (assuming $0 < t < t_0$), in order to find out what takes place for $t = t_0$, we can deduce from the state at $t = 0$ the state at $t = t'$ and, from the latter, the required state at $t = t_0$.

(B)(minor premise). If, at the instant $t = 0$ – or more exactly throughout a short interval $-\varepsilon \leq t \leq 0$ – we produce a luminous disturbance localized in the immediate neighbourhood of $O$, the effect of it will be, for $t = t'$, localized in the immediate neighbourhood of the surface of the sphere with centre $O$ and radius $\omega t'$: that is, will be localized in a very thin spherical shell with centre $O$ including the aforesaid sphere.

(C)(conclusion). In order to calculate the effect of our initial luminous phenomenon produced at $O$ at $t = 0$, we may replace it by a proper system of disturbances taking place at $t = t'$ and distributed over the surface of the sphere with centre $O$ and radius $\omega t'$.

Hadamard then adds: "Now it happens that, by 'Huygens' principle', different authors have meant indiscriminately any one of the above three statements: whereas we shall, in what follows, see that our opinion concerning each of them must be quite a different one".

Statement (A) is a philosophical principle of determinism, a universal law, expressing the existence of the group of transformations which represent the transition from the time $t$ to the time $t_0$ as a composition of the transformations from $t$ to $t'$ and then from $t'$ to $t_0$.

Statement (C) is a physical law with a large domain of applications. Its mathematical description is equivalent to a general property of hyperbolic equations.

Statement (B) has another character. In Hadamard's words "...it is quite a special property of certain special equations". He refers to statements (A) and (B) as "Huygens' major and minor premises", respectively, in order to distinguish them from Statement (C).

In the 1920s and 1930s Hadamard dedicated several papers to Huygens' major premise for general linear second order hyperbolic equations. He was interested in the integral equations which express the group property of solutions.[6] In the earlier papers [I.236], [I.240], [I.254], and [I.281], his study

---

[6]In this respect, note M. Reed and B. Simon's statement: "The connection between partial differential equations and semigroup theory goes back to J. Hadamard who noticed that the solution of the Cauchy problem has the semigroup property with respect to $t$: *Sur un problème mixte aux dérivées partielles*, Bull. Soc. Math. France **31** (1903), 208-

was facilitated by the assumption that there are no singularities on the family
of wave fronts corresponding to the time interval under consideration. We
recall that the wave front propagating from a curve is the locus of points
equidistant from this curve. If the time of propagation is small and the
initial curve is smooth, then the wave front is also smooth, but, from a
certain moment, singular points may immerge on it. They are situated on
curves called caustics, which are envelopes of rays normal to the initial curve.
In two subsequent articles on the subject [I.287], [I.310] Hadamard restricts
himself to the case of three independent variables but admits intersection
of wave fronts and caustics which makes the justification of Huygens' major
premise a more delicate problem.

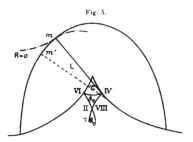

**Figure 13.** *The wave fronts and caustics (The figure
is borrowed from Hadamard's paper [I.310].)*

The geometrical theory of equidistants, envelopes, caustics, and the like
goes back to the seventeenth century and has a rich history (see [III.62,
p. 301]). Recently it got a new life in "catastrophe theory" [III.396], [III.13]
and found new applications to partial differential equations in the frame of
"microlocal analysis" (see, for example, [III.117]). It seems interesting that
Hadamard paid tribute to the difficult subject of singularities of wave fronts
as early as 1930.

Huygens' principle is understood nowadays in the sense of Hadamard's
minor premise, and, to be more precise, as the statement that the wave
produced by a perturbation localized in time and space has a 'backward'
front. We shall use this version of Huygens' principle from now on.

We illustrate the mathematical content of Huygens' principle by the
example of the Cauchy problem for the wave equation. Let the initial data

---

224, and *Principe de Huygens et prolongement analytique*, Bull. Soc. Math. France **52**
(1924), 241-278. But semigroup theory was not applied systematically to partial differential equations until Hille and Yosida developed the analytical tools in the late 1940s."
[III.339]

$f$ and $g$ in (14.16) vanish for $x^2 + y^2 + z^2 > R^2$, where $R$ is some positive number. From the Poisson formula (15.15) it follows immediately that the solution $u$ at a point $P$ at a distance $|P| > R$ from the origin, equals zero until the time $t = (|P| - R)/\omega$ and again vanishes identically for $t > (|P| + R)/\omega$. The point $P$ oscillates only during this time interval, and the wave has both a forward and a backward front. Thus Huygens' principle is valid.

It is interesting to compare this result with the case of two space variables. Let us turn to formula (15.16). Suppose, as in the three-dimensional case, that the functions $f$ and $g$ vanish for $|P| > R$ i.e. that the initial perturbation is concentrated in the cylinder $|P| \leq R$. Then, by (15.16), as for three space variables, $u(P, t) = 0$ for $t \leq (|P| - R)/\omega$. However, the perturbation never vanishes after that time: the wave has a forward front but no backward front. The property of two-dimensional wave motion of having an infinite tail is called wave diffusion. Obviously, the presence of diffusion is equivalent to the violation of Huygens' principle.

From an explicit formula for the solution of the Cauchy problem for the wave equation with $n$ space variables, obtained by Tedone in 1898, it follows that for $n = 3, 5, 7, \ldots$ there is no diffusion, i.e. that for these dimensions Huygens' principle is valid; whereas for $n = 1, 2, 4, \ldots$ diffusion takes place. Here is what L. Schwartz writes about this effect:

> While a spherical wave propagates in our three-dimensional space starting from a point and spreads out in spheres of increasing radius, in the case of even dimensions a wave which began at a certain point fills out the whole ball and not just the sphere, so that if we were to hear a concert in a space of even dimension there would be a residual sound after the passage of the wave front. Given the precision which we demand of music, I think that the difference would be quite significant [II.5, p. 16].

Huygens' principle for the general linear second-order hyperbolic equation (15.18) with variable coefficients was first investigated by Hadamard [I.223]. From the representation of the solution of the Cauchy problem he concluded that this principle is invalid for all even $n$. Later, in the article [I.324], Hadamard paid special attention to the case $n = 1$ and showed the presence of diffusion. For all other dimensions, the problem turned out to be more complicated. Hadamard established that a necessary and sufficient condition for the absence of diffusion for odd $n \geq 3$ is the identical vanishing of a function of $2n + 2$ variables which occurs in the decomposition of the elementary solution of the formally adjoint equation (15.19) (in other words, the decomposition (15.13) should not contain a logarithmic term). As regards this answer to the question about the conditions of validity of Huygens' principle, Hadamard expressed his desire that it "be '*plus résolu*' than it has been in the above. We have enunciated the necessary and sufficient condition, but we do not know how equations satisfying it can be found, or even whether any exist except $(e_{2m_1-1})^7$ (and of course, those which are deduced from $(e_{2m_1-1})$ by trivial transformations)" [I.223, p. 236].

The obvious transformations which are mentioned here are the change of independent variables as well as the multiplication of the solution and of the equation by given functions. Equations reducible to each other under such transformations are called equivalent.

A complete solution of the problem, posed by Hadamard, of the description of the whole class of equations (15.18) for which Huygens' principle is valid, has not been found to this day. However, the question of the existence of equations without diffusion which are not equivalent to the wave equation has been satisfactorily answered by the works of a large number of mathematicians. At the end of the 1930s, Myron Mathisson announced that there are no such equations in the case of three space variables. The proof was published in 1939 [III.272], but treated only equations with constant coefficients in the principal part of the operator. A continuation of this work did not appear, and it turned out later that the statement is not valid for variable coefficients $a_{ij}$. Independently, and using another method, Asgeirsson obtained a negative answer to Hadamard's question for $n = 3$ also under the assumption that the coefficients $a_{ij}$ are constant.[8] In the paper [I.359] dedicated to Mathisson's memory, Hadamard gave one more proof of the same theorem.

The first example of an equation without diffusion and not equivalent to the wave equation was constructed by Stellmacher for $n = 5$ [III.377]. He showed that equations of the form

$$\frac{\partial^2 u}{\partial t^2} - \sum_{i=1}^{5} \frac{\partial^2 u}{\partial x_i^2} - c(t, x)u = 0$$

satisfy Huygens' principle in the following three cases (up to equivalence):

$$c = 0, \qquad c = -2t^{-2}, \qquad c = 2x_1^{-2}.$$

Later he extended his example to arbitrary odd $n > 5$ [III.378]. For the remaining dimension $n = 3$ a class of equations (15.18) without diffusion

---

[7]The wave equation with an odd number of independent variables.

[8]Asgeirsson's paper [III.14] was published in 1956, but he obtained this result independently and about the same time as Mathisson. Douglis [III.109] mentions the unpublished manuscript of Asgeirsson as existing in 1936 (see also [III.90]).

which are not equivalent to the wave equation was constructed by Günther
[III.160]:

$$\frac{\partial^2 u}{\partial t^2} - \frac{\partial^2 u}{\partial x^2} - f(x-t)\frac{\partial^2 u}{\partial y^2} - \frac{\partial^2 u}{\partial z^2} = 0, \qquad (15.22)$$

where $f$ is an arbitrary positive function. The proof of the validity of Huygens' principle for equation (15.22) given by Günther consists in verifying the necessary and sufficient conditions of Hadamard.

Ibragimov soon found [III.186] (see also [III.189], [III.187], and [III.188]) that Huygens' principle is closely connected with the existence of the so-called non-trivial group of conformal mappings in an $(n+1)$-dimensional Riemannian space $V_{n+1}$ with metric

$$dl^2 = \sum_{i,j} b_{ij}(x)dx_i dx_j,$$

where $[b_{ij}] = [a_{ij}]^{-1}$, which was often used by Hadamard. Examples with non-trivial conformal group are the plane, spaces of constant curvature, any space conformal with the plane, as well as the space $V_4$ generated by equation (15.22). Assuming the existence of a non-trivial conformal group, Ibragimov gave a complete solution of Hadamard's problem for equation (15.18) in the case $n = 3$. The following is true: let $n = 3$ and let the Riemannian space $V_4$ have non-trivial conformal group, then equation (15.18) satisfies Huygens' principle in $V_4$ if and only if it is invariant with respect to the full group of conformal transformations of $V_4$; moreover, such a conformally invariant equation is unique up to equivalence transformations.

These results are only few in the list of known counterexamples and positive assertions concerning Huygens' principle. The reader who is interested in the history and the present state of this problem is recommended to read the books by Günther [III.161] and Ibragimov [III.188], and the papers [III.162], [III.28].

Let us note in conclusion, that a theory of lacunas of fundamental solutions of hyperbolic equations is closely connected with Huygens' principle. By a lacuna we mean, essentially, a connected domain in which the fundamental solution vanishes identically. A result of principal importance obtained by Petrovskiĭ in the theory of lacunas is his criterion of their existence for homogeneous, strictly hyperbolic equations of arbitrary order with constant coefficients. This result gave rise to a series of fundamental studies (see [III.15] and a survey in the commentary to Petrovskii's selected works [III.310]).

# Chapter 16

# Hadamard's Last Works

## 16.1   The book on the psychology of invention

We have already said that in 1945, Hadamard's book *The Psychology of Invention in the Mathematical Field* appeared in the USA. It became famous very quickly, and even today it is cited in books by mathematicians and psychologists. How does a mathematician guess new principles, how does he find proofs of new theorems? Such questions interested Hadamard all his life, but it was only when he was eighty that he decided to summarize his thoughts on these questions. In the foreword to his book, Hadamard writes:

> This study, like everything which could be written on mathematical invention, was first inspired by Henri Poincaré's famous lecture before the *Société de Psychologie* in Paris. I first came back to the subject in a meeting at the *Centre de Synthèse* in Paris (1937). But a more thorough treatment of it has been given in an extensive course of lectures delivered (1943) at the *École Libre des Hautes Etudes*, New York City [I.372].

Before his departure for Europe from the USA, Hadamard gave his lectures, in English, at Princeton University. The French translation of these lectures, made by his daughter Jacqueline Hadamard, appeared in 1945. In his review of this book, Hardy wrote:

> This is, apart from one famous lecture of Poincaré, the first attempt by a mathematician of the first rank to give a picture of his own modes of thought and those of other mathematicians [...] We must therefore all be grateful to Professor Hadamard for this

> stimulating little volume, written with all the authority of one
> of the greatest mathematicians of the last fifty years, and with
> the most charming frankness concerning his own achievements,
> triumphs and failures alike. Indeed it is the most personal parts
> of the book which seem to me the most attractive [II.16, p. 60].

Hadamard's book is written quite intelligibly, it contains almost no special philosophical or mathematical terms and reflects the author's versatility very clearly. He comments on the views of philosophers, psychologists, artists, and poets such as Abelard, Saint Augustine, Bergson, Spencer, Freud, Schopenhauer, Rodin, and Valéry. In his way of arguing respect for the opponent is a *sine qua non*.

Hadamard pays much attention to the phenomenon of the unconscious and its role in scientific discovery. Emphasizing that the study of the subconscious is in its infancy (this is true even now), he claims that it plays the main role in the creative process, that illumination (a subconscious burst of inspiration) is the result of a hidden, protracted process in the brain. Poincaré's account of one discovery made by him in the theory of automorphic functions [I.372, p. 37] is a good illustration of this, which can, in the words of Hardy from the article cited above, be confirmed by the experience of any other scientist who works intensively on some problem. Paul Valéry, who was an original thinker, and who was interested in the natural sciences, said similar things about illumination in poetical creativity. However, other views have been presented. For example, Gauss, who was a deeply religious man, explains the illumination which he experienced during one discovery in number theory as an inspiration sent from heaven, whereas the French biologist Nicolle considered illumination as a chance phenomenon similar to mutations in genetics. Without denying the role of chance, Hadamard argues that to explain the act of discovery as pure chance is "equivalent to no explanation at all and to asserting that there are effects without causes" [I.372, p. 19].

Hadamard emphasizes diversity as a special feature of the subconscious. First of all, it is hidden deep inside, beginning with the absolutely unconscious and finishing with a process close to conscious. Secondly, its diversity is manifested in the many variants which are possible in the process of solving a problem. "Discovery is discernment, selection," writes Poincaré [III.321, p. 51]. This aphorism resembles the words ascribed to Michelangelo: "In order to create a statue from a piece of marble, I only have to eliminate everything that is superfluous". Hadamard cites Valéry's words: "It takes two to invent anything. The one makes up combinations; the other one chooses, recognizes what he wishes and what is important to him in the mass of the things which the former has imparted to him" [I.372, p. 30].

An important criterion of choice is the esthetic factor of beauty. In mathematics it can be manifested in different ways: the shortest logical way leading to the result, various associations and analogies, generalizations, a desire to complete a gap in the edifice of some discipline. As an example of the last criterion, Hadamard describes the idea which led Volterra to the discovery of functional analysis: "Why was the great Italian geometer led to operate on functions as infinitesimal calculus had operated on numbers, that is to consider a function as a continuously variable element? Only because he realized that this was a harmonious way of completing the architecture of the mathematical building, just as the architect sees that the building will be better poised by the addition of a new wing" [I.372, p. 129].

In several places in his book, Hadamard persistently emphasizes the role of the esthetic factor. He maintains that the examples he gives refute the doubts expressed by Wallas[1] about the value of the sense of beauty as motive power in scientific discovery. He writes: "On the contrary, in our mathematical field, it seems to be almost the only useful one" [I.372, p. 130]. However, the notion of beauty is subjective in mathematics. For instance, Euler and Gauss enjoyed extremely long calculations and undoubtedly found them beautiful, whereas Poincaré said that he did not like long calculations. Invoking his own experience, as well as that of other mathematicians, Hadamard notes that sometimes direct ways of searching do not lead to a solution of the problem at hand, so that it is worthwhile to find an indirect approach by turning to some adjoining field. Following the psychologist Souriau, he calls this "thinking aside".

A very interesting question is posed in the subsection *Attempts to Govern our Unconscious.* Is it possible to understand the unconscious and to control it? It is known that in a hypnotized condition, a man is capable of actions which are impossible in an unhypnotized state: to walk on a tight-rope, to show artistic abilities, for example. Hadamard writes: "...it would be worth the trouble to try the experiment on good mathematicians in a state of hypnosis, to whom one would give questions whose solution they do not know (for example those we cite in Chapter 8). If they were to discover the solution in this way, one would have proved both the theorem in question and the fact of hypnosis" [I.372, p. 58].

In general, according to Hadamard, the creative process consists of four stages:

1. the choice of a problem and conscious preliminary work;

---

[1] Graham Wallas (1858-1932) was an English political scientist and psychologist, the author of *The Life of Francis Place* (1898), *Human Nature in Politics* (1908), *The Great Society* (1914), etc. His book *The Art of Thought* (1926) is quoted by Hadamard.

2. "incubation" – the period of unconscious thinking;

3. illumination – the moment of discovery;

4. analysis of the result obtained, its verification, and formulation.

The catalyst of creative inquiry is some kind of intuition – geometrical, physical. The significance of intuition was often stressed by Hilbert. However, in his works on mathematical logic and realization of the axiomatic method in concrete disciplines he suggested the possibility of deductive constructions without turning to intuition. Hadamard's point of view on intuition was shared by Pólya [III.323, p. 127]:

> I think that everybody prefers intuitive insight to formal logical arguments, including professional mathematicians. Jacques Hadamard, an eminent French mathematician of our times, expressed it so: "The object of mathematical rigour is to sanction and legitimize the conquests of intuition, and there never was any other object for it".

One more question discussed in the book is whether words are necessary for thinking. Hadamard disagrees in this with Müller,[2] who argued that any thinking, scientific or non-scientific, is somehow connected with words. Hadamard gives playing chess and communication by people who are deaf and dumb as examples of wordless thinking. Since he was mostly interested in mathematical creativity, Hadamard, during the preparation of his book, asked Einstein, Wiener, Pólya, and other eminent scientists living in the USA at that time, about their mental processes at work, in particular, the role of words in their creative activity. The answers they gave, which appear in the book, confirmed Hadamard's view.

Louis de Broglie, the famous physicist, was one of the recipients of Hadamard's questions. His replies are contained in the following note, dated August 26, 1945, when the book had already been published (there is no mention of this reply in the second edition of the book):

Replies to the questionnaire:

A. My thought is for the most of the time accompanied by words or other precise signs. If ideas come to me without this accompaniment, I immediately try to make them precise with the aid of words.

---

[2]Max Müller (1823-1900) was an orientalist, psychologist and philologist. Hadamard refers to his book *The Science of Thought* (1887).

B. I am in general guided by intuition, which most often leads me to evoke visual images or sometimes muscular ones. Auditory images do not appear to me to play an important role.

C. In ordinary life, I belong to the visual type, although I do not have a good visual memory (I easily forget the look of places I have visited or the faces of people I have met).

D. The question seems difficult. I believe that usually, when I evoke an image, it appears, somewhat schematized, in my full conscious. However, it may be that in certain cases it remains at the frontier of the clear conscious.

My reasonings are usually done in the full awareness. However there are times when a line of reasoning which seems difficult to establish, appears very much easier when I have spent some time without fixing my attention on it. I think that this is the result of work which is done in the more or less subconscious parts of the mind.

*The answers of Louis de Broglie to Hadamard's questionnaire*

E. Yes, I have experienced intimation. It is rarely false: however, it can sometimes be so.

    F. During sleepless nights or rather during nights when I sleep badly and insufficiently, I have often felt an activity of the mind which is very different from the waking state. But if this activity is the same as subconscious activity of the mind in the waking state, I cannot say. [IV.20]

The ability to distinguish between intuitionists and logicians among mathematicians is the theme of Chapter 7, *Different Kinds of Mathematical Minds*. Hadamard cites Klein's words from a lecture he gave in the USA in 1893: "It would seem as if a strong naive space intuition were an attribute of the Teutonic race, while the critical, purely logical sense is more developed in the Latin and Hebrew races" [I.372, p. 107].[3] But Hadamard clearly states his opposition to such statements, drawing a parallel with the racist views of the nazis. He writes: "...Klein implicitly considers intuition, with its mysterious character, as being superior to the prosaic way of logic...and is evidently happy to claim that superiority for his countrymen" [I.372, p. 107]. Giving several examples contradicting Klein's assertions and emphasizing the inconsistency of any criteria based on national characteristics, Hadamard even criticizes his fellow countryman Duhem:

    One will find such tendentious interpretations of facts whenever nationalistic passions enter into play. At the beginning of the First World War, one of our greatest scientists and historians of sciences, the physicist Duhem, was misled by them just as Klein had been, only in the opposite sense. In a rather detailed article, he depicts German scientists, especially mathematicians, as lacking intuition or even deliberately setting it aside.... If one or the other were right, the reader will realize by all that precedes that either Frenchmen or Germans would never have made any significant discovery [I.372, p. 107].

In discussing Poincaré's views on this subject, Hadamard agrees with him on the principle of not connecting it with politics, but disagrees with Poincaré's description of Hermite (who had been their teacher) as being a logician: "But to call Hermite a logician! Nothing can appear to me as more directly contrary to the truth. Methods always seemed to be born in his mind in some mysterious way" [I.372, p. 109].

---

[3]A more recent specimen of similar views can be found in Arnold's reminiscences about Kolmogorov [III.363]: "Andreĭ Nikolayevich tried to avoid technical work on the generalization of a developed theory (he said, incidentally, that at this stage Jews are especially successful, rather with admiration, since Andreĭ Nikolayevich considered his instinctive repugnance to this type of task as a shortcoming)".

In the section *The Choice of Subjects* in Chapter 9, Hadamard cites Renan:[4] "...there is a scientific taste just as there is a literary or artistic one" [I.372, p. 127], and adds that in choosing the subject of research, one is not necessarily led by the possibility of immediate applications; these usually come later. Keeping in mind his theorem on the multiplication of singularities of analytic functions, he recalls that, when he described it to his friend Duhem, he was asked about its applications:

> When I answered that so far I had not thought of that, Duhem, who was a remarkable artist as well as a prominent physicist, compared me to a painter who would begin by painting a landscape without leaving his studio and only then start on a walk to find in nature some landscape suiting his picture. This argument seemed to be correct, but as a matter of fact, I was right in not worrying about applications: they did come afterwards [I.372, p. 128].

Further on in the book, Hadamard illustrates the same point of view by the example of his inequality for determinants. While proving it, he did not think about its usefulness; he only had a feeling that it was interesting. Some years later, Fredholm published his theory, in which Hadamard's result turned out to be essential. He then mentions É. Cartan's discovery of spinors, connected with geometric transformations in the theory of groups, for which "no reason was seen, at that time, for special consideration of those transformations except just their esthetic character" [I.372, p. 129]. Volterra's theory of functionals was another example of what "...seemed to be an essentially and completely abstract creation of mathematicians" [I.372, p. 130]. After some time, both spinors and functionals found important applications in theoretical physics.

Thus, according to Hadamard, it is not the usefulness in some sense of a result, but the beauty of a problem and the feeling of its inner value that should lead a scientist in the choice of subject. One of the appendices in the book is "An inquiry into the working methods of mathematicians", published in *L'Enseignement Mathématique*, Volume 4, 1902, and Volume 6, 1904, which Hadamard often cites. It is a list of questions about the working habits of mathematicians. A modern version of such an inquiry appeared in 1988 [III.286].

---

[4]Ernest Renan (1823-1892) was a French philosopher and historian, the author of *La Vie de Jésus* (1863), five volumes of *Histoire du Peuple d'Israël* (1887-1894) and many other works.

## 16.2 The papers from the fifties

Hadamard's last mathematical papers concerned non-Euclidean geometry, history of mathematics and some classical problems in the theory of parabolic and elliptic equations. We also mention his book of 1951 on non-Euclidean geometry in the theory of automorphic functions [I.383] presently available only in Russian, where he gave an exposition of Poincaré's results in the domain. In 1954, he published an article [I.394] on parabolic equations without classical solutions which was inspired by A. Wintner's paper [III.424] of 1950. Wintner has shown that there exists a continuous function $f$ on the plane, such that there does not exist any domain $\mathcal{D}$ on which the equation $\Delta u = f$ possesses some non-zero solution. By a solution Wintner means a continuous function for which the derivatives $u_{xx}$ and $u_{yy}$ exist and which satisfy the equation in every point of the domain $\mathcal{D}$.

Hadamard established the same result for the heat equation

$$\frac{\partial^2 u}{\partial x^2} - \frac{\partial u}{\partial y} = f(x, y), \qquad (16.1)$$

speaking about continuous solutions of (16.1) which have the derivatives $u_{xx}$ and $u_y$. Wintner's construction relied heavily upon the work of H. Petrini [III.306] of 1909 on the existence of first and second derivatives of the logarithmic potential. Hadamard modified Petrini's argument in order to find a necessary condition for the existence of the derivative $v_{xx}$ of the double integral

$$v(x, y) = \int \int_{\mathcal{D}_y} f(\xi, \eta) \frac{1}{\sqrt{y - \eta}} \exp\left(-\frac{(\xi - x)^2}{4(\eta - y)}\right) d\xi d\eta,$$

where $\mathcal{D}_y$ is that part of a given domain $\mathcal{D}$ which lies below the line $\eta = y$.

The same year, Hadamard's paper [I.395] appeared, in which he gave the following analogue of Harnack's second theorem on harmonic functions and then applied it to solutions of the heat equation. Let $\{u_n\}$ be a monotone sequence of solutions of the equation $u_{xx} - u_y = 0$ in a domain $\mathcal{D}$, and suppose that for some $P \in \mathcal{D}$ the sequence $\{u_n(P)\}$ converges. Then the sequence $\{u_n\}$ is uniformly convergent on any proper subdomain. The greater part of this work concerns a non-trivial derivation of the result which asserts that Green's function $G(P, Q)$ and one of its derivatives is estimated from below by the distance from $P$ to the boundary of $\mathcal{D}$. Harnack's second theorem for the heat equation was obtained independently and simultaneously by Pini [III.312].

Hadamard's last mathematical article [I.400] appeared in 1957. It was published in a volume dedicated to Paul Lévy on the occasion of his seventieth birthday. In this paper, Hadamard gives a simple proof of Harnack's second theorem for a general second-order elliptic equation with smooth coefficients, based on the theory of boundary integral equations.

*The first page of the last mathematical article by Hadamard*

His interest in Harnack's theorems for the classical equations of mathematical physics might be explained by the fact that the seemingly indefatigable ninety-two-year-old Hadamard was now working on a new book, which we proceed to discuss.

## 16.3   The book on partial differential equations

Hadamard's book *La théorie des équations aux dérivées partielles* [I.405], was published in Beijing in 1964, in a luxurious edition, but he did not live to see his last book appear. Recalling the preparation and writing of this book, Laurent Schwartz writes:

> To the very end of his life he felt 'responsible' for partial differential equations. This feeling caused him anxiety and uneasiness. After the war, he worked on a book, which has just appeared in China, written in French, and which summarized his life experience with these equations. He had been to China before 1930,[5] and the Chinese academics had asked him to publish his lectures in French. Because of the war, he did not begin publication until afterwards, and the editing was intricate since he really felt the need to say everything he could say and even to read all the new memoirs! Clearly, this did not make sense, given the enormous number of articles which were appearing on the subject, and I believe that this situation worried him. He began this publishing after the war, and, by the age of 92 or 93, he had only prepared eight chapters out of twenty-two. Nevertheless, he wanted to publish everything! It is a remarkable fact which shows exceptional courage and spirit of enterprise: to have spent more than ten years in preparing eight chapters and not giving up the idea of publishing the remainder! He stopped, having been persuaded by his family and me, to finish the work at a given moment. Obviously the enterprise was no longer feasible, but I believe that this aspect of his behaviour shows the extent to which he felt himself, to the end of his life, connected with mathematics [II.5, p. 17].

Hadamard's last book resembles a course of lectures in the classical theory of partial differential equations, dealing with numerous problems: the theory of the Cauchy problem for ordinary and partial differential equations, the Dirichlet problem for second-order elliptic equations, 'elementary' solutions of elliptic and hyperbolic equations, the Cauchy problem and the mixed problem for linear, second-order hyperbolic equations, 'singular' equations, that is, equations with the operator

$$\frac{\partial^2}{\partial t^2} + \frac{k}{t}\frac{\partial}{\partial t}, \qquad k = \text{const},$$

equations which change type in a domain, as well as second-order parabolic equations. The major part of the book represents Hadamard's own research concerning elementary solutions of the Cauchy problem and the mixed problem for hyperbolic equations, as well as many other topics. The book also

---

[5]Hadamard's trip to China took place in 1936, and not before 1930 as Schwartz writes.

contains some questions which are rarely included in general courses, such as Wiener's theory of the regularity of a boundary point for the Dirichlet problem, the generalized potentials of Marcel Riesz and their applications to hyperbolic equations, the axially symmetric theory of harmonic potentials, second-order equations of mixed type. The text is rich in interesting notes, motivations, material of a review character, as well as numerous references to the original sources.

The presentation is most original. Its literary style is strikingly different from that of contemporary monographs on this subject: it contains more 'philosophy' and there are far fewer formulas. A special feature is the large number of references to the literature from the end of the last century and the beginning of the the present one. Hadamard shows a preference for constructive, analytic methods in the study of equations. The reader will see neither the function space $L^2$ nor a Sobolev space, although corresponding notions are sometimes used.

Was Hadamard's monograph of interest in the 1960s, and is it of interest now? These are not simple questions, and the answers are ambiguous. At the moment of its appearance, the theory of partial differential equations was a very diverse domain, enriched by its close connections with functional analysis, which was becoming increasingly powerful. Connections with algebra and topology were also coming into play. The theory of general linear differential operators with constant coefficients was already well developed, and the powerful apparatus of pseudo-differential operators was about to appear. So, at the beginning of the 1960s, Hadamard's *Theory of partial differential equations* could hardly be considered as a book for the study of the subject. One can not deny that the book already bore the stamp of archaism when it appeared.

However, Hadamard was one of the great mathematicians of our century who studied partial differential equations over a period of sixty years. This remarkable fact gives the book a timeless value. In the choice of material and the depth of exposition, the reader feels the touch of a great master. In spite of the diversity of the topics covered, Hadamard achieved a surprising unity in the presentation, which can undoubtedly be explained by the unifying role of the idea of well-posedness throughout the whole of the text. While reading Hadamard's book, one experiences a clear feeling of the living connection of the mathematics of today with the studies of the great masters of the past – a feeling which one rarely has when reading contemporary monographs.

# Epilogue

We conclude this book with the words of Laurent Schwartz: "Many have read his works. Certainly, young mathematicians read them less, partly because the language has changed a lot, and new books express the same ideas in modern language. It is easier to go directly and search in these new publications. However, it is still very profitable for every mathematician to go to the sources!"[II.5]

More than thirty years have elapsed since Hadamard died. The following graph, constructed with data obtained from the Science Citation Index,[6] shows how Hadamard's works are still used and referred to, and that his influence has not diminished with the years.

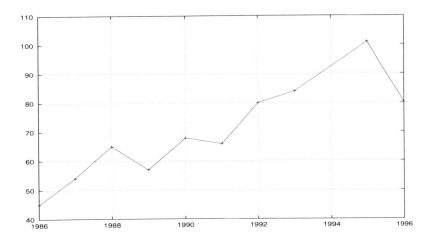

**Figure 14.**

[6]Hadamard's name is given in S.C.I. as J. Hadamard, J. Hadamand, F. Hadamard, G. Hadamard, H. Hadamard, I. Hadamard, J.M.J. Hadamard, J.S. Hadamard, J.W. Hadamard, L. Hadamard, M. Hadamard, M.J. Hadamard, P. Hadamard, T. Hadamard, J. Hadammard, I.S. Adamar, Z. Adamar. One can identify him by references to his works.

# Principal Dates in Hadamard's Life

The main source of the following data is the biographical dictionary [III.78].

1865 Born on December 8 in Versailles, to Claire-Marie-Jeanne Hadamard and Amédée Hadamard.

1876 Entered the *Lycée Louis le Grand*.

1882 *Bachelier ès lettres et ès sciences*.

1884 - 1888 Student of the *École Normale Supérieure*.

1886 Passed the *licence* examination.

1887 Passed the *agrégation* examination, spent one more academic year in the *École Normale Supérieure*.

1888 Continued education as a *boursier* of the city of Paris and of the *Collège de France*.

1888 - 1889 Teacher at the *Lycée de Caen*, free of duties.

1889 Appointed as *suppléant* at the *Lycée Saint-Louis*.

1890 - 1893 Mathematics teacher at the *Lycée Buffon*.

1892 Defended the thesis *Éssai sur l'étude des fonctions données par leur développement de Taylor* and received the degree of *docteur ès sciences mathématiques*.
Awarded the *Grand Prix* of the *Académie des Sciences*.
Married Louise-Anna Trénel on June 30.

1893 - 1896 *Chargé de cours* at the *Faculté des Sciences* of the University of Bordeaux.

1894 His first son Pierre was born in Bordeaux on October 5.

1896 - 1897 *Professeur* in Astronomy and Rational Mechanics at the *Faculté des Sciences* of the University of Bordeaux.

1896 Awarded *Prix Bordin* of the *Académie des Sciences*.

1897 His second son Étienne was born on July 26 in Cenon (Gironde).

1897 - 1900 *Maître de conférences* in differential and integral calculus at the *Faculté des Sciences* of the University of Paris.

1897 Appointed as *Professeur suppléant* in analytical and celestial mechanics at the *Collège de France*.

1898 Awarded *Prix Poncelet* of the *Académie des Sciences*.
Published the book *Leçons de géométrie élémentaire: Géométrie plane*.

1899 His third son Mathieu was born on February 27 in Paris.

1900 - 1909 *Professeur adjoint* at the *Faculté des Sciences* of the University of Paris.

1901 His first daughter Cécile was born on February 6 in Paris.
Published the book *La série de Taylor et son prolongement analytique*.
Published the book *Leçons de géométrie élémentaire: Géométrie dans l'espace*.

1902 His second daughter Jacqueline was born on November 29 in Paris.

1903 Published the book *Leçons sur la propagation des ondes et les équations de l'hydrodynamique*.
Awarded *Prix Petit d'Ormoy* of the *Académie des Sciences*.

1906 Elected as President of the French Mathematical Society.

1907 Awarded *Prix Vaillant* of the *Académie des Sciences*.

1908 Awarded *Prix Estrade Delcros* of the *Académie des Sciences*.

1909 Appointed *Professeur* in Mechanics at the *Collège de France*.

1910 Published the book *Leçons sur le calcul des variations*.

1912 Appointed as *Professeur d'Analyse* at the *École Polytechnique*.
Elected to the *Académie des Sciences* to succeed Poincaré.

1913 Started his seminar at the *Collège de France*.

1916 His sons Pierre and Étienne were killed at Verdun.

1920 Appointed as *Professeur* at the *École Centrale*.

1922 Published the book *Lectures on Cauchy's problem in linear partial differential equations*.

1926 Published the first volume of his *Cours d'analyse de l'École Polytechnique*.

1930 Published the second volume of his *Cours d'analyse de l'École Polytechnique*.

1937 Retired from the *Collège de France*, the *École Polytechnique*, and the *École Centrale*.

1941 Exile to the U.S.A.

1944 His son Mathieu died on July 1 at the front in Tripolitaine.
     Move to London.

1945 Returned to Paris.
     Published the book *The psychology of invention in the mathematical field*.

1955 Awarded Feltrinelli Prize by the *Accademia dei Lincei* in Rome.

1957 Promoted to the grade *Grand-Croix de la Légion d'honneur*.
     Awarded the Gold medal of the CNRS.

1960 His wife Louise died on July 6.

1962 Awarded the Gold medal of the *Académie des Sciences*.

1963 Died on October 17 in Paris.

1965 His book *La théorie des équations aux dérivées partielles* was published in China.

1966 His centenary was celebrated in Paris at the *École Polytechnique* on January 13.

1968 Four volumes of his *Œuvres* appeared in Paris.

# A Hadamard Collection

There are mathematical terms which are habitually attached with the names of the persons who discovered them: Pythagoras theorem, Euclidean geometry, Abel summation, Borel set, Euler constant, Laplace operator, Hilbert space, etc.

Hadamard's name is connected with numerous notions and results. Mostly this is fair and widely accepted. In some cases Hadamard's priority is doubtful or erroneous in spite of the common use of his name.

Here is a collection of Hadamard's mathematical objects. We supply each of them with a reference to the section of the present book where it is mentioned and/or to a work, chosen mostly by chance, where it can be found.

The Cauchy-Hadamard formula for the radius of convergence of the power series - Sections 1.9, 9.1.

Hadamard's test for singular points - Section 9.1, [III.296].

Hadamard's gaps - Section 9.1, [III.36].

Hadamard's gap theorem - Section 9.1, [III.36], [III.398].

Hadamard's determinants in the theory of meromorphic functions - Section 9.1, [III.313].

Hadamard's fractional integrals and derivatives - Section 9.1, [III.355].

Hadamard's three circles theorem - Sections 2.3, 9.3, [III.398].

Hadamard's theorem on multiplication of singularities - Sections 2.4, 9.1, [III.398].

Hadamard's algebra - [III.27].

Hadamard's quotient - [III.26].

Hadamard's theory of polar singularities - [III.155].

Hadamard's composition of power series - Section 9.1.

Hadamard's order - Section 9.1, [III.36].

Hadamard's factorization theorem - Section 9.2, [III.177], [III.398].

Hadamard-Cartan theorem for Riemannian manifolds of non-positive curvature (also called Hadamard's theorem) - Section 11.2, [III.40], [III.30].

Hadamard's conjecture on surfaces of negative Gaussian curvature - Section 11.2, [III.349].

Hadamard's graph transform (a method to prove the existence of invariant manifolds in dynamical systems theory) - [III.244].

Hadamard's representation of a linear functional on the space $C[a, b]$ - Section 12.5.

Hadamard's derivative of the functional - [III.83].

Hadamard's multiplicative inequality - Section 13.5.

The Hadamard-Kneser lemma - Section 13.5.

The Landau-Hadamard inequality - Section 13.5, [III.213].

Hadamard's inequality for convex functions (also called Hermite-Hadamard inequality) - [III.303], [III.280].

Hadamard's inequality for determinants - Sections 2.3, 13.1, [III.22].

Hadamard's approximation theorem for determinants - [III.288].

Hadamard's evanescence theorem for determinants - [III.288].

Hadamard matrices - Section 13.1, [III.5], [III.413].

Hadamard matrices of quaternion type - [III.20].

Hadamard matrices of the Williamson type - [III.20].

The Hadamard transform - [III.388].

The Walsh-Hadamard transform - [III.388].

Hadamard codes - Section 13.1, [III.261].

Hadamard designs - Section 13.1, [III.33].

Hadamard transform optics - Section 13.1, [III.168].

Hadamard transform computer - Section 13.1, [III.168].

Hadamard chemical balance weighing design - Section 13.1, [III.168].

Hadamard imaging spectrometer - Section 13.1, [III.168].

Hadamard dominant matrices - Section 13.1, [III.25].

Hadamard's product of matrices (also called Schur product or direct product ) - [III.181].

Hadamard's problem of the uniformization of sets - Section 5.5, [III.253].

The Legendre-Hadamard condition of strong ellipticity for systems of partial differential equations - Section 14.2, [III.148].

The Fresnel-Hadamard condition of wave propagation, the Fresnel-Hadamard theorem - Section 14.2, [III.401], [III.131].

Hadamard's paradox in the theory of wave propagation - [III.37, p. 42].

Hadamard's lemma on singular surfaces - [III.402].

Hadamard's variational formula for Green's functions - Section 14.3, [III.414], [III.140].

Hadamard's equation with functional derivatives - Section 14.3, [III.239].

Hadamard's hypothesis in the theory of plates - Section 14.3, [III.335], [III.112], [III.85].

Hadamard's integro-differential equation - Section 14.4, [III.56].

Well-posedness in the sense of Hadamard - Section 15.1.

Hadamard's differential - [III.135].

Hadamard's example of an ill-posed Cauchy problem for the Laplace equation - Section 15.1, [III.305].

Hadamard's counterexample to the Dirichlet principle - Section 12.3.

Hadamard's construction of the fundamental solution of the second order equation - Section 15.2.

Hadamard's construction of the solution of the Cauchy problem for second order hyperbolic equations - Section 15.3.

Hadamard's Ansatz - [III.16].

finite part of a divergent integral in the sense of Hadamard - Sections 3.1, 15.3.

Hadamard's principal value of the improper integral - [III.403].

Hadamard's criterion for the validity of Huygens' principle - Section 15.5, [III.161], [III.162].

Hadamard's problem in the theory of wave diffusion - Section 15.5, [III.272], [III.189].

Hadamard's estimate for the solution of a parabolic differential equation - [III.425].

Hadamard's formula for the increment of a function - [III.327].

Hadamard's inequality for the composition of two M.Riesz kernels - [III.23, p. 7]

# I. Bibliography of Jacques Hadamard

The following list extends the bibliography of Hadamard's works published in [I.333], [I.406], [II.37], and [II.4]. References are ordered according to the year of publication. Each of 180 papers included in the collection *Œuvres de Jacques Hadamard* [I.406] is supplied with the reference to the volume and pages in this collection.

*A part of the list of Hadamard's papers written by him for Notice sur les travaux scientifiques de M. Jacques Hadamard, 1909 [I.158]*

**1882**

[I.1] *Concours général de philosophie de 1881. Solution par M. Hadamard, élève au Lycée Louis-le-Grand*, J. Math. Élémentaires Sér. 2, **6**, 199-207.

**1883**

[I.2] *Concours général de 1883. Mathématiques élémentaires. Solution par M. Hadamard, élève au Lycée Louis-le-Grand (copie couronnée)*, J. Math. Élémentaires Sér. 2, No 1, 203-206.

## 1884

[I.3] *Sur le limaçon de Pascal* , J. Math. Spéc. Sér. 2, **3**, 80-83.

[I.4] *Sur l'hypocycloïde à trois rebroussements*, J. Math. Spéc. Sér. 2, **3**, 226-232.

[I.5] *Questions proposées: No 150*, J. Math. Spéc. Sér. 2, **3**, 240.

## 1885

[I.6] *Sur l'hypocycloïde à trois rebroussements*, J. Math. Spéc. Sér. 2, **4**, 41-42.

[I.7] *Questions proposées: No 138*, J. Math. Spéc. Sér. 2, **4**, 72.

[I.8] *Questions proposées: No 166*, J. Math. Élémentaires, Sér. 2, no. 3, 22-23.

[I.9] *Question 1528 proposée par J. Hadamard*, Nouvelles Annales de Mathématiques, Sér. 3, **4**, 151-152.

## 1888

[I.10] *Sur le rayon de convergence des séries ordonnées suivant les puissances d'une variable*, C.R. Acad. Sci. Paris **106**, 259-262 (Œuvres **1**, 3-6).

[I.11] *Recherche des surfaces anallagmatiques par rapport à une infinité de pôles d'inversion*, Bull. Sci. Math. Ser. 2, **12**, 118-121 (Œuvres **2**, 697-700).

## 1889

[I.12] *Sur la recherche des discontinuités polaires*, C.R. Acad. Sci. Paris **108**, 722-724.

## 1892

[I.13] *Essai sur l'étude des fonctions données par leur développement de Taylor*, Thèse de Doctorat de la Faculté des Sciences de Paris, J. Math. Sér. 4, **8**, 101-186 (Œuvres **1**, 792).

[I.14] *Sur les fonctions entières de la forme $e^{G(x)}$*, C.R. Acad. Sci. Paris **114**, 1053-1055.

## 1893

[I.15] *Étude sur les propriétés des fonctions entières et en particulier d'une fonction considérée par Riemann*. Mémoire couronné en 1892 par l'Académie: Grand Prix des Sciences mathématiques, J. Math. Sér. 4, **9**, 171-215 (Œuvres, **1**, 103-147).

[I.16] *Sur le module maximum que puisse atteindre un déterminant*, C.R. Acad. Sci. Paris **116**, 1500-1501 (Œuvres **1**, 237-238).

[I.17] *Résolution d'une question relative aux déterminants*, Bull. Sci. Math. Sér. 2, **17**, 240-246 (Œuvres **1**, 239-245).

[I.18] *Sur les caractères de convergence des séries à termes positifs*, C.R. Acad. Sci. Paris **117**, 844-845.

## 1894

[I.19] *Sur les caractères de convergence des séries à termes positifs et sur les fonctions indéfiniment croissantes* (avec note complémentaire), Acta Math. **18**, 319-336, 421 (Œuvres **1**, 249-271).

[I.20] *Remarque sur les centres de courbure des roulettes*, Proc. Verb. Soc. Sci. Phys. et Natur. Bordeaux, 19 avril, 37-38.

[I.21] *Sur les mouvements de roulement*, C.R. Acad. Sci. Paris **118**, 911-912.

[I.22] *Démonstration du théorème de Jacobi sur le mouvement d'un corps pesant de révolution fixé par un point de son axe*, Proc. Verb. Soc. Sci. Phys. et Natur. Bordeaux, 19 juillet, 46.

[I.23] *Sur l'élimination*, C.R. Acad. Sci. Paris **119**, 995-997 (Œuvres **4**, 2091-2092).

[I.24] *Réponse à question 128*, Intermediaire des Mathématiciens **1**, 127.

[I.25] *Sur les mouvements de roulement*, Mém. Soc. Sci. Phys. et Natur. Bordeaux Sér. 4, **5**, 397-417; reprinted in: P. Appel, *Les roulements en dynamique*, Coll. Sci. Carré et Naud, Paris, 1899, 47-68 (Œuvres **4**, 1725-1745).

**1895**

[I.26] *Sur le tautochronisme*, Proc. Verb. Soc. Sci. Phys. et Natur, Bordeaux 7 février, 16-17.

[I.27] *Sur l'expression du produit* $1 \cdot 2 \cdot 3 \ldots (n-1)$ *par une fonction entière*, Bull. Sci. Math. France Sér. 2, **19**, 69-71 (Œuvres **1**, 153-154).

[I.28] *Sur une congruence remarquable et sur un problème fonctionnel qui s'y rattache*, Proc. Verb. Soc. Sci. Phys. et Natur. Bordeaux, 14 février, 19-23 (Œuvres **2**, 701-706).

[I.29] *Sur la précession dans le mouvement d'un corps pesant de révolution fixé par un point de son axe*, Bull. Sci. Math. Sér. 2, **19**, 228-230 (Œuvres **4**, 1715-1717).

[I.30] *Sur la stabilité des rotations dans le mouvement d'un corps solide pesant autour d'un point fixe*, Assoc. Franç. pour l'Avancement des Sciences **24**, 175 (Œuvres **4**, 1719-1724).

[I.31] *Sur certains systèmes d'équations aux différentielles totales*, Proc.-Verb. Soc. Sci. Phys. et Natur. Bordeaux, 1894-1895, 17-18; reprinted in: P. Appel, *Les roulements en dynamique*, Coll. Sci. Carré et Naud, Paris, 1899, 69-70 (Œuvres **3**, 1051-1052).

[I.32] *Sur le mouvement d'un corps pesant de révolution fixé par un point de son axe*, Proc. Verb. Soc. Sci. Phys. et Natur. Bordeaux, 1894-1895, 61-62.

[I.33] *Sur la stabilité des rotations d'un corps solide*, Proc. Verb. Soc. Sci. Phys. et Natur. Bordeaux, 1894-1895, 70-71.

[I.34] *Sur les éléments infinitésimaux du second ordre dans les transformations ponctuelles*, Proc. Verb. Soc. Sci. Phys. et Natur. Bordeaux, 19 décembre, 11-12 (Œuvres **1**, 267-271).

**1896**

[I.35] *Mémoire sur l'élimination*, Acta Math. **20**, 201-238 (Œuvres **1**, 273-310).

[I.36] *Sur la géométrie non euclidienne*, Proc. Verb. Soc. Sci. Phys. et Natur. Bordeaux, 1895-1896, 24-25.

[I.37] *Une propriété des mouvements sur une surface*, Proc. Verb. Soc. Sci. Phys. et Natur. Bordeaux, 30 avril, 47-48.

[I.38] *Une propriété des mouvements sur une surface*, C.R. Acad. Sci. Paris **122**, 983-985 (Œuvres **4**, 1747-1748).

[I.39] *Sur l'instabilité de l'équilibre*, Proc. Verb. Soc. Sci. Phys. et Natur. Bordeaux, 21 mai, 48-50.

[I.40] *Sur les fonctions entières*, C.R. Acad. Sci. Paris **122**, 1257-1258 (Œuvres **1**, 149-150).

[I.41] *Sur les lignes géodésiques des surfaces spirales et les équations différentielles qui s'y rapportent*, Proc. Verb. Soc. Sci. Phys. et Natur. Bordeaux, 4 juin, 55-58.

[I.42] *Sur les zéros de la fonction $\zeta(s)$ de Riemann*, C.R. Acad. Sci. Paris **122**, 1470-1473 (Œuvres **1**, 183-186).

[I.43] *Sur la fonction $\zeta(s)$*, C.R. Acad. Sci. Paris **123**, 93 (Œuvres **1**, 187).

[I.44] *Sur les fonctions entières*, Bull. Soc. Math. France **24**, 186-187 (Œuvres **1**, 151-152).

[I.45] *Sur la distribution des zéros de la fonction $\zeta(s)$ et ses conséquences arithmétiques*, Bull. Soc. Math. France **24**, 199-200 (Œuvres **1**, 189-210).

[I.46] *Sur une forme de l'intégrale de l'équation d'Euler*, Bull. Sci. Math. Sér.2, **20**, 263-266 (Œuvres **3**, 1015-1017).

[I.47] *Sur la décomposition de deux figures géométriques équivalentes en un nombre fini d'éléments superposables chacun à chacun*, Proc. Verb. Soc. Sci. Phys. et Natur. Bordeaux, 24 décembre, 18-21.

### 1897

[I.48] *Sur certaines propriétés des trajectoires en dynamique*. Mémoire couronné en 1896 par l'Académie : Prix Bordin, J. Math. Sér. 5, **3**, 331-387 (Œuvres **4**, 1749-1805).

[I.49] *Sur les notions d'aire et de volume*, Proc. Verb. Soc. Sci. Phys. et Natur. Bordeaux, 21 janvier, 25-27 (Œuvres **4**, 2179-2180).

[I.50] *Sur les séries de Dirichlet*, Proc. Verb. Soc. Sci. Phys. et Natur. Bordeaux, 18 février, 41-45 (Œuvres **1**, 211-214).

[I.51] *Théorème sur les séries entières*, C.R. Acad. Sci. Paris **124**, 492.

[I.52] *Sur les lignes géodésiques des surfaces à courbures opposées*, C.R. Acad. Sci. Paris **124**, 1503-1505.

[I.53] *Sur les lignes géodésiques des surfaces à courbures opposées*, Proc. Verb. Soc. Sci. Phys. et Natur. Bordeaux, 1896-1897, 60-62.

[I.54] *Sur les principes fondamentaux de la mécanique*, Proc. Verb. Soc. Sci. Phys. et Natur. Bordeaux, 18 mars, 67-70 (Œuvres **4**, 1807-1809).

[I.55] *Sur la démonstration d'un théorème d'algèbre*, Proc. Verb. Soc. Sci. Phys. et Natur. Bordeaux, 1 avril, 84-86 (Œuvres **4**, 2093-2095).

[I.56] *Sur les conditions de décomposition d'une forme ternaire*, Proc. Verb. Soc. Sci. Phys. et Natur. Bordeaux, 13 mai, 100-102.

[I.57] *Sur les séries entières*, Proc. Verb. Soc. Sci. Phys. et Natur. Bordeaux, 3 juin, 110-111.

[I.58] *Sur les lignes géodésiques*, Proc. Verb. Soc. Sci. Phys. et Natur. Bordeaux, 17 juin, 115.

[I.59] *Sur les lignes géodésiques*, Proc. Verb. Soc. Sci. Phys. et Natur. Bordeaux, 1 juillet, 131.

[I.60] *Sur une surface à courbures opposées*, Proc. Verb. Soc. Sci. Phys. et Natur. Bordeaux, 22 juillet, 163-164.

### 1898

[I.61] *Sur la généralisation du théorème de Guldin*, Bull. Soc. Math. France **26**, 264-265.

[I.62] *Les surfaces à courbures opposées et leurs lignes géodésiques*, J. Math. Sér. 5, **4**, 27-73 (Œuvres **2**, 729-775).

[I.63] *Sur la forme de l'espace*, Proc. Verb. Soc. Sci. Phys. et Natur. Bordeaux, 3 février, 83-85.

[I.64] *Sur la courbure dans les espaces à plus de deux dimensions*, Proc. Verb. Soc. Sci. Phys. et Natur. Bordeaux, 3 février, 85-86.

[I.65] *Les invariants intégraux et l'optique*, C.R. Acad. Sci. Paris **126**, 811-812.

[I.66] *Sur la forme des géodésiques à l'infini et sur les géodésiques des surfaces réglées du second ordre*, Bull. Soc. Math. France **26**, 195-216 (Œuvres **2**, 707-728).

[I.67] *Leçons de géométrie élémentaire. Géométrie plane*, Armand Colin, Paris, (reprinted by J. Gabay, Paris, 1988).

[I.68] *Sur certaines applications possibles de la théorie des ensembles*, Verhandlungen des Internationalen Mathematiker-Kongresses in Zürich, 1897; Leipzig, 201-202 (Œuvres **1**, 311-312).

[I.69] *Sur le billard non euclidien*, Proc. Verb. Soc. Sci. Phys. et Natur. Bordeaux, 5 mai, 147-149.

**1899**

[I.70] *Théorème sur les séries entières*, Acta Math. **22**, 55-64 (Œuvres **1**, 93-101).

[I.71] *Sur les conditions de décomposition des formes*, Bull. Soc. Math. France **27**, 34-47 (Œuvres **1**, 313-326).

**1900**

[I.72] *Sur les points doubles des contours fermés*, Proc. Verb. Soc. Sci. Phys. et Natur. Bordeaux, 12 janvier, 4-7 (Œuvres **2**, 783-786).

[I.73] *Sur les intégrales d'un système d'équations différentielles ordinaires, considérées comme fonctions des données initiales*, Bull. Soc. Math. France **28**, 64-66 (Œuvres **3**, 1019-1021).

[I.74] *Sur l'intégrale résiduelle*, Bull. Soc. Math. France **28**, 69-90 (Œuvres **3**, 1065-1086).

[I.75] *Sur les singularités d'une certaine série*, Intermédiaire des Mathématiciens, **7**, 32.

[I.76] Review of the thesis of H. Fehr, *Application de la méthode vectorielle de Grassmann à la géométrie infinitésimale*, Rev. Gén. Sci. Pures et Appl. **11**, 556.

[I.77] Review of the thesis of E. Cahen, *Eléments de la théorie des nombres*, Rev. Gén. Sci. Pures et Appl. **11**, 807.

[I.78] *La bosse des mathématiques*, Rev. Gén. Sci. Pures et Appl. **11**, 913-914 (the paper does not bear any name, but it is encluded in the list of Hadamard's publications in [I.406]).

**1901**

[I.79] *Note sur l'induction et la généralisation en mathématiques*, Congrés International de Philosophie, Paris, 1900, Proc. Verb., Sommaires, Imprimerie Nationale, Paris,, 45 (Œuvres **4**, 2123-2126).

[I.80] *La série de Taylor et son prolongement analytique*, Gauthier-Villars, Paris.

[I.81] *Sur la propagation des ondes*, Bull. Soc. Math. France **29**, 50-60 (Œuvres **3**, 1087-1097).

[I.82] *Sur les réseaux de coniques*, Bull. Sci. Math. **25**, 27-30 (Œuvres **2**, 787-790).

[I.83] *Sur les éléments linéaires à plus de deux dimensions*, Bull. Sci. Math. **25**, 37-60 (Œuvres **2**, 791-794).

[I.84]  *Leçons de géométrie élémentaire. Géométrie dans l'espace*, Armand Colin, Paris, (reprinted by J. Gabay, Paris, 1988).

[I.85]  *Sur l'équilibre des plaques élastiques circulaires libres ou appuyées et sur celui de la sphère isotrope*, Ann. Éc. Norm. Sup. Sér. 3, **18**, 313-342 (Œuvres **4**, 1811-1840).

[I.86]  *Sur l'itération et les solutions asymptotiques des équations différentielles*, Bull. Soc. Math. France **29**, 224-228 (Œuvres **3**, 1023-1039).

[I.87]  *Notice sur les travaux scientifiques de M. Jacques Hadamard*, Gauthier-Villars, Paris.

### 1902

[I.88]  *Sur équations aux dérivées partielles à caractéristiques réelles*, Comptes Rendus du Deuxième Congrès International des Mathématiciens tenu à Paris, 1900, Gauthier-Villars, 373-375 (Œuvres **3**, 1061-1063).

[I.89]  *Sur les problèmes aux dérivées partielles et leur signification physique*, Bull. Univ. Princeton **13**, 49-52 (Œuvres **3**, 1099-1105).

[I.90]  *La théorie des plaques élastiques planes*, Trans. Amer. Math. Soc. **3**, 401-422 (Œuvres **4**, 1841-1862).

[I.91]  *Deux théorèmes d'Abel sur la convergence des séries*, Acta Math. **26**, 177-183 (Œuvres **1**, 327-333).

[I.92]  *Sur certaines surfaces minima*, Bull. Sci. Math. Sér. 2, **26**, 357-361 (Œuvres **2**, 777-780).

[I.93]  *Sur les dérivées des fonctions de lignes*, Bull. Soc. Math. France **30**, 40-43 (Œuvres **1**, 401-404).

[I.94]  *Sur une classe d'équations différentielles*, Bull. Soc. Math. France **30**, 208-220 (Œuvres **3**, 1027-1039).

[I.95]  *Sur une question de calcul des variations*, Bull. Soc. Math. France **30**, 253-256 (Œuvres **2**, 467-470).

[I.96]  *Sur une condition qu'on peut imposer à une surface*, Bull. Soc. Math. France **30**, 111.

[I.97]  Review of A. Larmor, *Aether and matter* (Cambridge Univ. Press, 1900), Bull. Sci. Math. Sér. 2, **26**, 319-328.

[I.98]  Review of A. Hatzfeld, *Pascal* (Alcan, Paris, 1901), Rev. Gén. Sci. Pures et Appl. **13**, 111.

[I.99]  Review of E. Goursat, *Cours d'analyse mathématique* (Gauthier-Villars, Paris, 1902), Rev. Gén. Sci. Pures et Appl. **13**, 694.

[I.100]  Review of E. Czuber, *Probabilités et moyennes géométriques* (Hermann, Paris, 1902), Rev. Gén. Sci. Pures et Appl. **13**, 787.

[I.101]  Review of L. Kronecker, *Vorlesungen über Mathematik* (Teubner, Leipzig, 1902), Rev. Gén. Sci. Pures et Appl. **13**, 836.

[I.102]  Review of E. Bouvier, *La méthode mathématique en économie politique* (Paris, 1902), Rev. Gén. Sci. Pures et Appl. **13**, 890-891.

[I.103]  *Sur les fonctions entières*, C.R. Acad. Sci. Paris **135**, 1309-1311 (Œuvres **1**, 155-157).

### 1903

[I.104]  *Sur les glissements dans les fluides*, C.R. Acad. Sci. Paris **136**, 299-301 (Œuvres **3**, 1107-1108).

[I.105] *Sur les glissements dans les fluides : Note complémentaire*, C.R. Acad. Sci. Paris **136**, 545 (Œuvres **3**, 1109).

[I.106] *Sur les opérations fonctionnelles*, C.R. Acad. Sci. Paris **136**, 351-354 (Œuvres **1**, 405-408).

[I.107] *Sur un problème mixte aux dérivées partielles*, Bull. Soc. Math. France **31**, 208-224 (Œuvres **3**, 1115-1131).

[I.108] *Sur les surfaces à courbure positive*, Bull. Soc. Math. France **31**, 300-301 (Œuvres **2**, 781-782).

[I.109] *Leçons sur la propagation des ondes et les équations de l'hydrodynamique*, Hermann, Paris.

[I.110] *Sur les équations aux dérivées partielles linéaires du second ordre*, C.R. Acad. Sci. Paris **137**, 1028-1030 (Œuvres **3**, 1111-1113).

[I.111] *Les sciences dans l'enseignement secondaire*, Conférence faite à l'École des Hautes Études Sociales, Alcan, Paris.

### 1904

[I.112] *Résolution d'un problème aux limites pour les équations linéaires du type hyperbolique*, Bull. Soc. Math. France **32**, 242-268 (Œuvres **3**, 1133-1159).

[I.113] *Sur un point de la théorie des percussions*, Nouv. Ann. Math. Sér. 4, **4**, 533-535.

[I.114] *Recherches sur les solutions fondamentales et l'intégration des équations linéaires aux dérivées partielles*, Ann. Éc. Norm. Sup. Sér. 3, **21**, 535-556 (Œuvres **3**, 1173-1194).

[I.115] *Sur les séries de la forme $\sum a_n e^{-\lambda_n z}$*, Nouv. Ann. Math. Sér. 4, **4**, 529-533 (Œuvres **1**, 215-219).

[I.116] *Le Troisième Congrès International des Mathématiciens*, Rev. Gen. Sci. **15**, 961-962 (the paper does not bear any name, but Hadamard acknowledged his authorship in [I.123, p. 270]).

### 1905

[I.117] *Sur les solutions fondamentales des équations linéaires aux dérivées partielles*, Verhandlungen des Dritten Internationalen Mathematiker-Kongresses in Heidelberg, 1904, Teubner, 265-271 (Œuvres **3**, 1165-1171).

[I.118] *Sur les données aux limites dans les équations aux dérivées partielles de la physique mathématique*, Verhandlungen des Dritten Internationalen Mathematiker-Kongresses in Heidelberg, 1904, Teubner, 414-416 (Œuvres **3**, 1161-1163).

[I.119] *Recherches sur les solutions fondamentales et l'intégration des équations linéaires aux dérivées partielles* (deuxième mémoire), Ann. Éc. Norm. Sup. Sér. 3, **22**, 101-141 (Œuvres **3**, 1195-1235).

[I.120] *Sur les équations linéaires aux dérivées partielles*, C.R. Acad. Sci. Paris **140**, 425-427 (Œuvres **3**, 1237-1238).

[I.121] *Sur quelques questions de calcul des variations*, Bull. Soc. Math. France **33**, 73-80 (Œuvres **2**, 485-513).

[I.122] *Sur la théorie des coniques*, Nouv. Ann. Math. Sér. 4, **3**, 145-152 (Œuvres **2**, 795-803).

[I.123] *Cinq lettres sur la théorie des ensembles* (Correspondance avec Borel, Baire et Lebesgue), Bull. Soc. Math. France **33**, 261-273; English translation in: G.H.

Moore, *Zermelo's axiom of choice. Its origins, developement, and influence*, Springer Verlag, New York-Heidelberg-Berlin, 1982 (Œuvres **1**, 335-348).

[I.124] *Remarque au sujet d'une note de M. Gyözö-Zemplen*, C.R. Acad. Sci. Paris **141**, 713.

[I.125] *À propos d'enseignement*, Rev. Gén. Sci. Pures et Appl. **16**, 192-194.

[I.126] *La théorie des ensembles*, Rev. Gén. Sci. Pures et Appl. **16**, 241-242.

[I.127] *Réflexions sur la méthode heuristique*, Rev. Gén. Sci. Pures et Appl. **16**, 499-504 (Œuvres **4**, 2127-2144).

[I.128] *Les principes des mathématiques et le problème des ensembles*, Rev. Gén. Sci. Pures et Appl. **16**, 541-543.

[I.129] Review of E. Czuber, *Wahrscheinlichkeitsrechnung und ihre Anwendung auf Fehlerausgleichung: Statistik und Lebensversicherung* (Teubner, Leipzig, 1905), Rev. Gén. Sci. Pures et Appl. **16**, 784-785.

**1906**

[I.130] *Sur un théorème de M. Osgood relatif au calcul des variations*, Bull. Soc. Math. France **34**, 61.

[I.131] *Sur la mise en équation des problèmes de mécanique*, Nouv. Ann. Math. Sér. 4, **6**, 97-100 (Œuvres **4**, 2193-2196).

[I.132] *Sur les transformations planes*, C.R. Acad. Sci. Paris **142**, 74-77.

[I.133] *Sur les caractéristiques des systèmes aux derivées partielles*, Bull. Soc. Math. France **34**, 48-52 (Œuvres **3**, 1239-1243).

[I.134] Review of J. Gibbs, *Elementary principles in statistical mechanics*, Bull. Amer. Math. Soc. **12**, 194-210; reprinted in Bull. Sci. Math. Sér. 2, **30**, 161-179.

[I.135] *Sur une méthode de calcul des variations*, C.R. Acad. Sci. Paris **143**, 1127-1129 (Œuvres **2**, 479-481).

[I.136] *La logistique et l'induction complète. La notion de correspondance*, Rev. Gén. Sci. Pures et Appl. **17**, 161-162 (the paper does not bear any name, but it is included in the list of Hadamard's papers in [I.406]) (Œuvres **4**, 2157-2160).

[I.137] *Les principes de la théorie des ensembles*, Rev. Gén. Sci. Pures et Appl. **17**, 5.

[I.138] *Sur les transformations ponctuelles*, Bull. Soc. Math. France **34**, 71-84 (Œuvres **1**, 349-363).

[I.139] *Sur le principe de Dirichlet*, Bull. Soc. Math. France **34**, 135-138 (Œuvres **3**, 1245-1248).

[I.140] *La logistique et la notion de nombre entier*, Rev. Gén. Sci. Pures et Appl. **17**, 906-909 (Œuvres **4**, 2145-2155).

**1907**

[I.141] *Les problèmes aux limites dans la théorie des équations aux dérivées partielles*, Conférences faites à la Société Mathématique de France et à la Société Française de Physique, J. Phys. Théor. et Appl. **6**, 202-241.

[I.142] *Sur quelques questions de calcul des variations*, Ann. Éc. Norm. Sup. Sér. 3, **24**, 203-231 (Œuvres **2**, 485-513).

[I.143] *Sur l'interprétation théorique des raies spectrales*, Bull. Soc. Franç. Phys.

[I.144] *Sur la variation des intégrales doubles*, C.R. Acad. Sci. Paris **144**, 1092-1093 (Œuvres **2**, 483-484).

**1908**

[I.145] *Sur le problème d'analyse relatif à l'équilibre des plaques élastiques encastrées.* Mémoire couronné en 1907 par l'Académie : Prix Vaillant, Mémoires présentés par divers savants à l'Académie des Sciences **33**, No 4 (Œuvres **2**, 515-629).

[I.146] *Sur les séries de Dirichlet*, Rend. Circolo Mat. Palermo **25**, 326-330 (Œuvres **1**, 221-225).

[I.147] *Rectification à la note "Sur les séries de Dirichlet"*, Rend. Circolo Mat. Palermo **25**, 395-396 (Œuvres **1**, 227-228).

[I.148] *Théorie des équations aux dérivées partielles linéaires hyperboliques et du problème de Cauchy*, Acta Math. **31**, 333-380 (Œuvres **3**, 1249-1296).

[I.149] *Sur l'expression asymptotique de la fonction de Bessel*, Bull. Soc. Math. France **36**, 77-85 (Œuvres **1**, 365-373).

[I.150] *Les paradoxes de la théorie des ensembles*, Rev. Gén. Sci. Pures et Appl. **19**, 681.

**1909**

[I.151] *Sur certains cas intéressants du problème biharmonique*, Atti del IV Congresso Internazionale dei Matematici, Roma, 1908; Accad. dei Lincei, Roma, 12-14 (Œuvres **3**, 1297-1303).

[I.152] *Sur certaines particularités du calcul des variations*, Atti del IV Congresso Internazionale dei Matematici, Roma, 1908; Accad. dei Lincei, Roma, 61-63 (Œuvres **2**, 643-645).

[I.153] *Sur les lignes géodésiques, à propos de la récente note de M. Drach*, C.R. Acad. Sci. Paris **148**, 272-274 (Œuvres **2**, 873-874).

[I.154] *Sur une propriété fonctionnelle de la fonction $\zeta(s)$ de Riemann*, Bull. Soc. Math. France **37**, 59-60 (Œuvres **1**, 229-230).

[I.155] *Détermination d'un champ électrique*, Ann. Chim. et Phys. Sér. 8, **16**, 403-432.

[I.156] *Notions élémentaires sur la géométrie de situation*, Nouv. Ann. Math. Sér. 4, **9**, 193-235 (Œuvres **2**, 829-871).

[I.157] *La géométrie de situation et son rôle en mathématiques* : Leçon d'ouverture professée au Collège de France, Rev. du Mois **8**, 38-60 (Œuvres **2**, 805-827).

[I.158] *Notice sur les travaux scientifiques de M. Jacques Hadamard* (kept at the *Archives de l'École Polytechnique*).

**1910**

[I.159] *Leçons sur le calcul des variations*, Hermann, Paris.

[I.160] *Sur les ondes liquides*, C.R. Acad. Sci. Paris **150**, 609-611 (Œuvres **3**, 1301-1303).

[I.161] *Sur les ondes liquides*, C.R. Acad. Sci. Paris **150**, 772-774 (Œuvres **3**, 1317-1320).

[I.162] *Quelques propriétés des fonctions de Green*, C.R. Acad. Sci. Paris **150**, 1664-1666 (Œuvres **3**, 1055-1057).

[I.163] *Sur quelques applications de l'indice de Kronecker*, Supplement to the 2nd edition of J. Tannery, *Introduction à la théorie des fonctions d'une variable*, Hermann, Paris (Œuvres **2**, 875-915).

[I.164] *Sur un problème de cinématique navale*, Nouv. Ann. Math. Sér. 4, **10**, 337-361, reprinted in Revue Maritime, Avril, 1911 (Œuvres **4**, 1863-1887).

**1911**

[I.165]  *Sur les trajectoires de Liouville*, Bull. Sci. Math. Sér. 2, **35**, 106-113 (Œuvres **4**, 1889-1895).

[I.166]  *Sur l'inégalité*

$$\left[\delta g_A^A \delta g_B^B - \left(\delta g_B^A\right)^2\right]\left[\delta g_C^C \delta g_D^D - \left(\delta g_D^C\right)^2\right] > \left(\delta g_C^A \delta g_D^B - \delta g_D^A \delta g_C^B\right)^2,$$

*à laquelle satisfont les variations de la fonction de Green quand on passe d'un contour à un contour voisin*, Bull. Soc. Math. France, **39**, 482.

[I.167]  *Sur la solution fondamentale des équations linéaires aux dérivées partielles de type parabolique*, C.R. Acad. Sci. Paris **152**, 1148-1149 (Œuvres **3**, 1309-1310).

[I.168]  *Mouvement permanent lent d'une sphère liquide et visqueuse dans un liquide visqueux*, C.R. Acad. Sci. Paris **152**, 1735-1738 (Œuvres **3**, 1311-1314).

[I.169]  *Relation entre les solutions des équations aux dérivées partielles des types parabolique et hyperbolique*, Bull. Soc. Math. France **39**, 14-15 (Œuvres **3**, 1059-1060).

[I.170]  *Sur les propriétés des fonctions de Green dans le plan*, Bull. Soc. Math. France **39**, 14-15.

[I.171]  *Propriétés générales des corps et domaines algébriques* (with M. Kürschak), in Encycl. Sci. Math., Édition française, **1** : 2, 233-385.

[I.172]  *Maurice Lévy*, Rev. Gén. Sci. Pures et Appl. **22**, 141-143.

[I.173]  Review of E. Barbette, *Les sommes de $p^{\text{èmes}}$ puissances distinctes égales à une $p^{\text{ème}}$ puissance* (Liège, 1911) and *Le dernier théorème de Fermat* (Liège, 1911), Rev. Gén. Sci. Pures et Appl. **22**, 541.

**1912**

[I.174]  *Le calcul fonctionnel*, Enseign. Math. **14**, 1-18 (Œuvres **4**, 2253-2266).

[I.175]  *Sur une question relative aux liquides visqueux. Note rectificative*, C.R. Acad. Sci. Paris **154**, 109 (Œuvres **3**, 1315).

[I.176]  *Sur les variations unilatérales et les principes du calcul des variations*, Bull. Soc. Math. France, **40**, 20.

[I.177]  *Sur les extrémales du problème isopérimétrique dans le cas des intégrales doubles*, Bull. Soc. Math. France, **40**, 20-23.

[I.178]  *Sur la généralisation de la notion de fonction analytique*, Bull. Soc. Math. France, **40**, 28-29 (Œuvres **1**, 175-176).

[I.179]  *Sur la loi d'inertie des formes quadratiques*, Bull. Soc. Math. France, **40**, 29-30.

[I.180]  *Propositions transcendantes de la théorie des nombres* (with M. Maillet), in Encycl. Sci. Math., Édition française, **1** : 3, 215-387.

[I.181]  *Itération des noyaux infinis dans le cas des intégrales doubles*, Supplement to the book of M. Fréchet and H.B. Heywood, *L'équation de Fredholm et ses applications à la physique mathématique*, Hermann, Paris (Œuvres **1**, 409-413).

[I.182]  *Propriétés de la résolvante de l'équation de Fredholm*, Supplement to the book of M. Fréchet and H.B. Heywood, *L'équation de Fredholm et ses applications à la physique mathématique*, Hermann, Paris (Œuvres **1**, 415-426).

**1913**

[I.183]  *Observations à propos de la communication de M. Borel "Remarque sur la théorie des résonateurs"*, Soc. Franç. Phys. Proc. Verb. et Résumé des Comm. faites pendant l'année 1912. Séance du 21 juin 1912, 79.

[I.184] *Sur la série de Stirling*, Proceedings of the Fifth International Congress of Mathematicians, Cambridge, 1912, Cambridge University Press, 303-305 (Œuvres **1**, 375-377).

[I.185] *Henri Poincaré et le problème des trois corps*, Rev. Métaphys. et Morale **21** : 5, 617-658; Rev. du Mois **16**, 385-418 (Œuvres **4**, 2007-2041).

[I.186] *La construction de Weierstrass et l'existence de l'extremum dans le problème isopérimétrique*, Ann. Mat. Sér. 3, **21**, 251-287 (Œuvres **2**, 647-682).

[I.187] *Observations à propos d'une note de M. Bouligand*, C.R. Acad. Sci. Paris **156**, 1364.

**1914**

[I.188] *Points pincés, arêtes de rebroussement et représentation paramétrique des surfaces*, Enseign. Math. **16**, 356-359 (Œuvres **2**, 917-920).

[I.189] *L'infini mathématique et la réalité*, Rev. du Mois.

[I.190] *Sur le module maximum d'une fonction et de ses dérivées*, Bull. Soc. Math. France **42**, 68-72 (Œuvres **1**, 379-382).

[I.191] *Observations au sujet d'une note de M. Paul Lévy*, C.R. Acad. Sci. Paris, **158**, 1010-1011 (Œuvres **1**, 427-428).

[I.192] *Henri Poincaré : L'œuvre scientifique, L'œuvre philosophique* (with V. Volterra, P. Langevin, P. Boutroux), Alcan, Paris.

**1915**

[I.193] *Sur un mémoire de M. Sundman*, Bull. Sci. Math. Sér. 2, **39**, 249-264 (Œuvres **4**, 1897-1912).

[I.194] *Four lectures on mathematics delivered at Columbia University in 1911*, Columbia University Press, New York.

**1916**

[I.195] *Sur les ondes liquides*, Rend. Accad. dei Lincei Sér. 5, **25**, 716-719 (Œuvres **3**, 1317-1320).

[I.196] *Sur l'élimination entre équations différentielles*, Nouv. Ann. Math. Sér. 4, **17**, 81-84 (Œuvres **4**, 2197-2200).

**1919**

[I.197] *Remarques sur l'intégrale résiduelle*, C.R. Acad. Sci. Paris **168**, 533-534 (Œuvres **3**, 1321-1322).

[I.198] *Sur les correspondances ponctuelles*, Bull. Soc. Math. France **47**, 28-29 (Œuvres **1**, 383-384).

[I.199] *Sur les singularités des séries entières*, Bull. Soc. Math. France **47**, 40.

[I.200] *Sur un théorème fondamental de la théorie des fonctions analytiques de plusieurs variables*, Bull. Soc. Math. France **47**, 44-46 (Œuvres **1**, 159-161).

[I.201] *Démonstration directe d'un théorème de Poincaré sur les périodes des intégrales abéliennes attachées à une courbe algébrique qui satisfait à une équation différentielle linéaire*, Bull. Soc. Math. France **47**, 46 (Œuvres **1**, 385).

[I.202] *Recherche du balourd dynamique des obus*, Travaux du Laboratoire d'Essais des Arts et Métiers.

**1920**

[I.203] *La solution élémentaire des équations aux dérivées partielles linéaires hyperboliques non analytiques*, C.R. Acad. Sci. Paris **170**, 149-154 (Œuvres **3**, 1323-1328).

[I.204] *Sur certaines solutions d'une équation aux dérivées fonctionnelles*, C.R. Acad. Sci. Paris **170**, 355-359 (Œuvres **1**, 429-433).

[I.205] *Rapport sur les travaux examinés et retenus par la Commission de balistique pendant la durée de la guerre*, C.R. Acad. Sci. Paris **170**, 436-445.

### 1921

[I.206] *Sur la solution élémentaire des équations linéaires aux dérivées partielles et sur les propriétés des géodésiques*, C. R. du Congrés International des Mathématiciens, Strasbourg, 1920 ; Édouard Privat, Toulouse, 179-184 (Œuvres **3**, 1329-1334).

[I.207] *Sur le problème mixte pour les équations linéaires aux dérivées partielles*, Comptes Rendus du Congrés International des Mathématiciens, Strasbourg, 1920; Édouard Privat, Toulouse, 499-503 (Œuvres **3**, 1335-1339).

[I.208] *L'œuvre mathématique de Poincaré*, Acta Math. **38**, 203-287 (Œuvres **4**, 1921-2005).

[I.209] *On some topics connected with linear partial differential equations*, Proc. Benares Math. Soc. **3**, 39-48.

[I.210] *A propos d'enseignement secondaire*, Rev. Internat. de l'Enseign. **75**, 289-294 (Œuvres **4**, 2201-2206).

[I.211] *Sur la comparaison des problèmes aux limites pour les deux principaux types d'équations aux dérivées partielles*, Bull. Soc. Math. France **49**, 28.

### 1922

[I.212] *L'enseignement secondaire et l'esprit scientifique*, Revue de France, Avril.

[I.213] *Einstein en France*, Rev. Internat. de l'Enseign. **76**, 129-137.

[I.214] *Les principes du calcul des probabilités*, Rev. de Métaphys. et de Morale **29**, 289-293 (Œuvres **4**, 2161-2165).

[I.215] *À propos des notions de dimension et d'homogénéité*, J. de Physique et Radium Sér. 6, **3**:5, 149-153.

[I.216] *Sur un théorème de géométrie élémentaire*, Bulletin Officiel de la Direction des Recherches Scientifiques et Industrielles et des Inventions, **38**.

[I.217] *Sur la fonction harmonique la plus voisine d'une fonction donnée*, Association Française pour l'Avancement des Sciences **46**, 108-109.

[I.218] *Sur une formule de calcul des probabilités*, Association Française pour l'Avancement des Sciences **46**, 109-110.

[I.219] *Les responsabilités des guerres: Comment déterminer l'agresseur?* Cahiers des Droits de l'Homme, avril, 184.

[I.220] *The early scientific work of H. Poincaré*, The Rice Institute Pamphlet **9**:3, 111-183.

[I.221] Preface to *Leçons d'analyse fonctionnelle*, by P. Lévy, Gauthier-Villars, Paris.

[I.222] Preface to *Introduction au calcul tensoriel et au calcul différentiel*, by G. Juvet, Albert Blanchard, Paris.

### 1923

[I.223] *Lectures on Cauchy's problem in linear partial differential equations*, New Haven, Yale Univ. Press; reprinted by Dover Publications, New York, 1954.

[I.224] *La notion de différentielle dans l'enseignement*, Scripta Univ. Jerusalem **1**:4; reprinted in Enseign. Sci. **35** (1931), 136-137.

[I.225] *Poincaré i la teoria de les ecuaciones differentiales*, Conférences prononcées à l'Institut d'Études Catalanes de Barcelone.

[I.226] *La réforme de l'enseignement secondaire*, Bull. Sci. des Étudiants de Paris **9**, 2-13.

[I.227] *Sur les points doubles des lieux géométriques et sur la construction par régions*, Nouv. Ann. Math. Sér. 5, **1**, 364-379 (Œuvres **2**, 921-933).

[I.228] *La pensée française dans l'evolution des sciences exactes*, France et Monde **92**, 321-343.

[I.229] *Sur une formule déduite de la théorie des cubiques*, Bull. Soc. Math. France **51**, 295-296 (Œuvres **2**, 935).

[I.230] *Observations à propos d'une communication de Mordoukhay-Boltovskoy*, C.R. Acad. Sci. Paris **176**,727-728.

[I.231] *Remarque sur une communication de P. Noaillon*, C.R. Acad. Sci. Paris **176**, 1059.

[I.232] *Sur les tourbillons et les surfaces de glissement dans les fluides*, C.R. Acad. Sci. Paris **177**, 505-506 (Œuvres **3**, 1341-1342).

[I.233] *Déclare qu'une expérience a donné raison aux hypothèses exposées dans sa note*, C.R. Acad. Sci. Paris **177**, 568.

[I.234] *Observations à propos d'une note de A. Bloch "Sur les cercles paratactiques et la cyclide de Dupin"*, C.R. Acad. Sci. Paris **177**, 734.

[I.235] *La vrai culture générale*, Pour et contre, novembre.

**1924**

[I.236] *Principe de Huygens et prolongement analytique*, Bull. Soc. Math. France **52**, 241-278 (Œuvres **3**, 1375-1413).

[I.237] *Quelques conséquences analytiques du principe de Huygens* (13ème Réunion de la Société Italienne pour l'Avancement des Sciences à Naples), Atti della Soc. Ital. per il Progresso della Sci. **16**, 164-168.

[I.238] *Sobre la representación gráfica del espacio de cuatro dimensiones*, Rev. Matem. Hisp.-Amer. **6**, 265-269.

[I.239] *Comment je n'ai pas découvert la relativité*, Atti Congr. Intern. Philos. Naples, 441-453; abridged English translation: *How I did not discover relativity*, translated by I.H. Rose and edited by B.B. Mandelbrot, Math. Intelligencer **10**:2 (1988), 65-67.

[I.240] *Le principe de Huygens* (Conférence pour le cinquantenaire de la Société Mathématique de France), Bull. Soc. Math. France **52**, 610-640 (Œuvres **3**, 1343-1373).

**1925**

[I.241] *On quasi-analytic functions*, Proc. Nat. Acad. Sci. USA **11**, 447-448 (Œuvres **1**, 177-178).

[I.242] *Sur le calcul approché des intégrales définies*, Proc. Nat. Acad. Sci. USA **11**, 448-450 (Œuvres **1**, 387-389).

[I.243] *Itération et fonctions quasi analytiques*, Rev. Gén. Sci. Pures et Appl.

[I.244] *Sobre un tipo de ecuaciones integrales singulares*, Rev. Acad. Madrid **22**, 187-191.

**1926**

[I.245] *Développement de la notion de fonction* (in Portuguese), Conférences à l'École polytechnique de Rio de Janeiro, le 23 sept. 1924, rédigées par J. Nicoletis, Revista da Academia Brasileira de Sciencias **1**, 82-111.

[I.246] *Sur une série entière en relation avec le dernier théorème de Fermat*, Bull. Soc. Math. France **54**, 21-22 (Œuvres **1**, 231-232).

[I.247] *Sur les équations intégrables par la méthode de Laplace*, Bull. Soc. Math. France **54**, 33-35.

[I.248] *Sur la géométrie anallagmatique*, Bull. Soc. Math. France **54**, 35-39.

[I.249] *Sur la géometrie anallagmatique (Addition à l'article précédent)* (Œuvres **2**, 975-981).

[I.250] *Quelques cas d'impossibilité du problème de Cauchy*, In memoriam N.I. Lobachevskii, Glavnauka, Kazan, vol. 2, 163-176 (Œuvres **3**, 1457-1470).

[I.251] *Sur la théorie des séries entières*, Nouv. Ann. Math. Sér. 6, **1**, 161-164 (Œuvres **4**, 2227-2230).

[I.252] *A propos du nouveau programme de mathématiques spéciales*, Nouv. Ann. Math. Sér. 6, **1**, 257-276, 391-393 (Œuvres **4**, 2207-2226).

[I.253] *La série de Taylor et son prolongement analytique* (with S. Mandelbrojt), second edition, revised and completed, Gauthier-Villars, Paris.

[I.254] *Le principe de Huyghens dans le cas de quatre variables indépendantes*, Acta Math. **49**, 203-344 (Œuvres **3**, 1415-1456).

[I.255] Preface to *Les fondements des Mathématiques*, by F. Gonseth (Blanchard, Paris, 1926), Bull. Sci. Math. Sér. 2, **51**, 66-73.

[I.256] *Remarque au sujet d'une communication de Leonida Tonelli "Sur la méthode d'adjonction dans le calcul des variation"*, C.R. Acad. Sci. Paris **182**, 679.

[I.257] *Observations sur les communications de I. Karamata "Sur certaines limites rattachées aux intégrales de Stieltjes" et de P. Lévy "Remarques sur les precédés de sommation des séries divergentes"*, C.R. Acad. Sci. Paris **182**, 838.

**1927**

[I.258] *Cours d'analyse de l'École Polytechnique*, Hermann, Paris, vol. 1.

[I.259] *Récents progrès de la géométrie anallagmatique*, Rev. Mat. Hisp.-Amer. **2**; Nouv. Ann. Math. Sér. 6, **2**, 257-273, 289-320 (Œuvres **2**, 937-974).

[I.260] *Sur les éléments riemanniens et le déplacement parallèle*, Bull. Soc. Math. France **55**, 30-31.

[I.261] *Sur la théorie des fonctions entières*, Bull. Soc. Math. France **55**, 135-137 (Œuvres **1**, 163-165).

[I.262] *Sulle funzioni intere di genere finito. Nota dei Soci Stranieri J. Hadamard e E. Landau*, Rend. Accad. Lincei **6**, 3-9 (Œuvres **1**, 167-173).

[I.263] *L'œuvre de Duhem dans son aspect mathématique*, Mém. Soc. Sci. Phys. et Natur. Bordeaux **1**, 637-665.

[I.264] *Observation sur la note de P. Lévy "Sur un théorème de M. Hadamard relatif à la multiplication des singularités"*, C.R. Acad. Sci. Paris **184**, 581.

[I.265] *Sur le battage des cartes*, C.R. Acad. Sci. Paris **185**, 5-9 (Œuvres **4**, 2065-2069).

[I.266] *Observation sur la note précédente de M. Ragnar Frisch*, C.R. Acad. Sci. Paris **185**, 1245-1246.

[I.267] *Les méthodes d'enseignement des sciences expérimentales*, Rev. Internat. de l'Enseign. **81**, 355-356.

**1928**

[I.268]  *Observation sur une note de M.B. Hostinsky,* C.R. Acad. Sci. Paris **186**, 62.

[I.269]  *Sur les opérations itérées en calcul des probabilités,* C.R. Acad. Sci. Paris **186**, 189-192 (Œuvres **4**, 2079-2082).

[I.270]  *Sur le principe ergodique,* C.R. Acad. Sci. Paris **186**, 275-276.

[I.271]  *Observation sur une note de A. Haar,* C.R. Acad. Sci. Paris **187**, 25-26.

[I.272]  *Deux exercices de mécanique,* Enseign. Sci. **1**, 173-177.

[I.273]  *A propos de géométrie anallagmatique,* Enseign. Sci. **1**, 296-298.

[I.274]  *Une propriété de la fonction $\zeta(s)$ et des séries de Dirichlet,* Association Française pour l'Avancement des Sciences **52**, 29-30 (Œuvres **1**, 233-235).

[I.275]  *Quelques remarques sur l'enseignement de la mécanique,* Association Française pour l'Avancement des Sciences **52**, 77-79 (Œuvres **4**, 2231-2233).

[I.276]  *La peine de mort et le Code pénal,* Cahiers des Droits de l'Homme **30**, 715.

[I.277]  *Une application d'une formule intégrale relative aux séries de Dirichlet,* Bull. Soc. Math. France **56**, 43-44.

### 1929

[I.278]  *Le développement et le rôle scientifique du calcul fonctionnel,* Atti del Congresso Internazionale dei Matematici, Bologna, 1928, Nicola Zanichelli, Bologna, vol. 1, 143-161 (Œuvres **1**, 435-453).

[I.279]  *Les responsabilités de la guerre,* Cahiers des Droits de l'Homme **31**, 729-732.

[I.280]  *Princip d'Huyghens* (in Czech), Časopis pro Pěstování Matematiky a Fysiky **58**, 346-366.

[I.281]  *Le principe de Huyghens pour les équations à trois variables independantes,* J. Math. Pures et Appl. Sér. 9, **8**, 197-228 (Œuvres **3**, 1471-1502).

[I.282]  *On ordinary restricted extrema in connection with point transformations,* Bull. Amer. Math. Soc. **35**, 823-828 (Œuvres **1**, 391-397).

[I.283]  *Analyse du livre de E. Landau "Vorlesungen über Zahlentheorie"* (with S. Mandelbrojt), Bull. Sci. Math. **53**, 164-182.

[I.284]  *Observation sur une note d'Étienne Halphen,* C.R. Acad. Sci. Paris **188**, 846.

[I.285]  *Observation sur une note de M. de Franchis,* C.R. Acad. Sci. Paris **188**, 1026.

### 1930

[I.286]  *Sur les arêtes de rebroussement des certaines enveloppes,* 1er Congrés des Mathématiciens des Pays Slaves, Warszawa, 1929, Sprawozdanie, Warszawa, 318-322 (Œuvres **2**, 987-991).

[I.287]  *Remarques géométriques sur les enveloppes et la propagation des ondes,* Acta Math. **54**, 247-261 (Œuvres **3**, 1503-1519).

[I.288]  *La physique et la culture générale,* Œuvre, 30 janvier.

[I.289]  *La question de la physique,* Œuvre, 16 février.

[I.290]  *Un nouveau pas à faire dans la voie de la Paix: les manuels scolaires,* La Paix par le Droit **1**, 1-4.

[I.291]  *Cours d'analyse de l'École Polytechnique,* Hermann, Paris, vol. 2.

### 1931

[I.292]  *Sur le battage des cartes et ses relations avec la mécanique statistique,* Atti del Congresso Internazionale dei Matematici, Bologna, 1928; Nicola Zanichelli, Bologna, **5**, 133-139 (Œuvres **4**, 2071-2077).

[I.293] *Parlons encore culture générale*, Œuvre, 13 janvier.

[I.294] *Formation ou déformation intellectuelle*, Œuvre, 19 janvier.

[I.295] *Une culture qu'il ne faudrait pas détruire*, Œuvre, 24 janvier.

[I.296] *La question de la physique*, Œuvre, 16 février.

[I.297] *La formation des agrégés de mathématiques*, Enseign. Sci. **4**:35, 135-136.

[I.298] *Multiplication et division*, Enseign. Sci. **4**:39, 267-269.

[I.299] *A propos des lettres de M. L. Blum. Réflexions générales sur un cas particulier*, Enseign. Sci. **4**:40, 309-311.

[I.300] *Remarques sur la note de E.O. Lovett "Sur un problème de M. Gambier dans la déformation des surfaces"*, C.R. Acad. Sci. Paris **193**, 567-568.

## 1932

[I.301] *Sur la théorie des équations aux dérivées partielles du premier ordre*, Enseign. Sci. **5**:46, 161-163 (Œuvres **3**, 1521-1523).

[I.302] *Sur les équations aux dérivées partielles d'ordre supérieur*, Verhandlungen des Internationalen Mathematiker-Kongresses, Zürich, 1932, Teubner, Zürich-Leipzig, vol. 2, 78-80 (Œuvres **3**, 1567-1569).

[I.303] *Réponse à une enquête sur l'histoire des sciences dans l'enseignement*, Enseign. Sci. **5**:47, 309-211.

[I.304] *Coordination d'enseignements*, Enseign. Sci. **5**:51, 2-3.

[I.305] *Le problème de Cauchy et les équations aux dérivées partielles linéaires hyperboliques*, translation of the revised book [I.223], Hermann, Paris.

[I.306] *Réponse à une enquête sur la revision des traités*, Paix Mondiale **2**.

[I.307] *Équations aux dérivées partielles et variables réelles* (in Ukrainian), Zapiski Kharkiv. Mat. Tovar. ta Ukraïn. Inst. Matem. Nauk Ser. 4, **5**, 11-20.

[I.308] *Sur certaines propriétés des trajectoires en dynamique*, J. Math. Pures et Appl. **11**, 207.

## 1933

[I.309] *Propriétés d'une équation linéaire aux dérivées partielles du quatrième ordre*, Tôhoku Math. J. **37**, 133-150 (Œuvres **3**, 1571-1588).

[I.310] *La propagation des ondes et les caustiques*, Comment. Math. Helvetici **5**, 137-173 (Œuvres **3**, 1525-1565).

[I.311] *The later scientific work of Henri Poincaré*, The Rice Institute Pamphlet **20**:1, 1-85.

[I.312] *Painlevé, le savant*, Vu.

[I.313] *Sur les probabilités discontinues des événements "en chaîne"* (with M. Fréchet), Zeitschr. Angew. Math. Mech. **13**, 92-97 (Œuvres **4**, 2083-2088).

[I.314] *Observation sur une note de A. Przeborski*, C.R. Acad. Sci. Paris **197**, 302.

[I.315] *Observations sur une note récente de M. Sixto Rios*, C.R. Acad. Sci. Paris **197**, 1374.

[I.316] *Observation sur une note de N. Adamoff*, C.R. Acad. Sci. Paris **198**, 218.

[I.317] Preface to *Leçons sur les progrés récents de la théorie des séries de Dirichlet*, by V. Bernstein, Paris, Gauthier-Villars.

## 1934

[I.318] *Remarques sur deux résultats dus à M. Demoulin et à M. Perron*, Bull. Soc. Math. France **62**, 25.

[I.319] *Sur une question relative aux congruences de sphères*, Bull. Soc. Math. France. **62**, 25-26.

[I.320] *L'œuvre scientifique de Paul Painlevé*, Rev. Métaphys. et Morale, **41**, 289-325.

[I.321] *Un terme à effacer de l'enseignement mathématique : "effectuer"*, Enseign. Sci. **7**:66, 167-168.

[I.322] *Réponses à l'enquête sur les bases de l'Enseignement des Mathématiques*, Enseign. Sci. **7**:66, 175-177, **7**:68, 247.

[I.323] *La non-résolubilité de l'équation du cinquième degré*, Enseign. Sci. **7**:68, 225-235, **7**:69, 257-260 (Œuvres **4**, 2097-2107, 2109-2119).

[I.324] *Un cas simple de diffusion des ondes*, Matem. Sbornik **41**, 402-404 (Œuvres **3**, 1589-1592).

[I.325] Preface to *Über gewisse Ideale in einer einfachen Algebra*, by H. Hasse, Actualités Sci. Indust., **109**, Hermann, Paris.

[I.326] *Observations au sujet de la note de M. Mursi*, C.R. Acad. Sci. Paris, **199**, 179-180.

[I.327] Preface to *Méthodes topologiques dans les problèmes variationnels*, by L. Lusternik and L. Schnirelmann, Hermann, Paris.

**1935**

[I.328] *Polynomes linéaires adjoints*, Enseign. Sci. **8**:74, 97-100.

[I.329] *Les développables circonscrites à la sphère*, Enseign. Sci. **8**:75, 129-130 (Œuvres **2**, 993-994).

[I.330] *Réponse à l'enquête sur l'enseignement de la mécanique*, Enseign. Sci. **8**:77, 193.

[I.331] *La théorie des équations du premier degré*, Enseign. Sci. **8**:79, 257-262.

[I.332] *Extrait d'une lettre à M.T. Kubota*, Tôhoku Math. J. **40**, 198.

[I.333] *Selecta: Jubilé Scientifique de M. Jacques Hadamard*, Gauthier-Villars, Paris.

[I.334] *La notion de différentielle dans l'enseignement*, Math. Gaz. **19**, 341-342.

**1936**

[I.335] *Sur les caustiques des enveloppes à deux paramètres*, Soc. Math. de France, C.R. des Séances, 29-30.

[I.336] *Équations aux dérivées partielles et fonctions de variables réeles*, Proc. of the First Congress of Mathematicians of the U.S.S.R. (Kharkov, 1930), ONTI, Moscow-Leningrad, 106-118.

[I.337] *Principe de Huyghens et théorie d'Hugoniot*, Proc. of the First Congress of Mathematicians of the U.S.S.R. (Kharkov, 1930), ONTI, Moscow-Leningrad, 276-279.

[I.338] *Équations aux dérivées partielles. Les conditions définies en général. Las cas hyperbolique*, Conférence Internationale sur les Équations aux Dérivées Partielles, Genève, 17-20 juin 1935, Enseign. Math. **35**, 5-42 (Œuvres **3**, 1593-1630).

[I.339] *La caustique des enveloppes à deux paramètres*, J. Math. Pures et Appl. **15**, 333-337 (Œuvres **2**, 995-999).

**1937**

[I.340] *Un problème topologique sur les équations différentielles*, Prace Mat. Fiz. **44**, 1-7 (Œuvres **3**, 1041-1047).

[I.341] *Calcul des variations et différentiation des intégrales*, Trudy Tbilisskogo Mat. Inst. **1**, 55-63 (Œuvres **2**, 685-693).

[I.342] *Le problème de Dirichlet pour les équations hyperboliques*, J. Chin. Math. Soc. **2**, 6-20 (Œuvres **3**, 1631-1645).

[I.343] *Observations sur la note de M. Krasner et B. Ranulac*, C.R. Acad. Sci. Paris **204**, 399.

[I.344] *Observations sur les notes précédentes de J.-L. Destouches et A. Appert*, C.R. Acad. Sci. Paris **204**, 458.

[I.345] *Observations sur la note de M. Mandelbrojt*, C.R. Acad. Sci. Paris **204**, 1458-1459 (Œuvres **1**, 179).

[I.346] *La science mathématique*, Encyclopédie française, **1**, 1.52.1-1.58.7.

[I.347] *Les équations différentielles*, Encyclopédie française, **1**, 1.76.1-1.76.7.

[I.348] *L'existence et le domaine de validité des solutions*, Encyclopédie française, **1**, 1.76.8-1.76.14.

[I.349] *Étude directe des solutions* (with J. Chazy), Encyclopédie française, **1**, 1.78.5-1.80.8.

[I.350] *Divers types de conditions définies et d'équations aux dérivées partielles*, Encyclopédie française, **1**, 1.82.1-1.82.13.

**1938**

[I.351] *Remarque sur l'intégration approchée des équations différentielles: Extrait d'une lettre*, Ann. Soc. Polon. Math. **16**, 126 (Œuvres **3**, 1049-1050).

[I.352] *Sur certaines questions de calcul intégral*, Ann. Soc. Ci. Argentina **125**, 1-18 (Œuvres **4**, 2235-2252).

[I.353] *L'homogénéité en mécanique*, Bull. Sci. Math. Sér. 2, **62**, 6-10 (Œuvres **4**, 1913-1917).

[I.354] *Un problème de géométrie*, Enseign. Sci. **11**:108, 225-229.

[I.355] *A propos de la non-résolubilité par radicaux de l'équation du 5ème degré*, Enseign. Sci. **11**:108, 229.

[I.356] *La science mathématique*, Enseign. Sci. **11**:110, 290-296.

**1940**

[I.357] *Les mathématiques dans l'Encyclopédie française*, Mathematica (Cluj) **16**, 1-5.

[I.358] *Les diverses formes et les diverses étapes de l'esprit scientifique*, Thalès **4**, 23-27.

**1942**

[I.359] *The problem of diffusion of waves*, Ann. Math. **43**, 510-522 (Œuvres **3**, 1647-1659).

[I.360] *Le problème de Dirichlet dans le cas hyperbolique*, included in Œuvres **3**, 1661-1662, without reference.

[I.361] *On the Dirichlet problem for the hyperbolic case*, Proc. Nat. Acad. Sci. USA **28**, 258-263 (Œuvres **3**, 1663-1668).

**1943**

[I.362] *Émile Picard*, J. London Math. Soc. **18**, 114-128 (Œuvres **4**, 2043-2057).

[I.363] *La science et le monde moderne*, Renaissance (Revue trimestrielle publiée par l'École Libre des Hautes Études, New York) **1**:4, 523-558.

**1944**

[I.364] *Émile Picard, 1856-1941*, Obit. Not. Roy. Soc. London **4**, 129-140.

[I.365] *Two works on iteration and related questions*, Bull. Amer. Math. Soc. **50**, 67-75.

[I.366] *A known problem of geometry and its cases of indetermination*, Bull. Amer. Math. Soc. **50**, 520-528 (Œuvres **2**, 1001-1009).

[I.367] *An open letter addressed by Professor Jacques Hadamard to the chairman of the American Jewish Joint Distribution Committee and of the United Jewish Appeal*, Pamphlet published by the Union for the Protection of the Human Person and Committee for the Defense of the Rights of Jews of Central and Eastern Europe, Inc., New York, 1-15.

**1945**

[I.368] *Problèmes à apparence difficile*, Mat. Sbornik **17**, 3-8.

[I.369] *Remarques sur le cas parabolique des équations aux dérivées partielles*, Publ. Inst. Math. Univ. Nac. Littoral **5**, 3-11 (Œuvres **3**, 1669-1677).

[I.370] *Notice nécrologique sur George David Birkhoff*, C.R. Acad. Sci. Paris **220**, 719-721.

[I.371] *On the three-cusped hypocycloid*, Math. Gaz. **29**, 66-67 (Œuvres **2**, 1011-1012).

[I.372] *The psychology of invention in the mathematical field*, Princeton Univ. Press, Princeton, N.J.

[I.373] *Subconscient, intuition et logique dans la recherche scientifique*, Conférence faite au Palais de la Découverte le 8 décembre 1945, Ed. Université de Paris, Paris.

**1947**

[I.374] *Observation sur la note de F. Bureau*, C.R. Acad Sci. Paris **225**, 854.

[I.375] *Newton and the infinitesimal calculus*, Roy Soc. Newton Tercentenary Celebrations, July 15-19, 1946, Cambridge Univ. Press, Cambridge, 35-42.

**1948**

[I.376] *Sur le cas anormal du problème de Cauchy pour l'équation des ondes*, Studies and essays presented to R. Courant on his 60th birthday, January 8, Intersci. Publ., New York, 161-165 (Œuvres **3**, 1679-1683).

[I.377] *Le rôle de l'inconscient dans la recherche scientifique*, Atomes, tous les aspects scientifiques d'un nouvel age **26**, 166-169.

[I.378] *Le cinquantenaire de la Ligue. Souvenirs*, Cahiers des Droits de l'Homme, **49-52**, 376-377.

**1949**

[I.379] *An essay on the psychology of invention in the mathematical field* Princeton Univ. Press, Princeton, N.J., revised edition of [I.372] (reprinted by Dover Publications, New York, 1954).

**1950**

[I.380] *Célébration du deuxième centenaire de la naissance de P.S. Laplace*, Arch. Internat. Hist. Sci. **29**, 287-290 (Œuvres **4**, 2059-2062).

**1951**

[I.381] *Quelques résultats accessoires de la théorie des équations aux dérivée partielles*, Dodatek Roczn. Polsk. Towarz. Mat. **22**, 17-21.

[I.382] *Les fonctions de classe supérieure dans l'équation de Volterra*, J. Analyse Math. **1**, 1-10 (Œuvres **1**, 455-464).

[I.383] *Non-Euclidean geometry in the theory of automorphic functions* (in Russian), GTTI, Moscow-Leningrad.

[I.384] *Partial differential equations and functions of real variable* (in Portuguese), Gaz. Mat. (Lisboa), **12**:50, 3-6.

**1952**

[I.385] *Remarque sur la note de Florent Bureau*, C.R. Acad. Sci. Paris **234**, 792.

[I.386] *Are we lacking words ?* Proceedings of the International Congress of Mathematicians, Cambridge, Massachusetts, August 30 - September 6, 1950. AMS, Providence, R.I., 726-727.

[I.387] *Lettre ouverte au professeur Einstein*, Action, **389**, semaine du 14 au 20 mars.

**1953**

[I.388] *Histoire des sciences et psychologie de l'invention*, Actes du 7ème Congrès International d'Histoire des Sciences, Jérusalem, Collection des travaux de l'Académie Internationale d'Histoire des Sciences, **8**, 350-357.

[I.389] *L'origine, l'esprit et le role de la science moderne*, Cahiers Rationalistes, **132**, 1-16.

**1954**

[I.390] *La géométrie non-euclidienne et les définitions axiomatiques*, Acta Math. Acad. Sci. Hung. **5**, suppl., 95-104; Hungarian translation: Magyar Tud. Akad. Mat. Fiz. Oszt. Közlemenyi **3** (1953), 199-208 (Œuvres **4**, 2167-2176).

[I.391] *Histoire de la science et la psychologie de l'invention*, Mathematica **1**, 1-3; Hungarian translation: Mat. Lapok. **9** (1958), 64-66.

[I.392] *Sur des questions d'histoire des sciences: La naissance du calcul infinitésimal*, An. Acad. Brasil. Sci. **26**, 19-23 (Œuvres **4**, 2267-2271).

[I.393] *Le centenaire d'Henri Poincaré*, Rev. Hist. Sci. Appl. **7**, 101-108.

[I.394] *Équations du type parabolique dépourvues de solutions*, J. Rational Mech. Anal. **3**, 3-12 (Œuvres **3**, 1685-1694).

[I.395] *Extension à l'équation de la chaleur d'un théorème de A. Harnack*, Rend. Circ. Mat. Palermo **3**, 337-346 (Œuvres **3**, 1695-1704).

**1955**

[I.396] *Échanges culturels France-Israël*, Amitiés France-Israël, **1**, 9-10.

[I.397] *À propos des humanités classiques*, La Pensée **64**, 78-79.

**1956**

[I.398] *Henri Poincaré et les mathématiques*, in: *Œuvres d'Henri Poincaré*, tome 11, *Livre du centenaire de la naissance d'Henri Poincaré (1854-1954)*, 50-57, Gauthier-Villars, Paris.

[I.399] *L'Affaire Dreyfus*, La Pensée **68**, 77-88.

**1957**

[I.400] *Sur le théorème de A. Harnack*, Publ. Inst. Statist. Univ. Paris **6**, 177-181; reprinted in: Bull. Inst. Politech. Iasi, **3**, 1-6 (Œuvres **3**, 1705-1709).

[I.401] *Un point obscur de plus dans l'affaire Dreyfus*, La Pensée, **71**, 3-4.

**1958**

[I.402] *Soixante ans d'activité au service de la justice*, Cahiers des Droits de l'Homme **4**, 47-49.

**1959**

[I.403] *De la Renaissance à l'époque actuelle: deux conceptions opposées*, in: The Golden Jubilee Commemoration volume, 1958-1959, Calcutta Mat. Soc. **1**, 11-14.

[I.404] *Essai sur la psychologie de l'invention dans le domaine mathématique*, first French edition of [I.372], revised and extended, Albert Blanchard, Paris.

**1964**

[I.405] *La théorie des équations aux dérivées partielles*, Éd. Sci. Pékin.

**1968**

[I.406] *Œuvres de Jacques Hadamard*, vol.1-4, Centre National de la Recherche Scientifique, Paris.

# II. Publications about Jacques Hadamard and his Work

[II.1] Amerio, L., *Jacques Hadamard*, Rend. Ist. Lombardo. Sci. e Lettere, Parte gen. e Atti uffic. **98** (1965), 88-89.

[II.2] Bouligand, J., *Introduction à la pensée créatrice de Jacques Hadamard*, Rev. d'Histoire des Sciences **19** (1966), 247-265.

[II.3] Bouligand, J., *Centenaire de la naissance de Jacques Hadamard*, Rev. Gén. Sci. Pures et Appl. **123**:5-6 (1966), 133-138.

[II.4] Cartwright, M., *Jacques Hadamard*, Biogr. Memoirs of Fellows of the Roy. Soc. **11** (1965), 75-98.

[II.5] *Centenaire de Jacques Hadamard, Mathématicien (1865-1963)*, Plaquette éditée par la Société Amicale des Anciens Élèves de l'École Polytechnique (1966), 1-35; extrait de La Jaune et la Rouge, **204**, mai 1966.

[II.6] Chapelon, J., *Hommage au professeur Jacques Hadamard*, La Pensée, **67** (1956), 106-108.

[II.7] Chatelet, A., *Hommage à Jacques Hadamard*, Le Courrier Rationaliste, **6** (1957), 118-119; reproduced in the same journal, **6** (1964), 167-169.

[II.8] Denjoy, A., *Notices sur les membres décédes: Hadamard*, Association Amicale des Anciens Élèves de l'École Normale Supérieure (1965), 33-35.

[II.9] Desforge, J., Iliovici, G., Robert, P., *L'œuvre de M. Jacques Hadamard et l'enseignement secondaire*, Enseig. Scien. **9** (1936), 97-117.

[II.10] *En hommage à la mémoire de J. Hadamard. Délégations et messages au domicile mortuaire*, L'Humanité, 19 octobre 1963.

[II.11] Fréchet, M., Review of the book of J. Hadamard, *Cours d'analyse professé à l'École Polytechnique*, Rev. Gén. Sci. Pures et Appl. **36** (1925), 547.

[II.12] Fréchet, M., *Jacques Hadamard*, Pensée, no. 112 (1963), 102-104.

[II.13] Fréchet, M., *Notice nécrologique sur Jacques Hadamard*, Acad. Sci. Paris, Notices et Discours, no. 34 (1963), 1-16; an abridged variant in C.R. Acad. Sci. Paris **257**:26 (1963), 4081-4086.

[II.14] Gelfond, A.O., Shnirel'man, L.G., *On the works of Academician Jacques Hadamard in the theory of functions of complex variable and in number theory* (in Russian) Uspehi Mat. Nauk, **2** (1936), 92-117.

[II.15] George, A., *Jacques Hadamard*, Nouvelles Littéraires, octobre 1963.

[II.16] Hardy, G.H., Review of the book of J. Hadamard, *The psychology of invention in the mathematical field*, Math. Gaz., **30** (1946), 111-115; reprinted in Math. Intelligencer **5**:2 (1983), 60-63.

[II.17] Heilbrown, H., Howarth, L., *Jacques Hadamard*, Nature, **200**, (1963), 937-938.

[II.18] *Hommage à Jacques Hadamard*, Nouvelles Littéraires, 2 février 1956.

[II.19] *Hommage italien au mathématicien Jacques Hadamard*, Le Monde, 21 novembre 1965.

[II.20] Itard, J., *Jacques Hadamard (1865-1963)*, Bull. de l'association des professeurs de mathématiques de l'enseignement public, **233** (1963), 107-111.

[II.21] *Jacques Hadamard a reçu hier une grand médaille d'or frappée à son nom*, L'Humanité, 22 décembre 1962.

[II.22] *Jacques Hadamard. Une vie de science et de conscience*, L'Humanité, 18 octobre 1963.

[II.23] *Jacques Hadamard*, France Nouvelle, 22 octobre 1963.

[II.24] *Jacques Hadamard : un grand savant, un homme de progrès*, L'Humanité Dimanche, 20 octobre 1963.

[II.25] *Jacques Hadamard, mathématicien*, Figaro Littéraire, 26 octobre 1963.

[II.26] *Jacques Hadamard n'est plus*, Le Courrier Rationaliste, **11** (1963), 237.

[II.27] *Jubilé scientifique de M. Jacques Hadamard*, Gauthier-Villars, Paris, 1937.

[II.28] Kahane, J.-P., *Un grand savant, un grand témoin de notre temps : Jacques Hadamard*, L'Humanité, 21 décembre 1962.

[II.29] Kahane, J.-P., *Jacques Hadamard*, Math. Intelligencer **13**:1 (1991), 23-29.

[II.30] *La médaille d'or de l'Académie remise vendredi au professeur Hadamard*, L'Humanité, 18 décembre 1962.

[II.31] *La médaille d'or de l'Académie des Sciences au professeur Hadamard*, Le Monde, 19 décembre 1962.

[II.32] *La semaine de la science française en URSS. Les savants de l'URSS saluent leurs confrères français. Jacques Hadamard*, Le Journal de Moscou, mai 1934.

[II.33] *Le professeur Jacques Hadamard a aujourd'hui 90 ans*, L'Humanité, 8 décembre 1955.

[II.34] *Le professeur Jacques Hadamard a 90 ans*, Droit et Liberté, 20 décembre 1955.

[II.35] *L'illustre mathématicien Jacques Hadamard*, L'Humanité, 22 octobre 1963.

[II.36] Lévy, P., *Jacques Hadamard*, La Jaune et la Rouge, janvier 1964, 3-9.

[II.37] Lévy, P., Malgrange, B., Malliavin, P., Mandelbrojt, S., *La vie et l'œuvre de Jacques Hadamard (1865-1963)*, Monographie de l'Enseign. Math. **16** (1967).

[II.38] Maeder, A.M., *Jacques Hadamard* (in Portuguese), Boletim Soc. Paranaense Mat. **7** (1963), 5-7.

[II.39] Malgrange, B., *Jacques Hadamard : un grand savant, un homme de progrès*, L'Humanité Dimanche, 20 octobre 1963.

[II.40] Mandelbrojt, S., *The mathematical work of Jacques Hadamard*, Amer. Math. Month. **60** (1953), 599-603.

[II.41] Mandelbrojt, S., *Jacques Hadamard au Collège de France*, La Jaune et la Rouge, mai 1966, 23-25.

[II.42] Mandelbrojt, S., *Hadamard Jacques*, Dictionary of Scientific Biography, New York, vol. 6, 1972, 3-5.

[II.43] Mandelbrojt, S., Schwartz, L., *Jacques Hadamard*, Bull. Amer. Math. Soc. **71** (1965), 107-129.

[II.44] Margulis, A.Ya., Yushkevich, A.P., *Jacques Hadamard* (in Russian), Mat. v shkole, **2** (1964), 77-80.

[II.45] *Mort hier à l'âge de 98 ans, le mathématicien Jacques Hadamard fut célèbre des son entrée à Polytechnique : jamais un élève ne fut admis avec un plus grand nombre de points*, Liberté, Clermont-Ferrand, 18 octobre 1963.

[II.46] *Mort du mathématicien Jacques Hadamard de l'Académie des Sciences*, Le Figaro, 18 octobre 1963.

[II.47] *Mort du mathématicien Jacques Hadamard, doyen de l'Institut de France*, Le Monde, 19 octobre 1963.

[II.48] *Mort d'un grand mathématicien*, Les Lettres Françaises, 24 octobre 1963.

[II.49] Nicoletis, J., *Souvenirs sur Jacques Hadamard*, La Jaune et la Rouge, janvier 1964, 10-12.

[II.50] Painlevé, P., *Rapport sur le mémoire de M. J. Hadamard*, in *Prix Vaillant*, C.R. Acad. Sci. Paris **145** (1907), 984-986.

[II.51] Petrovskiĭ, I.G., Sobolev, S.L., *Works of Jacques Hadamard on partial differential equations* (in Russian), Uspehi Mat. Nauk, **2** (1936), 82-91.

[II.52] Polyshchuk, E.M., Shaposhnikova, T.O., *Jacques Hadamard*, Nauka, Leningrad, 1990.

[II.53] *Prix Albert $I^{er}$ de Monaco*, C.R. Acad. Sci. Paris **227** (1948), 1302.

[II.54] *Recepción del profesor doctor Jacobo Hadamard el 13 de mayo de 1930*, Anales de la Academia Nacional de Ciencias Exactas, Físicas y Naturales de Buenos Aires **110** (1930), 66-80.

[II.55] *Relazione sul Concorso al Premio Internazionale Antonio Feltrinelli per la Matematica e l'Astronomia per il 1951*, Accademia Nazionale dei Lincei, Rendiconti delle Adunanze solenni **5**:6 (1951), 293-295.

[II.56] *Remise à M. Jacques Hadamard d'une médaille d'or à l'occasion du cinquantième anniversaire de son élection à l'Académie*, Acad. Sci. Paris, Notices et Discours, **4** (1962), 730-739.

[II.57] Renteln, M. von, *Geschichte der Analysis im 20 Jahrhundert von Hilbert bis J.v. Neumann*, Skriptum zur Vorlesung, Universität Karlsruhe, 1987.

[II.58] Rossat-Mignod, S., Rossat-Mignod, A., *Jacques Hadamard*, Les Cahiers Rationalistes, **269** (1969), 306-358.

[II.59] Steklov, V.A., Uspenskiĭ, Ya.V., Ioffe, A.F., *A note on scientific works of Jacques Hadamard* (in Russian), Izv. Rossiĭskoĭ Akad. Nauk, Ser. 6, **16** (1922), 33-37.

[II.60] Shilov, G.E., *Jacques Hadamard and the formation of functional analysis* (in Russian), Uspehi Mat. Nauk **19**:3 (1964), 183-185.

[II.61] Tricomi, F.G., *Commemorazione del Socio straniero Jacques Hadamard*, Accademia Nazionale dei Lincei, Rendiconti della Classe di Scienze fisiche, matematiche e naturali, Ser. 8, **39**:5 (1965), 375-379.

[II.62] Tsortsis, A., *On the occasion of the Jubilee of Jacques Hadamard* (in Greek), Hellenike Mathematike Hetaireia **18** (1938), 203-207.

# III. General Bibliography

[III.1] Abragam, A., *Time reversal. An autobiography*, Clarendon Press, Oxford, 1989.

[III.2] Accadi, L., *Vito Volterra and the development of functional analysis*, Convegno Internazionale in Memoria di Vito Volterra (Roma, 8-11 ottobre 1990), Accademia Nazionale dei Lincei, Roma, 1992, 151-181.

[III.3] Adhemar, R. d' *Sur une classe d'équations aux dérivées partielles de second ordre, du type hyperbolique*, J. Math. Pures et Appl. Sér. 5, **10** (1904), 131-207.

[III.4] Adhemar, R. d' *Sur l'intégration des équations aux dérivées partielles du second ordre du type hyperbolique*, J. Math. Pures et Appl. Sér. 6, **2** (1906), 357-379.

[III.5] Agaian, S.S., *Hadamard matrices and their applications*, Springer-Verlag, Berlin-Heidelberg-New York, 1985.

[III.6] Albers, D.J., Alexanderson, G.L., Reid, C., *International Mathematical Congresses. An illustrated history 1893-1986*, Springer-Verlag, Berlin-Heidelberg-New York, 1987.

[III.7] Albers, D.J., Alexanderson, G.L., Reid, C., (Eds.) *More mathematical people*, Harcourt Brace Jovanovich, Boston-San Diego-New York, 1990.

[III.8] Aleksandrov, V.A., *Embedding of locally Euclidean and conformally Euclidean metrics* (in Russian), Mat. Sbornik **182**:8 (1991), 1105-1117; English translation: Math. USSR Sbornik **73**:2 (1992), 467-478.

[III.9] Almansi, E., *Sull' integrazione dell' equazione differenziale*, Atti Accad. Sci. Torino **31** (1896), 527-534.

[III.10] Andersson, K.G, *Poincaré's discovery of homoclinic points*, Arch. Hist. Exact Sci. **48**:2 (1994), 133-147.

[III.11] Anosov, D.V., *Geodesic flows on closed Riemannian manifolds of negative curvature* (in Russian). Trudy Mat. Inst. Steklov **90** (1967), 3-209; English translation: Proceedings of the Steklov Institute of Mathematics **20** (1967), 3-239.

[III.12] Arnold, V.I., *Mathematical methods of classical mechanics*, Springer-Verlag, New York-Heidelberg-Berlin, 1978.

[III.13] Arnold, V.I., *Singularities of caustics and wave fronts*, Kluwer Academic Publishers, Dordrecht-Boston-London, 1990.

[III.14] Asgeirsson, L., *Some hints in Huygens' principle and Hadamard's conjecture*, Commun. Pure Appl. Math. **9**:3 (1956), 307-237.

[III.15] Atiyah, M.F., Bott, R., Gårding, L., *Lacunas for hyperbolic differential operators with constant coefficients*, I. Acta Math. **124** (1970), 109-189; II. ibid. **131** (1973), 145-206.

[III.16] Babich, V.M., *The Hadamard Ansatz, its analogues, generalizations and applications* (in Russian) Algebra and Analysis, **3**:5 (1991), 1-37; English translation: St. Petersburg Math. J. **3**:5 (1992), 937-972.

[III.17] Babich, V.M., Buldyrev, V.S., and Molotkov, I.A., *Space-time ray method. Linear and non-linear waves* (in Russian), Leningrad Univ. Press, Leningrad, 1985.

[III.18] Bamberger, A., Ha Duong, T., *Formulation variationnelle espace-temps pour le calcul par potentiel retardé de la diffraction d'une onde acoustique*, I. Math. Methods Appl. Sci. **8**:3 (1986), 405-435.

[III.19] Barrow-Green, J., *Poincaré and the three body problem*, AMS, Providence, R.I., 1997.

[III.20] Baumert, L.D., Hall, M., Jr., *Hadamard matrices of the Williamson type*, Math. of Computation **19** (1965), 442-447.

[III.21] Beaulieu, L., *A Parisan café and ten proto-Bourbaki meetings (1934-1935)*, The Math. Intelligencer **15**:1 (1993), 27-35.

[III.22] Beckenbach, E.F., Bellman, R., *Inequalities*, Springer-Verlag, Berlin-Göttingen-Heidelberg, 1961.

[III.23] Begehr, H., Zhenyuan, Xu, *Nonlinear Half-Dirichlet problems for first order elliptic equations in the unit ball of $R^m (m \geq 3)$*, Applicable Analysis **45** (1992), 3-18.

[III.24] Belhoste, B., Dalmedico, A.D., Picon, A., *La formation polytechnicienne 1794-1994*, Dunod, Paris, 1994.

[III.25] Bellman, R., *Introduction to matrix analysis*, Tata McGraw-Hil, New Delhi, 1979.

[III.26] Benzaghou, B., *Sur le quotient de Hadamard de deux fractions rationnelles*, C.R. Acad. Sci. Paris **267** (1968), 212-214.

[III.27] Benzaghou, B., *Algèbres de Hadamard*, Bull. Soc. Math. France, **98** (1970), 209-252.

[III.28] Berest, Yu.Yu., Veselov, A.P., *Huygens' principle and integrability*, Uspekhi Mat, Nauk **49**:6 (1994), 7-78.

[III.29] Berezanskiĭ, Yu. M., *Expansion in eigenfunctions of selfadjoint operators*, Naukova Dumka, Kiev, 1965; English translation: AMS, Providence. R.I. 1965.

[III.30] Berger, M., Gostiaux, B., *Differential geometry: manifolds, curves and surfaces*, Springer-Verlag, New York-Berlin-Heidelberg, 1988.

[III.31] Berger, M.S., Berger, M.S., *Perspectives in nonlinearity*, Benjamin, New York, 1970.

[III.32] Bernstein, S.N., *Sur la nature analytique des solutions des équations aux dérivées partielles du second ordre*, Thèse Fac. Sci. Paris, Teubner, Leipzig, 1904.

[III.33] Beth, Th., Jungnickel, D., Lenz, H., *Design theory*, Bibliographisches Institut, Mannheim-Wien-Zürich, 1985.

[III.34] Beurling, A., *Analyse de la loi asymptotique de la distribution des nombres premiers généralisés* I, Acta Math. **68** (1937), 255-291.

[III.35] Beurling, A., *Ensembles exceptionnels*, Acta Math. **72** (1940), 1-13.

[III.36] Bieberbach, L., *Analytische Fortsetzung*, Springer-Verlag, Berlin, 1955.

[III.37] Birkhoff, G., *Hydrodynamics, A study in logic, fact, and similitude*, Dover, New York, 1955.

[III.38] Birkhoff, G.D., *Dynamical systems*, AMS Publications, Providence, 1927; reprinted 1966.

[III.39] Birkhoff, G., (Ed.) with the assistance of Merzbach, U., *A source book in classical analysis*, Harvard University Press, Cambridge, Massachusetts, 1973.

[III.40] Bishop, R.L., Crittenden, R.J., *Geometry of manifolds*, Academic Press, New York-London, 1964.

[III.41] Blagoveščenskii, A.S., *The characteristic problem for the ultra hyperbolic equation* (in Russian), Mat. Sb. **63** (1964), 137-168.

[III.42] Block, H., *Sur les équations aux derivées partielles du type parabolique*, Arkiv för Matem., Astronomi, Fysik **6**:31 (1911), 1-42.

[III.43] Blum, L., *Souvenirs sur l'Affaire*, Gallimard, Paris, 1936.

[III.44] Boggio, T., *Sull' equilibrio delle piastre elastiche incastrate*, Rend. Accad. Lincei **10** (1901), 201-203.

[III.45] Boggio, T., *Determinazione della deformazione di un corpo elastico per date tensioni superficiali*, Atti Accad. Lincei **7** (1907), 441-449.

[III.46] Bolza, O., *Vorlesungen über Variationsrechnung*, Teubner, Leipzig-Berlin, 1909.

[III.47] Bombieri, E., *On the large sieve*, Mathematika **12** (1965), 201-225.

[III.48] Borel, É., *Démonstration élémentaire d'un théorème de M. Picard sur les fonctions entières*, C.R. Acad. Sci. Paris **122** (1896), 1045-1048 (see also *Œuvres de Émile Borel*, vol. 1, 571-574, CNRS, Paris, 1972).

[III.49] Borel, É., *Sur les zéros des fonctions entières*, Acta Math. **20** (1897), 357-396 (see also *Œuvres de Émile Borel*, vol. 1, 577-616, CNRS, Paris, 1972).

[III.50] Borel, É., *Leçons sur les fonctions méromorphes*, Gauthier-Villars, Paris, 1903.

[III.51] Borel, É., *Remarques sur les principes de la théorie des ensembles*, Math. Ann. **60** (1905), 194-195 (see also *Œuvres de Émile Borel*, vol. 3, CNRS, Paris, 1972).

[III.52] Borel, É., *Documents sur la psychologie de l'invention dans le domaine de la science*, Organon, Varsovie, **1** (1935), 33-42 (see also *Œuvres de Émile Borel*, vol. 4, 2093-2102, CNRS, Paris, 1972).

[III.53] Bottazzini, U., *The higher calculus: a history of real and complex analysis from Euler to Weierstrass*, Springer-Verlag, New York-Berlin-Heidelberg, 1986.

[III.54] Bottazzini, U., *Three traditions in complex analysis: Cauchy, Riemann and Weierstrass*, In: *Companion encyclopedia of the history and philosophy of the mathematical sciences*, Grattan-Guiness, I. (Ed.), Routledge, London-New York, 1994, 419-431.

[III.55] Bottazzini, U., Gray, J.J., *Complex function theory from Zurich (1897) to Zurich (1932)*, Rend. Circolo Matematico di Palermo, Ser. II Supplemento, **44** (1996), 85-111.

[III.56] Bouligand, G., *Sur divers problèmes de la dynamique des liquides*, Gauthier-Villars, Paris, 1930.

[III.57] Bourbaki, N., *Espaces vectoriels topologiques*, Hermann, Paris, 1953.

[III.58] Bourbaki, N., *Éléments d'histoire des mathématiques*, Hermann, Paris, 1960.

[III.59] Bourgin, D.G., Duffin, R., *The Dirichlet problem for the vibrating string equation*, Bull. Amer. Math. Soc. **45** (1939), 851-859.

[III.60] Bradley, S., (Ed.) *Archives biographiques françaises*, Bowker-Saur, London-Paris-Munich-New York, 1988-1990.

[III.61] Browder, F.E., *Nonlinear operators and nonlinear equations of evolution in Banach spaces*, Proc. Symposia in Pure Math. **18** (1976), part 2, AMS, Providence, R.I.

[III.62] Bruce, J.W., Giblin, P.J., *Curves and singularities*, Cambridge University Press, Cambridge, 1992.

[III.63] Bureau, F.J., *Divergent integrals and partial differential equations*, Comm. Pure Appl. Math. **8** (1955), 143-202.

[III.64] Bureau, F.J., *Sur la question du Prix Bordin 1933*, Cahiers du séminaire d'histoire des mathématiques **7** (1986), 1-13.

[III.65] Campbell, D.M., *Beauty and the beast: the strange case of André Bloch*, Math. Intelligencer **7**:4 (1985), 36-38.

[III.66] Cannell, D.M., *George Green. Mathematician and physicist, 1793-1841. The background to his life and work*, The Athlone Press, London and Atlantic Highlands, NJ, 1993.

[III.67] Carathéodory, C., Revue de Hadamard *Leçons sur le calcul des variations*, Bull. Sci. Math. **35** (1911), 124-142.

[III.68] Carleman, T., *Sur un théorème de M. Denjoy*, C.R. Acad. Sci. Paris **174** (1922), 373-376.

[III.69] Carleman, T., *Les fonctions quasi analytiques*, Hermann, Paris, 1926.

[III.70] Cartan, E. – Einstein, A., *Letters on absolute parallelism 1929-1932*, (R. Debever, Ed.), Princeton University Press, 1979.

[III.71] Cartan, H., Ferrand, J., *The case of André Bloch*, Math. Intelligencer **10**:1 (1988), 23-26.

[III.72] Cattaneo, C., *Su un teorema fondamentale nelle teoria delle onde di discontinuità*, Atti Accad. Naz. Lincei, Rend. cl. sci. fis., mat. e natur. **1**:1 (1946), 66-72, **1**:6 (1946), 728-734.

[III.73] Cauchy, A., *Cours d'analyse de l'École Royale Polytechnique. Analyse algébrique*, Imprimerie Royale, Paris, 1821.

[III.74] Cauchy, A., *Mémoire sur diverses formules relatives à la théorie des intégrales définies et sur la conversion des différences finies des puissances en intégrales de cette espèce*, J. École Polytechnique, **18**, 28ème Cahier (1844), 147-248.

[III.75] Cauchy, A., *Sur un nouveau genre d'intégrales : Exercices de mathématiques: 1826. Œuvres complètes*, Gauthier-Villars, Paris, 1887, Vol. 6, p. 78-88.

[III.76] *Chansons nationales*, Dumersan et Segur, Paris, 1851.

[III.77] Chauvin, A., *Louis le Grand, lycée scientifique*, in: *Louis le Grand, 1563-1963. Études. Souvenirs. Documents*, Paris, 1963.

[III.78] Charle, C., Telkes, E., *Les professeurs du Collège de France, Dictionnaire biographique*, Éditions du CNRS, Paris, 1988.

[III.79] Chen, J.T., Hong, H.K., *On the dual integral representation of boundary value problems in Laplace equation*, Boundary Elements Abstracts, **4**:3 (1993), 114-116.

[III.80] Christoffel, E.B., *Untersuchungen über die mit dem Fortbestehen li- nearer partieller Differentialgleichungen verträglichen Unstetigkeiten*, Ann. Math. **8**:2 (1877), 81-112.

[III.81] Christophe, *L'idée fixe du savant Cosinus*, Armand Colin, Paris, 1991.

[III.82] *Chronique*, Archives Israélites de France **6** (1845), 170.

[III.83] Clarke, F.H., *Optimization and nonsmooth analysis*, John Wiley and Sons, New York, 1983.

[III.84] Coffman, C.V., *On the structure of solutions to $\triangle^2 u = \lambda u$ which satisfy the clamped plate conditions on a right angle.* SIAM J. Math. Anal. **13** (1982), 746-757.

[III.85] Coffman, C.V., Duffin, R.J., *On the fundamental eigenfunctions of a clamped punctured disc*, Research Report Nr. 91-101, Dept. of Math., Carnegie Mellon Univ., 1991.

[III.86] *Compte rendu de la Conférence internationale de l'enseignement mathématique*, Paris, 1-4 avril 1914, Enseign. Math. **16** (1914), 174-177, 298-302.

[III.87] Cooke, R., *The mathematics of Sonya Kovalevskaya*, Springer-Verlag, New York-Berlin-Heidelberg, 1984.

[III.88] *Corpus Iuris Civilis*, vol. 2: *Codex Iustinianus*, Apud Weidmannos, Berolini, 1915.

[III.89] *Correspondance d'Hermite et de Stieltjes*, (publiée par les soins de B. Baillaud et H. Bourget), tome 2, Gauthier-Villars, Paris, 1905.

[III.90] Courant, R., *Partial differential equations*, Interscience publishers, New York, 1962.

[III.91] R. Courant, R., Hilbert, D., *Methods of mathematical physics*, vol. 2, Interscience Publishers, New York, 1962.

[III.92] Darboux, G., *Mémoire sur l'approximation des fonctions de très grands nombres et sur une classe étendue de développements en série*, J. Math. Pures Appl. **4**:3 (1878), 5-56.

[III.93] Dauben, J.W., *Mathematicians and World War I: the international diplomacy of G.H. Hardy and Gösta Mittag-Leffler as relected in their personal correspondence*, Historia Mathematica **7** (1980), 261-288.

[III.94] *David Hilbert on Poincaré, Klein, and the world of mathematics*, (translated from the German by D.E. Rowe), Math. Intelligencer **8**:1 (1986), 75-77.

[III.95] Delbourgo, D., Elliott, D., *On the approximate evaluation of Hadamard finite-part integrals*, IMA J. of Numerical Analysis **14** (1994), 485-500.

[III.96] Dell'Agnola, C.A., *Estensione di un teorema di Hadamard*, Ven. Ist. Atti **58** (1899), 525-539, 669-677.

[III.97] Demidov, S.S., *Création et développement de la théorie des équations différentielles aux dérivées partielles dans les travaux de J. d'Alembert*, Rev. His. Sci. **35**:1 (1982), 3-42.

[III.98] Demidov, S.S., *The study of partial differential equations of the first order in the 18th and 19th centuries*, Arch. for Hist. Ex. Sci. **26** (1982), 325-350.

[III.99] *Den exakta forskningen i Sverige. Fyra uttalanden för Svenska Dagbladets jubileumsenquête*, Svenska Dagbladet, October 27, 1939.

[III.100] Denjoy, A., *Sur les fonctions quasi-analytiques de variable réelle*, C.R. Acad. Sci. Paris **173** (1921), 1329-1331.

[III.101] Desplanques, J., *Théorème d'algèbre*, J. de Math. Spec. **9** (1887), 12-13.

[III.102] *Déterminisme et causalité dans la physique contemporaine, C.R. de séance du 12 novembre 1929*, Bull. de la Société française de philosophie **29** (1929), 141-160.

[III.103] Diacu, F., Holmes, Ph.,*Celestial encounters. The origins of chaos and stability.*, Princeton University Press, Princeton, New Jersey, 1996.

[III.104] Dieudonné, J., *The work of Nicolas Bourbaki*, Amer. Math. Monthly **77** (1970), 134-145.

[III.105] Dieudonné, J., *Abrégé d'histoire des mathématiques: 1700-1900*, vol. 1,2, Hermann, Paris, 1978.

[III.106] Dirichlet, P.C.L., *Beweis des Satzes, dass jede unbegrenzte arithmetische Progression, deren erstes Glied und Differenz ganze Zahlen ohne gemeinschaftlichen Faktor sind, unendlich viele Primzahlen enthält*, Abh. Akad. Berlin. Math. Abh. 1837-1839, 45-71.

[III.107] *Discours prononcé à la distribution solennelle des prix le 2 août 1878*, Donnaud, Paris, 1878.

[III.108] Domar, Y., *On the foundation of Acta Mathematica*, Acta Math. **148** (1982), 3-8.

[III.109] Douglis, A., *A criterion for the validity of Huygens' principle*, Commun. Pure Appl. Math. **9**:3 (1956), 391-402.

[III.110] Du Bois-Reymond, P., *Die allgemeine Funktionentheorie*, Verlag der H. Laupp'schen Buchhandlung, Tübingen, 1882.

[III.111] Du Bois-Reymond, P., *Über lineare partielle Differentialgleichungen zweiter Ordnung*, J. reine und angew. Math. **104** (1889), 241-301.

[III.112] Duffin, R.J., *On a question of Hadamard concerning superbiharmonic functions*, J. Math. and Phys. **27**:1 (1949), 253-258.

[III.113] Duffin, R.J., *Some problems of mathematics and science*, Bull. Amer. Math. Soc. **80**:6 (1974), 1053-1070.

[III.114] Duhem, P., *Hydrodynamique, élasticité, acoustique*, Hermann, Paris, 1891.

[III.115] Duhem, P., *Recherches sur l'élasticité. Troisième partie: la stabilité des milieux élastiques*, Ann. École Norm. Sup. **20**:3 (1905), 143-217.

[III.116] Duhem, P., *La théorie physique: son objet, sa structure*, Marcel Rivière et Cie, Paris, 1906; English translation: *The aim and structure of physical theory*, Atheneum, New York, 1974.

[III.117] Duistermaat, J.J., *Oscillatory integrals, Lagrange immersions and unfolding of singularities*, Comm. Pure Appl. Math. **27** (1974), 207-281.

[III.118] Edwards, H.M., *Riemann's zeta function*, Academic Press, New York-London, 1974.

[III.119] Egorov, D.F., *Letters to N.N. Luzin*, (in Russian), published by F.A. Medvedev and A.P. Yushkevich, Istor. Mat. Issled. **25** (1980), 335-361.

[III.120] Einstein, A., *Œuvres choisies*, tome 4: *Correspondances françaises*, Éditions du Seuil and Éditions du CNRS, Paris,1989.

[III.121] Ellison, W., Ellison, F., *Prime numbers*, John Wiley and Sons, New York-London-Sydney-Toronto and Hermann, Paris, 1985.

[III.122] Engelsman, S.B., *Lagrange's early contributions to the theory of first order partial differential equations*, Hist. Math. **7** (1980), 7-23.

[III.123] Engliš, M., Peetre, J., *Green's function for the annulus*, Annali di Matematica, **171**, 313-377.

[III.124] Erdös, P., *On a new method in elementary number theory which leads to an elementary proof of the prime number theorem*, Proc. Nat. Acad. Sci. (Washington) **35** (1949), 374-384.

[III.125] Euler, L., *Découverte d'une loi tout extraordinaire des nombres, par rapport à la somme de leurs diviseurs* (Presented to Berlin Acad. of Sci. on June 22, 1747), Comment. Arithm. Collect. Petropoli **2** (1849), 639-647.

[III.126] Euler, L., *Correspondance de Léonard Euler avec A.C. Clairaut, J. d'Alembert et J.L. Lagrange*, in: Euler, L., *Opera Omnia*, vol. 5, ser. 4, 1980.

[III.127] Fabry, E., *Sur les points singuliers d'une fonction donnée par son développement de Taylor*, Ann. Éc. Norm. Sup. Paris **13**:3 (1896), 367-399.

[III.128] Fantappiè, L., *I funzionali analitici*, Atti Accad. dei Lincei **3**:6 (1928-1929), 453-683.

[III.129] Fattorini, H.O., *The Cauchy problem*, Addison-Wesley, London-Amsterdam, 1983.

[III.130] Fichera, G., *Linear elliptic differential systems and eigenvalue problems*, Springer-Verlag, Berlin,1965.

[III.131] Fichera, G., *Sulla propagazione delle onde in un mezzo elastico*, Proc. Symp. on continuum mechanics and related problems of analysis (Tbilisi, September 23-29, 1971), Mezniereba, Tbilisi, 1973, vol. 2, 567-574.

[III.132] Fichera, G., *Vito Volterra and the birth of functional analysis*, in: Pier, J.-P., (Ed.) *Development of mathematics 1900-1950*, Birkhäuser Verlag, Basel-Boston-Berlin, 1994, 171-183.

[III.133] *Foreign scientists on their impressions* (in Russian), Vestnik Akad. Nauk SSSR, **7** (1945), 149-152.

[III.134] Frantz, E.G., (Ed.) *Die Chronik Hessens*, Harenberg Verlag, Dortmund, 1991.

[III.135] Fréchet, M., *Sur la notion de différentielle*, Journal de Mathématiques Pures et Appliquées, **16** (1937), 233-250.

[III.136] Fréchet, M., *Les principes de la théorie des probabilités. Second livre: Méthode des fonctions arbitraires. Théorie des événements en chaine dans le cas d'un nombre fini d'états possibles*, Gauthier-Villars, Paris, 1938.

[III.137] Fredholm, I., *Œuvres complètes*, Litos reprotryck, Malmö, 1955.

[III.138] Fredholm, I., *Sur une nouvelle méthode pour la résolution du problème de Dirichlet*, Öfversigt af Kongl. Vetenskaps-Akademiens Förhandlingar **1** (1900), 39-46 (see also *Œuvres complètes*, 61-68).

[III.139] Freudental, H., *The cradle of modern topology according to Brouwer's inedita*, Hist. Math. **2** (1975), 495-502.

[III.140] Fujiwara, D., Ozawa, S., *The Hadamard variational formula for the Green functions of some normal elliptic boundary value problems*, Proc. Japan Acad. **54**, Ser. A (1978), 215-220.

[III.141] Gagliardo, E., *Ulteriori proprietà di alcune classi di funzioni in più variabili*, Ric. Mat. **8**:1 (1959), 24-51.

[III.142] Gantmacher, F.G., *The theory of matrices*, vol.1, 2, Chelsea Publishing

Company, New York, 1977.

[III.143] Garabedian, P.R., *A partial differential equation arising in conformal mapping*, Pacific J. Math. **1** (1951), 485-524.

[III.144] Garipov, R.M., *On the linear theory of gravity waves: the theorem of existence and uniqueness*, Arch. Rat. Mech. Anal. **24**:5 (1967), 352-362.

[III.145] Gâteaux, R., *Sur la notion d'intégrale dans le domaine fonctionnel et sur la théorie du potentiel*, Bull. Soc. Math. France **97** (1919), 47-70.

[III.146] Gâteaux, R., *Fonctions d'une infinité de variables indépendantes*, Bull. Soc. Math. France **97** (1919), 70-96.

[III.147] Gauss, C.F., *Werke*, Leipzig, Bd.2, 1863.

[III.148] Giaquinta, M., Souček, J., *Caccioppoli's inequality and Legendre-Hadamard condition*, Math. Ann. **270** (1985), 105-107.

[III.149] Gispert, H., *La France mathématiques*, Société Française d'Histoire des Sciences et des Techniques. Société Mathématique de France, Paris, 1991.

[III.150] Glivenko, V.I., *The notion of differential according to Marx and Hadamard* (in Russian), Pod znamenem marksizma **5** (1934), 79-85.

[III.151] Goldstein, L.J., *A history of the prime number theorem*, Amer. Math. Monthly **80** (1973), 599-615.

[III.152] Goldstine, H.H., *A history of the calculus of variations from the 17th through the 19th century*, Springer-Verlag, New York-Heidelberg-Berlin, 1980.

[III.153] Goncourt, E. de, Goncourt, J. de, *Journal*, tome 9, Fasquelle et Flammarion, Monaco, 1956.

[III.154] Goursat, E., *A course in mathematical analysis*, Athenaeum Press, Boston, Mass., vol.1 1904, vol.2 1916, vol.3 1917.

[III.155] Gragg, W. B., *On Hadamard's theory of polar singularities*, in: Graves-Morris, P.R., (Ed.) *Padé approximants and their applications*, Proc. of the Conference held at the University of Kent 17-21 July, 1972, Academic Press, London and New York, 1973.

[III.156] Gray, J.D., *The shaping of the Riesz representation theorem: a chapter in the history of analysis*, Arch. Hist. Exact Sci. **31** (1984), 125-187.

[III.157] Gray, J.J., *Linear differential equations and group theory from Riemann to Poincaré*, Birkhäuser Verlag, Boston-Basel-Stuttgart, 1986.

[III.158] Gray, J.J., *Green and Green's functions*, Math. Intelligencer **16**:1 (1994), 45-47.

[III.159] Grunau, H.-Ch., Sweers, G., *Positivity for perturbations of polyharmonic operators with Dirichlet boundary conditions in two dimensions*, Math. Nachr. **179** (1996), 89-102.

[III.160] Günther, P., *Zur Gültigkeit des Huygensschen Prinzips bei partiellen Differentialgleichungen vom normalen hyperbolischen Typ*, Ber. Verh. Sächs. Akad. Wiss. Leipzig. Math.-Naturwiss. Kl. **100** (1952), H.2.

[III.161] Günther, P., *Huygens' principle and hyperbolic equations*, Academic Press, Boston, 1988.

[III.162] Günther, P., *Huygens' principle and Hadamard's conjecture*, Math. Intelligencer **13**:2 (1991), 56-63.

[III.163] Halmos, P.R., *I want to be a mathematician*, Springer-Verlag, New York-Berlin-Heidelberg-Tokyo, 1985.

[III.164] Hamel, G., *Über die Geometrien in denen die Geraden die Kürzesten sind*,

Göttingen, 1901.

[III.165] Hardy, G.H., *Sur les zéros de la foncton $\zeta(s)$ de Riemann*, C.R. Acad. Sci. Paris **158** (1914), 1012-1014.

[III.166] Hardy, G.H., Littlewood, J.E., *Contributions to the arithmetic theory of series*, Proc. Lond. Math. Soc. Ser. 2, **11** (1912-1913), 411-478.

[III.167] Hardy, G.H., Littlewood, J.E., Pólya, G., *Inequalities*, Cambridge University Press, London, 1934.

[III.168] Harwit, M., Sloane, N.J.A., *Hadamard transform optics*, Academic Press, New York, 1979.

[III.169] Hawkins, T., *Lebesgue's theory of integration. Its origins and development*, The Univ. of Wisconsin Press, Madison, Milwaukee, and London, 1970.

[III.170] Hayman, W.K., Korenblum, B., *Representation and uniqueness theorems for polyharmonic functions*, J. Anal. Math. **60** (1993), 113-133

[III.171] Hedayat, A., Wallis, W.D., *Hadamard matrices and their applications*, Annals of Statistics **6**:6 (1978), 1184-1238.

[III.172] Hedenmalm, H., *A computation of Green functions for the weighted biharmonic operators $\Delta|z|^{-2\alpha}\Delta$ with $\alpha > -1$*, Duke Math. J. **75**:1 (1994), 51-78.

[III.173] Heim, R., *Exposé sur la nécessité de protéger certaines ressources en voie de disparition*, C.R. Acad. Sci. Paris, Vie Académique, **265** (1967), 140-144.

[III.174] Hermite, C., *Estrait d'une lettre de M. Ch. Hermite à M. Mittag-Leffler sur quelques points de la théorie des fonctions*, J. für Mathem. **91** (1881), 53-78.

[III.175] Herz, H., *Die Prinzipien der Mechanik in neuen Zusammenhängen dargestellt*, Ges. Werke, Leipzig, Bd. 3, 1894-1895.

[III.176] Hilbert, D., *Sur les problèmes futurs des mathématiques*, Comptes Rendus du deuxième Congrès international des mathématiciens tenu à Paris du 6 au 12 août 1900, Gauthier-Villars, Paris, 1902, 58-114.

[III.177] Holland, A.S.B., *Introduction to the theory of entire functions*, Academic Press, New York-London, 1973.

[III.178] Holmgren, E., *Über Systeme von linearen partiellen Differentialgleichungen*, Öfversigt af Kongl. Vetenskapsakad. förh. **58** (1901), 91-105.

[III.179] Holmgren, E., *Sur l'extension de la méthode d'intégration de Riemann*, Ark. Mat., Astron., Fys. **1**:22 (1904), 317-326.

[III.180] Hörmander, L., *On the theory of general partial differential operators*, Acta Math. **54** (1955), 161-248.

[III.181] Horn, R.A., Johnson, C.R., *Topics in matrix analysis*, Cambrige University Press, Cambridge, 1991.

[III.182] Huber, A., *Die erste Randwertaufgabe für geschlossene Bereiche bei der Gleichung $\partial^2 z/\partial x \partial y = f(x, y)$*, Monatshefte Math. Phys. **39** (1932), 79-100.

[III.183] Hugoniot, H., *Mémoire sur la propagation du mouvement dans les corps et spécialement dans les gaz parfaits*, J. Éc. Polytechn. **57** (1887), 1-97; **58** (1889), 1-125.

[III.184] Hunt, L.R., Su, R., Meyer, G., *Global transormations of nonlinear systems*, IEEE Transactions on Automatic Control **28**:1 (1983), 24-30.

[III.185] Hurwitz, A., *Sur un théorème de M. Hadamard*. C.R. Acad. Sci. Paris **128** (1899), 350-353.

[III.186] Ibragimov, N.H., *Conformal invariance and Huygens' principle*, Dokl.

Akad. Nauk SSSR **194**:1 (1970), 24-27; English translation: Soviet Math. Dokl. **11**:5 (1970), 1153-1157.

[III.187] Ibragimov, N.H., *Huygens' principle* (in Russian), in: *Certain problems of mathematics and mechanics*, Nauka, Leningrad, 1970, p.159-170.

[III.188] Ibragimov, N.H., *Transformation groups applied to mathematical physics*, Reidel, Dordrecht, 1985.

[III.189] Ibragimov, N.H., Mamontov, E.V., *Sur le problème de J. Hadamard relatif à la diffusion des ondes*, C.R. Acad. Sci. Paris **270** (1970), 456-458.

[III.190] Ikehara, S., *An extension of Landau's theorem in the analytical theory of numbers*, J. Math. and Phys. **10** (1931), 1-12.

[III.191] Infeld, L., *My reminiscences about Einstein* (in Polish), Twórczość **9** (1955), 41-85.

[III.192] Ingham, A.E., *The distribution of prime numbers*, Cambridge University Press, Cambridge, 1932.

[III.193] Jacobi, C.G.J., *Vorlesungen über Dynamik*, Reimer, Berlin, 1866.

[III.194] Jaki, S.L., *Uneasy genius: the life and work of Pierre Duhem*, Martinus Nijhoff, Dordrecht-Boston-Lancaster, 1987.

[III.195] Jeannin, P., *Deux siècles à Normale Sup. Petite histoire d'une Grand École*, Larousse, Paris, 1994.

[III.196] Jensen, J.L.W., *Sur un nouvel et important théorème de la théorie des fonctions*, Acta Math. **21** (1897), 359-365.

[III.197] Jentsch, R., *Über Integralgleichungen mit positivem Kern*. J. für die reine und angew. Math. **141** (1912), 235-244.

[III.198] John, F., *The Dirichlet problem for a hyperbolic equation*, Amer. J. Math., **63**:1 (1941), 141-154.

[III.199] John, F., *On quasi-isometric mappings*, I, Comm. Pure and Appl. Math. **21** (1968), 77-110.

[III.200] Johnson, D.M., *The problem of the invariance of dimension in the growth of modern topology*, II, Archive for History of Exact Sciences **25** (1981), 85-266.

[III.201] Julia, G., *Variation de la fonction qui fournit la représentation conforme d'une aire sur un cercle, lorsque le contour de l'aire varie*, C.R. Acad. Sci. Paris **172** (1921), 568-570.

[III.202] Julia, G., *Deux conséquences de l'équation aux dérivées fonctionnelles qu'on tire de la représentation conforme*, C.R. Acad. Sci. Paris **172** (1921), 738-741.

[III.203] Julia, G., *Sur une équation aux dérivées fonctionnelles analogue à l'équation de M. Hadamard*, C.R. Acad. Sci. Paris **172** (1921), 831-833.

[III.204] Julia, G., *Les séries d'itérations et les fonctions quasi analytiques*, C.R. Acad. Sci. Paris **180** (1925), 720-723.

[III.205] Julia, G., *Sur un type de fonctions quasi analytiques*, C.R. Acad. Sci. Paris **180** (1925), 1150-1153.

[III.206] Julia, G., *Fonctions quasi analytiques et fonctions entières d'ordre nul*, C.R. Acad. Sci. Paris **180** (1925), 1240-1242.

[III.207] Kahane, J.-P., *Des séries de Taylor au mouvement brownien, avec un aperçu sur le retour*, in: *Development of Mathematics 1900–1950*, Jean-Paul Pier (Ed.), Birkhäuser Verlag, Basel-Boston-Berlin, 1994, 415-429.

[III.208] Kantorovich, L.V., *My journey in science* (in Russian), Uspekhi Mat. Nauk **42**:2 (1987), 183-213. English translation: Russian Math. Surveys **42**:2 (1987), 233-270.

[III.209] Kennedy, H.C., *Karl Marx and the foundations of differential calculus*, Istoriko-matematicheskie Issledovania **26** (1982), 17-39. Congress of Mathematicians, Cambridge, Massachusetts, U.S.A., August 30-September 6, 1950; AMS, Providence, R.I., 1952, p.121-123.

[III.210] Kneser, A., *Studien über die Bewegungsvorgänge in der Umgebung instabiler Gleichgewichtslagen*, J. für die reine und angew. Math. **118** (1897), 186-223.

[III.211] Kneser, A., *Lehrbuch der Variationsrechnung*, Braunschweig, 1900.

[III.212] Koch, H. von, *Sur la distribution des nombres premiers*, Acta Math. **24** (1901), 159-182.

[III.213] Kolmogorov, A.N., *Une généralisation de l'inégalité de M. J. Hadamard entre les bornes supérieures des dérivées successives d'une fonction*, C.R. Acad. Sci. Paris **207** (1938), 764-765.

[III.214] Kolmogorov, A.N., *On inequalities between least upper bounds of sequences of derivatives of an arbitrary function on an infinite interval*, (in Russian), Uch. Zap. MGU, Matematika **3**:3 (1939), 3-16; English translation in: V.M. Tikhomirov (Ed.), *Selected works of A.N. Kolmogorov*, vol. 1, Kluwer, Dordrecht-Boston-London, 1991, p.277-290.

[III.215] Kolmogorov, A.N., *On some asymptotic characteristics of totally bounded metric spaces* (in Russian), Dokl. Akad. Nauk SSSR **108**:2 (1956), 179-182.

[III.216] Kolmogorov, A.N., *Selected works* (V.M. Tikhomirov, Ed.), vol. 1, Kluwer, Dordrecht-Boston-London, 1991.

[III.217] Kőnig, G., *Ein allgemeiner Ausdruck für die ihrem absoluten Betrage nach kleinste Wurzel der Gleichung n-ten Grades*, Math. Annalen **9** (1876), 530-540.

[III.218] Kozlov, V.A., Kondrat'ev, V.A. and Maz'ya, V.G., *On sign variability and the absence of "strong" zeros of solutions of elliptic equations* (in Russian), Izv. Akad. Nauk SSSR Ser. Mat. **53**:2 (1989), 328-344; English translation: Math. USSR Izv. **34**:2 (1990), 337-353.

[III.219] Kronecker, L., *Zur Theorie der Elimination einer Variabeln aus zwei algebraische Gleichungen*, Berlin Ber. (1881), 535-600.

[III.220] *La création artistique. C.R. de séance du 28 janvier 1928*, Bull. de la Société française de Philosophie **28** (1928), 1-23.

[III.221] Lamb, H., *Hydrodynamics*, Cambridge University Press, London, (Sixth edition), 1932.

[III.222] Landau, E., *Über die Multiplikation Dirichlet'scher Reihen*, Rend. Circolo Mat. Palermo **24** (1907), 81-160.

[III.223] Landau, E., *Handbuch der Lehre von der Verteilung der Primzahlen*, Bd. 1, 2, Teubner, Leipzig, 1909.

[III.224] Landau, E., *Einige Ungleichungen für zweimal differenzierbare Funktionen*, Proc. London Math. Soc. **13** (1913), 43-49.

[III.225] Landau, E., *Vorlesungen über Zahlentheorie*, Bd. 1, 2, Verlag von S. Hirzel, Leipzig, 1927.

[III.226] Landis, E.M., Oleinik, O.A., *Generalized analyticity and some related properties of solutions of elliptic and parabolic equations* (in Russian), Uspehi Mat. Nauk **29**:2 (1974), 190-206.

[III.227]  *La théorie de la relativité. C.R. de séance du 6 avril 1922*, Bull. de la Société française de Philosophie **17** (1922), 91-113.

[III.228]  Lauricella, G., *Integrazione dell' equazione* $\triangle^2(\triangle^2 u) = 0$ *in un campo di forma circolare*, Atti Acad. Sci. Torino **31** (1896), 610-618.

[III.229]  Lebesgue, H., *Sur la définition de l'aire d'une surface*, C.R. Acad. Sci. Paris **129** (1899), 870-873.

[III.230]  Lecornu, L., *Sur les séries entières*, C.R. Acad. Sci. Paris, **104** (1887), 349-352.

[III.231]  Legendre, A.M., *Recherches d'analyse indéterminée*, Histoire de l'Académie Royale des Sciences, Année 1785, avec les Mémoires de Mathématiques et de Physique pour la même année, Imprimerie Royale, Paris, 1788, 465-559.

[III.232]  Legendre, A.M., *Essai sur la théorie des nombres*, Courcier, Paris, 1808.

[III.233]  *Le lycée de Versailles*, Revue de l'Histoire de Versailles et de Seine-et-Oise, Léon Bernard, Versailles, 1911.

[III.234]  *Les chevaux savants d'Elberfeld. C.R. de séance du 13 mars 1913*, Bull. de la Société française de Philosophie **13** (1913).

[III.235]  *Letter of N.N. Luzin to O.Ju. Schmidt*, (in Russian), Publication and preface of S.S. Demidov, Ist. Mat. Issled. **28** (1985), 278-285.

[III.236]  *Letters of D.F. Egorov to N.N. Luzin*, (in Russian), Publication of F.A. Medvedev in collaboration with A.P. Youschkevitch. Preface of P.S. Alexandrov, Ist. Mat. Issled. **25** (1980), 335-361.

[III.237]  Lévy, L., *Sur la possibilité de l'équilibre électrique*, C.R. Acad. Sci. Paris **93** (1881), 706-708.

[III.238]  Lévy, M., *Mémoire sur la théorie des plaques élastiques planes*, J. Math. Pures et Appl. Sér. 3, **3** (1877), 219-307.

[III.239]  Lévy, P., *Leçons d'analyse fonctionnelle*, Gauthier-Villars, Paris, 1922.

[III.240]  Lévy, P., *Problèmes concrets d'analyse fonctionnelle*, Gauthier-Villars, Paris, 1951.

[III.241]  Lévy, P., *Quelques aspects de la pensée d'un mathématicien*, Albert Blanchard, Paris, 1970.

[III.242]  Lévy, P., *Les noms des Israélites en France, histoire et dictionnaire*, Paris, 1960.

[III.243]  Lewy, H., *An example of a smooth linear partial differential equation without solution*, Ann. Mat. **66** (1957), 155-158.

[III.244]  Li, Y., McLaughlin, D.W., Shatah, J., Wiggins, S., *Persistent homoclinic orbits for a perturbed nonlinear Schrödinger equation*, Comm. Pure Appl. Math. **49**:11 (1996), 1175-1255.

[III.245]  Littlewood, J.E., *Sur la distribution des nombres premiers*, C.R. Acad. Sci. Paris **158** (1914), 1869-1872.

[III.246]  Littlewood, J.E., *A mathematician's miscellany*, Methuen, London, 1953.

[III.247]  Loewner, Ch., *On generation of solution of biharmonic equations in the plane by conformal maping*, Pacific J. Math. **3** (1953), 417-436.

[III.248]  Lorch, L., *Review of the book* by D.J. Albers, G.L. Alexanderson, C. Reid, *International Mathematical Congresses. An illustrated history, 1893-*

*1986*, Math. Intelligencer **10**:1 (1988), 65-69.

[III.249] Lützen, J., *The prehistory of the theory of distributions*, Springer-Verlag, New York-Heidelberg-Berlin, 1982.

[III.250] Lützen, J., *Joseph Liouville, 1809-1882, master of pure and applied mathematics*, Springer-Verlag, New York-Berlin-Heidelberg, 1990.

[III.251] Lützen, J., *Partial differential equations*. in: *Companion encyclopedia of the history and philosophy of the mathematical sciences*, I. Grattan-Guiness (Ed.), vol. 2, 1994, London-New York, 452-469.

[III.252] Lützen, J., *Interaction between mechanics and differential geometry in the 19th century*, Arch. Hist. Exact Sci. **49**:1 (1995), 1-72.

[III.253] Luzin, N., *Sur le problème de M. J. Hadamard d'uniformisation des ensembles*, C.R. Acad. Sci. Paris **190** (1930), 349-351.

[III.254] Luzin, N.N., *A letter to O.Yu. Schmidt* (in Russian), published by S.S. Demidov, Istor. Mat. Issled. **28** (1985), 278-285.

[III.255] *Lycée Charlemagne. Distribution des prix*, Seringe frères, Paris, 1874.

[III.256] *Lycée de Louis-le-Grand. Distribution des prix faite aux élèves du grand et du moyen collège le 7 aout 1877*, Donnaud, Paris, 1877.

[III.257] *Lycée de Louis-le-Grand. Distribution des prix faite aux élèves du grand et du moyen collège le 6 aout 1878*, Donnaud, Paris, 1978.

[III.258] *Lycée de Louis-le-Grand. Distribution des prix faite aux élèves du grand et du moyen collège le 5 aout 1879*, Donnaud, Paris, 1879.

[III.259] *Lycée de Louis-le-Grand. Distribution des prix*, Donnaud, Paris, 1880-1884.

[III.260] Lyusternik, L.A., *The early years of the Moscow mathematical school. I, II, III, IV* (in Russian), Uspehi Mat. Nauk **22**:1 (1967), 137-161, **22**:2 (1967), 199-239, **22**:4 (1967), 147-185, **25**:4 (1970), 189-196; English translation: Russian Math. Surveys **22**:1 (1967), 133-157, **22**:2 (1967), 171-211, **22**:4 (1967), 55-91, **25**:4 (1970), 167-174.

[III.261] MacWilliams, F.J., Sloane, N.J.A., *The theory of error-correcting codes*, North-Holland, Amsterdam, 1993.

[III.262] Malyshev, V.A., *Hadamard's conjecture and estimates of the Green function* (in Russian), Algebra and Analysis, **4**:4 (1992), 1-45; English translation: St. Petersburg Math. J. **4**:4 (1993), 633-666.

[III.263] Mandelbrojt, S., *Séries adhérentes, régularisation des suites, applications*, Gauthier-Villars, Paris, 1952.

[III.264] Mandelbrojt, S., *Selecta*, Gauthier-Villars, Paris, 1981.

[III.265] Mandelbrojt, S., *Souvenirs à bâtons rompus de Szolem Mandelbrojt recueillis en 1970 et préparés par Benoît Mandelbrot*, Cahiers du Séminaire d'Histoire des Mathématiques **6** (1985), 1-46.

[III.266] Mangoldt, J. von, *Zu Riemann's Abhandlung "Über die Anzahl..."*, J. reine angew. Math. **114** (1895), 255-305.

[III.267] Marguet, *Notice nécrologique sur Amédée Hadamard*, Association amicale des anciens élèves du lycée Charlemagne. Avril 1889, Seringe frères et Noailles, Paris, 1889.

[III.268] Markushevich, A.I., *Theory of functions of a complex variable*, vol. I, II, Prentice-Hall, Inc., Englewood Cliffs, N.J., 1965.

[III.269] Marx, K., *Mathematische Manuskripte*, Scriptor-Verlag, 1974.

[III.270] Mathieu, E.L., *Sur le mouvement vibratoire des plaques*, J. Math. Pures et Appl. Sér. 2, **14** (1869), 241-259.

[III.271] Mathisson, M., *Eine neue Lösungsmethode für Differentialgleichungen von normalem hyperbolischen Typus*, Math. Ann. **107** (1932), 400-419.

[III.272] Mathisson, M., *Le problème de M. Hadamard relatif à la diffusion des ondes*, Acta Math. **70** (1939), 249-282.

[III.273] Mazja, W.G., Nasarow, S.A. Plamenewski, B.A., *Asymptotische Theorie elliptischer Randwertaufgaben in singulär gestörten Gebieten*, Akademie-Verlag, Berlin, Bd. I, II. 1991.

[III.274] Menshov, D.E., *Impressions sur mon voyage à Paris en 1927*, Cahiers du Séminaire d'Histoire des Mathématiques **6** (1985), 55-59.

[III.275] Michell, J.H., *The flexture of circular plates*, Proc. Math. Soc. London **34** (1901), 223-228.

[III.276] Miller, D.G., *Pierre-Maurice-Marie Duhem*, Dictionary of scientific biography, New York, vol. 4, 1971, 225-233.

[III.277] Minkowski, H., *Zur Theorie der Einheiten in den algebraischen Zahlkörpern*, Nachr. Königlichen Ges. Wiss. Göttingen Math. Phys. Kl., 90-93, Gesammelte Abh. 1 (1900), 316-319.

[III.278] Mikhlin, S.G., *On uniform convergence of series of analytic functions*, Mat. Sbornik **39**:3 (1932), 88-96.

[III.279] Mirsky, L., *On a generalization of Hadamard's inequality due to Szasz*, Ark. Mat. **8** (1957), 274-275.

[III.280] Mitrinović, D.S., Lacković, I.B., *Hermite and convexity*, Aequationes Mathematicae **28** (1985), 229-232.

[III.281] Mittag-Leffler, G., *Une page de la vie de Weierstrass*, Comptes Rendus du Deuxième Congrès International des Mathématiciens tenu à Paris du 6 au 12 août 1900, Gauthier-Villars, Paris, 1902, 131-153.

[III.282] Mittag-Leffler, G., *Om en generalisering av potensserien*, Övers. Vet. Akad. Stockholm **55** (1898), 135-138.

[III.283] Moore, G.H., *Zermelo's axiom of choice. Its origins, development, and influence*, Springer-Verlag, New York-Heidelberg-Berlin, 1982.

[III.284] Mordell, L.J., *Reminiscences of an octogenarian mathematician*, Amer. Math. Monthly **78**:6 (1971), 952-961.

[III.285] Morrey, C.B., Jr., *Multiple integrals in the calculus of variations*, Springer, Berlin-Heidelberg-New York, 1966.

[III.286] Muir, A., *The psychology of mathematical creativity*, Math. Intelligencer, **10**:1 (1988), 33-37.

[III.287] Muir, T., *Hadamard's approximation theorem since 1900*, Trans. of the Royal Soc. of South Africa **13** (1926), 299-308.

[III.288] Muir, T., *Contributions to the history of determinants* 1900–1920, Blackie, London, 1930.

[III.289] Nadirashvili, N.S., *Rayleigh's conjecture on the principal frequency of the clamped plate*, Arch. Rational Mech. Anal. **129** (1995), 1-10.

[III.290] Nadirashvili, N.S., *Hadamard's and Calabi-Yau's conjectures on negatively curved and minimal surfaces*, Invent. math. **126** (1996), 457-465.

[III.291] Nathan, O., Norden, H., (Eds.), *Einstein on peace*, Simon and Schuster, New York, 1960.

[III.292]  *Nécrologie. Notice sur Madame Rebecca Hadamard, née Lambert, de Metz*, Archives Israélites de France **4** (1843), 220-223.

[III.293]  Neumann, C., *Untersuchungen über das logarithmische und Newton'sche Potential*, Teubner, Leipzig, 1877.

[III.294]  Nirenberg, L., *On elliptic partial differential equations: Lecture II*, Ann. Sc. Norm. Sup. Pisa, Ser. 3, **13** (1959), 115-162.

[III.295]  Nirenberg, L., *Partial differential equations in the first half of the century*, in: Pier, J.-P. (Ed.) *Development of mathematics 1900-1950*, Birkhäuser-Verlag, Basel-Boston-Berlin, 1994, 479-515.

[III.296]  Ostrowski, A., *On Hadamard's test for singular points*, J. London Math. Soc. **1** (1926), 236-239.

[III.297]  Ostrowski, A., *Über quasi-analytische Funktionen und Bestimmtheit asymptotischer Entwicklungen*, Acta Math. **53** (1929), 181-266.

[III.298]  Ostrowski, A., *Sur la détermination des bornes inférieures pour une classe de déterminants*, Bull. Sci. Math. **61**:2 (1937), 19-32.

[III.299]  Painlevé, P., *Sur les lignes singulières des fonctions analytiques*, Thése, 1887; Toulouse Ann., 1888.

[III.300]  Painlevé, P., *Rapport sur le mémoire de M. J. Hadamard*, C.R. Acad. Sci. Paris, **145** (1907), 984-986.

[III.301]  Painlevé, P., *Œuvres*, tome 3 : *La correspondance entre P. Painlevé et G. Mittag-Leffler*, CNRS, Paris, 1975.

[III.302]  Palais, R.S., *Natural operations on differential forms*, Trans. Amer. Math. Soc. **92** (1959), 125-141.

[III.303]  Pečarić, J.E., *Notes on convex functions*, Internat. Series of Numer. Math. **103** (1992), 449-454.

[III.304]  Pellegrino, F., *Una condizione necessaria e sufficiente perchè una serie di potenze abbia sulla circonferenza di convergenza un solo polo multiplo*, Pont. Acc. Sci. **6** (1942), 115-123.

[III.305]  Persson, J., *The Cauchy problem and Hadamard's example*, J. Math. Anal. Appl. **59** (1977), 522-530.

[III.306]  Petrini, H., *Les dérivées premières et second du potentiel logarithmique*, Journal de Mathématiques, ser. 6, **5** (1909), 127-223.

[III.307]  Petrowsky, I.G., *Sur le problème de Cauchy pour un système d'équations aux dérivées partielles dans le domaine réel*, C.R. Acad. Sci. Paris **202** (1936), 1010-1012.

[III.308]  Petrowsky, I.G., *Über das Cauchy'sche Problem für Systeme von partiellen Differentialgleichungen*, Mat. Sb. **2**:5 (1937), 815-870.

[III.309]  Petrovskii, I.G., *On the diffusion of waves and the lacunas for hyperbolic equations*, Mat. Sb. **17**:3 (1945), 289-370.

[III.310]  Petrovskii, I.G., *Selected works: Systems of partial differential equations. Algebraic geometry* (in Russian), Nauka, Moscow, 1986.

[III.311]  Peyrefitte, A., *Rue d'Ulm. Chroniques de la vie normalienne*, Fayard, Paris, 1994.

[III.312]  Pini, B., *Sulla soluzione generalizzata di Wiener per il primo problema di valori al contorno nel caso parabolico*, Rend. Sem. Mat. Univ. Padova **23** (1954), 422-434.

[III.313]  Piranian, G., *Algebraic-logarithmic singularities and Hadamard's determi-

*nants*, Duke Math. J. **2** (1944), 147-153.

[III.314] Plastock, R., *Homeomorphisms between Banach spaces*, Trans. Amer. Math. Soc. **200** (1974), 169-183.

[III.315] Poincaré, H., *Sur les fonctions entières*, Bull. Soc. Math. France **11**, (1883), 136-144.

[III.316] Poincaré, H., *Sur les fonctions à espaces lacunaires*, Acta Soc. Sci. Fennicae **12** (1883),343-350.

[III.317] Poincaré, H., *Sur les intégrales irrégulières des équations linéaires*, Acta Math. **8** (1886), 295-344.

[III.318] Poincaré, H., *Les méthodes nouvelles de la mécanique céleste*, Gauthier-Villars, Paris, vol. 1 1892, vol. 2 1893, vol. 3 1899; English translation: Dover, New York, 1957.

[III.319] Poincaré, H., *La notation différentielle et l'enseignement*, Enseignement Math. **1** (1899), 106-110.

[III.320] Poincaré, H., *Calcul des probabilités*, Red. de A. Quiquet, 2ème éd., Gauthier-Villars, Paris, 1912.

[III.321] Poincaré, H., *Science et méthode*, Flammarion, Paris, 1918; English translation: *Science and method*, Dover, New york, 1952.

[III.322] Pólya, D., *Über gewisse Determinantenkriterien für eine Potenzreihe*, Math. Ann. **99** (1928), 687-706.

[III.323] Pólya, D., *Mathematical discovery*, John Wiley and Sons, vol. 1 1962, vol. 2 1965.

[III.324] Pólya, G., *Picture album: encounters of a mathematician*, Birkhäuser, Boston-Basel, 1987.

[III.325] Polya, G., Szegö, G., *Isoperimetric inequalities in mathematical physics*, Princeton University Press, Princeton, 1951.

[III.326] Pont, J.C., *La topologie algébrique des origines à Poincaré*, Presses Univ. de France, Paris, 1974.

[III.327] Pontryagin, L.S., *Ordinary differential equations*, Addison-Wesley, Reading-Palo Alto-London, 1962.

[III.328] Posner, E.C., *Combinatorial structures in planetary reconnaissance*, in: Mann, H.B. (Ed.), *Error correcting codes*, Wiley, New York, 1969.

[III.329] Pourciau, B., *Global invertibility of nonsmooth mappings*, J. Math. Anal. and Appl. **131** (1988), 170-179.

[III.330] Poznanski, R., *Être juif en France pendant la Seconde Guerre mondiale*, Hachette, Paris, 1996.

[III.331] Pringsheim, A., *Vorlesungen über Funktionenlehre*, II, Teubner, Leipzig, 1932.

[III.332] *Proceedings of the International Congress of Mathematicians, Cambridge, Massachusetts, U.S.A., August 30 – September 6, 1950*, AMS, Providence, R.I., 1952.

[III.333] Protter, M.H., Weinberger, H.F., *Maximum principles in differential equations*, Prentice-Hall, Englewood Cliffs, N.J., 1967.

[III.334] *Prix Bordin*, C.R. Acad. Sci. Paris **123**, juillet-décembre (1896), 1109-1111.

[III.335] Prym, F., *Zur Integration der Differentialgleichung $\frac{\partial^2 u}{\partial x^2} + \frac{\partial^2 u}{\partial y^2} = 0$*, J. reine angew. Math. **73** (1871), 340-364.

[III.336] Raghavarao, D., *Constructions and combinatorial problems in design of experiments*, Wiley, New York, 1971.

[III.337] Rankine, N.J., *On the thermodynamic theory of waves of finite longitudinal disturbance*, Trans. Roy. Soc. London **160** (1870), 277-288.

[III.338] Rayleigh, J.W., (Strutt, J.W.), *The theory of sound*, vol. 1, 2, Macmillan, London, 1894-1896.

[III.339] Reed, M., Simon, B., *Methods of modern mathematical physics*, Academic Press, New York-San Francisco-London, 1975.

[III.340] Reid, C., *Hilbert*, Springer-Verlag, Berlin-Heidelberg-New-York, 1970.

[III.341] *Report on a trip abroad for research of N. Luzin, privat-docent of the Moscow University* (in Russian). Prepared for publication by L.E. Maisterov, Ist. Mat. Issled. **8** (1955), 57-70.

[III.342] Riemann, B., *Über die Anzahl der Primzahlen unter einer gegebenen Grösse*, Monatsber. Berlin. Akad. (1859), 671-680.

[III.343] Riemann, B., *Über die Fortpflanzung ebener Luftwellen von endlicher Schwinungsweite*, Göttinger Abh. **8** (1860).

[III.344] Riemann, B., *Gesammelte mathematische Werke*, Hrsg. von H. Weber und R. Dedekind, Leipzig, 1876.

[III.345] Riesz, F., *Sur les opérations fonctionnelles linéaires*, C.R. Acad. Sci. Paris **149** (1909), 974-977.

[III.346] Rivkind, V. Ja., *Steady-state motion of a viscous drop with account taken of its deformation*, (in Russian), Zap. Naučn. Sem. Leningrad. Otdel. Mat. Inst. Steklov (LOMI) **84** (1979), 220-242.

[III.347] Rowe, D.E., *David Hilbert on Poincaré, Klein, and the world of mathematics*, Math. Intelligencer **8**:1 (1986), 75-77.

[III.348] Rowe, D.E., *Klein, Mittag-Leffler, and the Klein-Poincaré correspondence of 1881-1882.*, in: Demidov,S.S., Folkerts, M., Rowe, D.E., Scriba, C.J., (Eds.), *Amphora*, Birkhäuser Verlag, Basel-Bostin-Berlin, 1992, 597-618.

[III.349] Rozendorn, E.R., *Surfaces of negative curvature*, in: Burago, Yu.D., Zalgaller, V.A. (Eds.), *Geometry III*, Springer-Verlag, Berlin-Heidelberg-New York, 1992, 89-178.

[III.350] Rybczynski, W., *Über die fortschreitende Bewegung einer Hussigen Kugel in einem zähen Medium*, Bull. Inst. Acad. Cracovia. Cl. sci. math. et natur. Ser. A (1911), 40-44.

[III.351] Saalschütz, L., *Bemerkungen über die Gammafunktionen mit negativen Argumenten*, Zeitschr. Math. u. Phys. **32** (1887), 246-250.

[III.352] Saalschütz, L., *Weitere Bemerkungen über die Gammafunktionen mit negativen Argumenten*, Zeitschr. Math. u. Phys. **33** (1888), 362-374.

[III.353] Saint-Venant, B. de, *Mémoire sur la torsion des prismes*, Mémoires présentés par divers savants à l'Acad. Sci. **14** (1856), 233-560.

[III.354] Salaff, S., *A biography of Hua Lo-keng*, in: Sivin, N. (Ed.), *Science and technology in East Asia* Science History Publications, New York, 1977, 207-247.

[III.355] Samko, S.G., Kilbas, A.A., Marichev, O.L., *Fractional integrals and derivatives. Theory and applications*, Nauka i Technika, Minsk, 1987; English translation: Gordon and Breach, 1993.

[III.356] Schauder, J., *Das Anfangswertproblem einer quasilinearen hyperboli-*

*schen Differentialgleichung zweiter Ordnung*, Fundamenta Mathematicae **24** (1935), 213-246.

[III.357] Schwartz, L., *Quelques réflexions et souvenirs sur Paul Lévy*, Colloque Paul Lévy sur les processus stochastiques (22-26 juin 1987, École Polytechnique, Palaiseau), Astérisque **157-158** (1988), 13-28.

[III.358] Schwartz, L., *Un mathématicien aux prises avec le siècle*, Odile Jacob, Paris, 1997.

[III.359] Schwartz, W., *Some remarks on the history of the prime number theorem from 1896 to 1960*, in: Pier, J.-P. (Ed.), *Development of mathematics 1900-1950*, Birkhäuser-Verlag, Basel-Boston-Berlin, 1994, 565-614.

[III.360] Selberg, A., *On the zeros of Riemann's zeta function*, Skr. Norske Vid. Akad. Oslo **10** (1943), 3-59.

[III.361] Selberg, A., *An elementary proof of the prime number theorem*, Ann. of Math. **50** (1949), 305-313.

[III.362] Shapiro, H.S., Tegmark, M., *An elementary proof that the biharmonic Green function of an eccentric ellipse changes sign*, SIAM Review **36**:1 (1994), 99-101.

[III.363] Shiriaev, A.N., (Ed.) *Kolmogorov in reminiscences*, (in Russian), Nauka, Moscow, 1993.

[III.364] Siegmund-Schultze, R., *Die Anfänge der Funktionalanalysis und ihr Platz im Umwälzungsprozess der Mathematik um 1900*, Archive for Hist. of Exact Sci. **26**, (1982), 13-71.

[III.365] Sjöstrand, O., *Sur une équation aux dérivées partielles du type composite*, (Note 1), Arkiv Mat., Astron., Fysik **25**A:21 (1937), 1-11.

[III.366] Sjöstrand, O., *Sur une équation aux dérivées partielles du type composite*, (Note 2), Arkiv Mat., Astron., Fysik **26**A:1 (1938), 1-10.

[III.367] Smirnov, V.I., *From the correspondence of P. Appell, J. Hadamard, J. Burckhardt, V. Volterra, P. Duhem, C. Jordan, A. Poincaré and N. Radau with academician A.M. Liapunov* (in Russian), Trudy Inst. Istorii Estestvozn. Tech. **19** (1957), 690-719.

[III.368] Sobolev, S.L., *On one generalization of the Kirchhoff formula*, (in Russian), Dokl. Akad. Nauk USSR **1** (1933), 256-262.

[III.369] Sobolev, S.L., *Le problème de Cauchy dans l'espace des fonctionnelles*, Dokl. Akad. Nauk USSR **7**:3 (1935), 291-294.

[III.370] Sobolev, S.L., *Sur quelques évaluations concernant les familles de fonctions ayant des dérivées de carré intégrable*, Dokl. Akad. Nauk USSR **1**:7 (1936), 279-282.

[III.371] Sobolev, S.L., *Méthode nouvelle pour résoudre le problème de Cauchy pour les équations linéaires hyperboliques*, Mat. Sb. **1**:1 (1936), 39-72.

[III.372] Sommerfeld, A., *Randwertaufgaben in der Theorie der partiellen Differentialgleichungen*, Enzykl. math. Wiss. Leipzig **2**:4 (1904), 504-560; **2**:5, 561-570.

[III.373] Speziali, P., *Tannery Jules*, Dictionary of scientific biography, New York, 249-251.

[III.374] Spivak, M., *A comprehensive introduction to differential geometry*, vol. 3, Publish or Perish, Berkeley, 1979.

[III.375] Steklov, V.A., *Les méthodes générales pour résoudre les problèmes fon-*

*damentaux de la physique mathématique*, Ann. Fac. Sci. Toulouse Sér. 2, **2** (1900), 207-272.

[III.376] Steklov, V.A., *Mathematics and its significance for the mankind* (in Russian), Gosizdat. Berlin, 1923.

[III.377] Stellmacher, K.L., *Ein Beispiel einer Huygensschen Differentialgleichung*, Nachr. Akad. Wiss. Göttingen. Math.-Phys, Kl., **11**a:10 (1953), 133-138.

[III.378] Stellmacher, K.L., *Eine Klasse Huygensscher Differentialgleichungen und ihre Integration*, Math. Ann. **130** (1955), 219-233.

[III.379] Stieltjes, T.J., *Sur une fonction uniforme*, C.R. Acad. Sci. Paris **101**:2 (1885), 153-154.

[III.380] Stone, H.A., *Dynamics of drop deformation and breakup in viscous fluids*, Ann. Rev. Fluid Mech. **26** (1994), 65-102.

[III.381] Sylvester, J.J., *Thoughts on inverse orthogonal matrices, simultaneous sign successions, and tessellated pavements in two or more colours, with applications to Newton's rule, ornamental tile-work, and the theory of numbers*, Ann. of Philosophy Ser.4, **34** (1867), 461-475.

[III.382] Sylvester, J.J., *On Tchebycheff's theorem on the totality of prime numbers comprised within given limits*, Amer. J. Math. **4** (1881), 230-247.

[III.383] Szász, O., *Über eine Verallgemeinerung des Hadamardschen Determinantensatzes*, Monat. Math. Phys. **28** (1917), 253-257.

[III.384] Szegö, G., *On membranes and plates*, Proc. Nat. Acad. of Sciences USA **36**:3 (1950), 210-216.

[III.385] Szegö, G., *Remark on the preceeding paper of Charles Loewner*, Pacific J. Math. **3** (1953), 437-446.

[III.386] Szénássy, B., *History of mathematics in Hungary until the 20th century*, Akadémiai Kiadó, Budapest, 1992.

[III.387] Szőkefalvy-Nagy, B., *Über Integralungleichungen zwischen einer Funktion und ihrer Ableitung*, Acta Sci. Math. Szeged **10** (1941), 64-74.

[III.388] Sutter, E.E., *The fast m-transform: a fast computation of cross-correlations with binary m-sequences*, SIAM J. Comput. **20**:4 (1991), 686-694.

[III.389] Tannery, J., *Introduction à la théorie des fonctions d'une variable*, A. Hermann, Librairie scientifique, Paris, 1886.

[III.390] Taussky, O., *A recurring theorem on determinants*, Amer. Math. Monthly, **56** (1949), 672-676.

[III.391] Taylor, A.E., *A study of Maurice Fréchet: I. His early work on point set theory and the theory of functionals*, Arch. History of Exact Sciences **37** (1982), 233-295.

[III.392] Taylor, S.J., *Paul Lévy*, Bull. London Math. Soc. **7** (1975), 300-320.

[III.393] Tchebycheff, P.L., *Sur les nombres premiers*, J. Math. **17** (1852), 366-390.

[III.394] Teissier, G.F., *Essai philologique sur les commencements de la typographie à Metz, et sur les imprimeurs de la ville*, Docquet, Metz, 1828.

[III.395] *The fire that has been burning for ten years*, The New York Times, December 22, 1943; The New York Herald Tribune, December 28, 1943.

[III.396] Thom, R., *Structural stability and morphogenesis*, W.A. Benjamin, Reading, Massachusetts, 1975.

[III.397] Titchmarsh, E.C., *The theory of the Riemann zeta function*, Clarendon Press, Oxford, 1951.

[III.398] Titchmarsh, E.C., *The theory of functions*, Oxford University Press, London, 1960.

[III.399] Tricomi, F.G., *La mia vita di matematico attraverso la cronistoria dei miei lavori*, Edizioni Cedam, Padova, 1967.

[III.400] Triebel, H., *Interpolation theory. Function spaces. Differential operators*, Deutscher Verlag der Wissenschaften, Berlin, 1978.

[III.401] Truesdell, C., *General and exact theory of waves in finite elastic strain*, Arch. Rat. Mech. Anal. **8**:4 (1961), 263-296.

[III.402] Truesdell, C., Topin, R., *The classical field theories*, in: Flügge, S. (Ed.) *Encyclopedia of Physics*, vol. III/1. *Principles of classical mechanics and field theory*, Springer-Verlag, Berlin-Göttingen-Heidelberg, 1960, 226-793.

[III.403] Tuck, E.O., *Application and solution of Cauchy singular integral equations*, in: *The application and numerical solution of integral equations*, Anderssen, R.S. et al. (Eds.), Sijthoff and Noordhoff, Alphen aan den Rijn, 1980.

[III.404] Valéry, P., *Lettres à quelques-uns*, Gallimard, Paris, 1952.

[III.405] Vallée Poussin, Ch. de la, *Recherches analytiques sur la théorie des nombres premiers* Ann. Soc. Sci. Bruxelles **20** (1896), 183-256, 281-297.

[III.406] Vallée Poussin, Ch. de la, *Sur la fonction $\zeta(s)$ de Riemann et le nombre des nombres premiers, inférieurs à une limite donnée*, Mém. Acad. Roy. Sci. Belgique **59**:1 (1899-1900), 7-38.

[III.407] Volterra, V., *Sur les vibrations des corps élastiques isotropes*, Acta Math. **18** (1894), 161-232.

[III.408] Volterra, V., *Betti, Brioschi, Casorati. Trois analystes italiens et trois manières d'envisager les questions d'analyse*, Comptes Rendus du Deuxième Congrès International des Mathématiciens tenu à Paris, 1900, Gauthier-Villars, Paris, 1902, 43-57.

[III.409] Volterra, V., *Sur les équations aux dérivées partielles*, C.R. du Deuxième Congrès International des Mathématiciens tenu à Paris, 1900; Gauthier-Villars, Paris, 1902, 377-378.

[III.410] Volterra, V., *Sull' applicazione del metodo della immagine alle equazioni di tipo iperbolico*, Atti IV Congr. Intern. Mat. **2** (1908), 90-93.

[III.411] Volterra, V., *Leçons sur les fonctions de lignes*, Gauthier-Villars, Paris, 1913.

[III.412] Walfisz, A., *Weylsche Exponentialsummen in der neueren Zahlentheorie*, Deutscher Verl. der Wiss., Berlin, 1963.

[III.413] Wallis, W.D., Street, A.P. and Wallis, J.S., *Combinatorics: Room squares, sum-free sets, Hadamard matrices*, Springer-Verlag, Berlin-Heidelberg-New York, 1972.

[III.414] Warschawski, S.E., *On Hadamard's variational formula for Green's function*, J. Math. Mech. **9** (1960), 497-511.

[III.415] Weierstrass, K., *Zur Theorie der eindeutigen analytischen Funktionen*, Math. Abh. Akad. Wiss. Berlin (1876), 11-60.

[III.416] Weierstrass, K., *Zur Funktionenlehre*, Monatsber. Acad. Wiss. Berlin (1880), 719-743.

[III.417] Weil, A., *Number Theory. An approach through history. From Hammurapi to Legendre*, Birkhäuser, Boston-Basel-Stuttgart, 1983.

[III.418] Weil, A., *Souvenirs d'apprentissage*, Birkhäuser, Basel, 1991; English translation: *The apprenticeship of a mathematician*, Birkhäuser, Basel-Boston-Berlin, 1992.

[III.419] Whitham, G.B., *Linear and non-linear waves*, Wiley, New York, 1974.

[III.420] Wiener, N., *A new method in Tauberian theorems*, J. Math. and Phys. Massachusetts Inst. Tech. **7** (1928), 161-184.

[III.421] Wiener, N., *Tauberian theorems*, Ann. Math. **33** (1932), 1-100.

[III.422] Wiener, N., *I am a mathematician*, Doubleday, New York, 1956.

[III.423] Wiener, N., *Collected works with commentaries*, (P. Masani, Ed.), vol. 4, The MIT Press, Cambridge, Massachusetts, and London, England, 1985.

[III.424] Wintner, A., *On the Hölder restrictions in the theory of partial differential equations*, Amer. Journ. of Math. **72** (1950), 731-738.

[III.425] Wloka, J., *Partial differential equations*, Cambridge Univ. Press, 1987.

[III.426] Young, L.C., *Lectures on calculus of variations and optimal control theory*, Saunders, Philadelphia, 1969.

[III.427] Zermelo, E., *Beweis, dass jede Menge wohlgeordnet werden kann (Aus einem an Herrn Hilbert gerichteten Briefe)*, Math. Annalen **59** (1904), 514-516.

[III.428] Zerner, M., *Le règne de J. Bertrand (1874-1900)*, in: Gispert, H., *La France mathématique*, Société Française d'Histoire des Sciences et des Techniques. Société Mathématique de France, Paris, 1991, 299-322.

[III.429] *Zprávy. Návštěva francouzského matematika* (in Czech), Časopis pro Pěstování Matematiky a Fysiky **57** (1928), 319-320.

[III.430] Zygmund, A., *Trigonometric series*, vol. 2, Cambridge Univ. Press, Cambridge-London-New York, 1959.

# IV. Archival Material

## Papers of Jacques Hadamard's family

[IV.1] Jacqueline Hadamard, *Enfant de grand homme*, (autobiographical manuscript).

[IV.2] Jacques Hadamard, *À nos petits-enfants*, (handwritten copybook).

[IV.3] Jacques Hadamard, *Congrès Scientifique Indien, février-mars* 1947, (4 typewritten pages).

[IV.4] *Tableau d'ascendance de Jacques Salomon Hadamard*, (composed by P.-A. Meyer on the basis of the registres *d'état civil de la communauté juive de Metz*).

[IV.5] Daniel Mayer's speech at J. Hadamard's funeral (5 typewritten pages).

[IV.6] Letter of Chu Chia-hua to J. Hadamard.

[IV.7] Letter of C.V. Raman to J. Hadamard.

[IV.8] Letter of Hua Loo-Keng to J. Hadamard.

[IV.9] Letter of Ou-Sing-Mo to J. Hadamard.

[IV.10] Letter of G. Birkhoff to J. Hadamard.

[IV.11] Letter of A. Ebin to J. Hadamard.

[IV.12] Letter of J. Hadamard to P. Mendès-France.

[IV.13] Reminiscences of E. Kahane.

[IV.14] Letter of P. Montel to J. Hadamard.

[IV.15] Greeting to Hadamard from the Congress on partial differential equations, Paris, June 1962.

[IV.16] Letter of H. Lebesgue to J. Hadamard.

[IV.17] Telegram from N.M. Butler to J. Hadamard.

[IV.18] Letter from P. Robeson to J. Hadamard.

[IV.19] Letter from A. Einstein to J. Hadamard.

[IV.20] Louis de Broglie's reply to Hadamard's questionnaire.

[IV.21] Letter from S. Mandelbrojt to Louise Hadamard.

## *Archives de l'Académie des Sciences*

[IV.22] J. Hadamard's personal file.

[IV.23] P. Lévy's personal file.

### Archives Nationales

[IV.24]  *Dossiers des anciens fonctionnaires des enseignements primaire, secondaire et supérieure, XIX siècle*, (Includes Amédée Hadamard's and David Hadamard's personal files), F17/20923.

[IV.25]  *Archives personnelles des professeurs du Collège de France*, (Jacques Hadamard's personal file), F17/24600.

[IV.26]  *Dossiers administratifs du rectorat de Paris*, (Jacques Hadamard's personal file), AJ16/6016.

[IV.27]  *Dossier d'entrée à l'École Normale Supérieure de Jacques Hadamard*, AJ61/229.

### Archives de l'École Polytechnique

[IV.28]  *Concours d'admission de 1884*, IIc, 3.

[IV.29]  *Extrait du Journal Officiel du 13 novembre 1937, page 12469. Pensions civiles*, VI, 1, Sect. b, no 2.

[IV.30]  Letter of A. Vasilesco to J. Hadamard, A3b/367.

### Archives of the Massachusetts Institute of Technology

[IV.31]  Letters of J. Hadamard to N. Wiener, MC22, boxes 1, 4.

### Woodson Research Center, Rice University

[IV.32]  Seven letters of J. Hadamard to E.O. Lovett.

[IV.33]  Newsclipping from *Town & Country*, a New York City newspaper, of April 1, 1920.

[IV.34]  The article *Noted Frenchman is guest of Institute* from the Rice student newspaper *The Thresher*.

### Central Archives of the Hebrew University of Jerusalem

[IV.35]  Minutes of the Second Conference of the Academic Council of the Hebrew University held on August 16th and 18th, 1929, at Zurich.

[IV.36]  Minutes from the meetings of the Board of Governors, 1925-1931.

### Archives of the Russian Academy of Sciences (St.-Petersburg)

[IV.37]  Two letters of J. Hadamard to A.M. Liapunov, fond 186, op. 1.

[IV.38]  Letter of J. Hadamard to V.A. Steklov, fond 162, op. 2.

[IV.39]  Letter of V.A. Steklov to A.M. Liapunov (in Russian) fond 257, op. 1.

[IV.40]  Letter of J. Hadamard to A.P. Karpinskiĭ, fond 265, op. 6.

[IV.41]  Letters of J. Hadamard to A.N. Krylov, fond 759, op. 3.

[IV.42] A.N. Krylov's typewritten speech in Hadamard's honour (in Russian), fond 759, op. 2.

### Bibliothèque Nationale

[IV.43] *Factum pour le Sieur Paul Guerre Maître de Forge à Moyeuvre, Appellant d'une Sentence renduë au Bailliage de Metz le vingt-troisième Août dix-sept cent quinze & Demandeur aux fins de la commission du vingt & un Décembre suivant contre Natan Hodomard, Juif, résidant à Metz, Intimé & anticipant. Et contre Isaac Spire L évy, autre Juif, Marchand de Fer, demeurant audit Metz, Deffendeur. Metz, 1715, B.N. naf 22705.*

### Library of the *Accademia dei Lincei*

[IV.44] Correspondence of J. Hadamard and V. Volterra.

### Library of the Mittag-Leffle Institute

[IV.45] Letters of G. Mittag-Leffler to J. Hadamard.
[IV.46] Letters of J. Hadamard to G. Mittag-Leffler.
[IV.47] T. Carleman, Obituary of G.M. Mittag-Leffler.
[IV.48] Letter of J. Hadamard to I. Fredholm.

### Niedersächsische Staats- und Universitätsbibliothek Göttingen. Abteilung für Handschriften und seltene Drucke

[IV.49] Two letters of J. Hadamard to A. Hurwitz, Math. Arch 76: 183-184.
[IV.50] Three letters of J. Hadamard to D. Hilbert, Cod. Ms. D. Hilbert 125.

# Index

## M